D1769354

Selected Titles in This Series

For a complete list of titles in this series, visit the AMS Bookstore at **www.ams.org/bookstore/**.

CONTEMPORARY
MATHEMATICS

279

Topology, Geometry, and Algebra: Interactions and New Directions

Conference on Algebraic Topology
in Honor of R. James Milgram
August 17–21, 1999
Stanford University

Alejandro Adem
Gunnar Carlsson
Ralph Cohen
Editors

American Mathematical Society
Providence, Rhode Island

This volume contains the proceedings of a conference on "Topology, Geometry, and Algebra: Interactions and New Directions," which was held at Stanford University, August 17–21, 1999.

2000 *Mathematics Subject Classification.* Primary 55Pxx, 55Rxx, 55Txx, 55Uxx, 57Nxx, 57Rxx.

Library of Congress Cataloging-in-Publication Data

Conference on Algebraic Topology (1999 : Stanford, Calif.)

Topology, geometry, and algebra : interactions and new directions : Conference on Algebraic Topology in honor of R. James Milgram, August 17–21, 1999, Stanford, California / Alejandro Adem, Gunnar Carlsson, Ralph Cohen, editors.

p. cm. — (Contemporary mathematics, ISSN 0271-4132 ; 279)

Includes bibliographical references.

ISBN 0-8218-2063-X (alk. paper)

1. Algebraic topology—Congresses. I. Milgram, R. James. II. Adem, Alejandro. III. Carlsson, G.(Gunnar), 1952– IV. Cohen, Ralph L., 1952– V. Title. VI. Contemporary mathematics (American Mathematical Society) ; v. 279.

QA612.A67 1999
514′.2—dc21

2001037329
CIP

∞ The paper used in this book is acid-free and falls within the guidelines
established to ensure permanence and durability.
Visit the AMS home page at URL: http://www.ams.org/

10 9 8 7 6 5 4 3 2 1 06 05 04 03 02 01

Contents

PREFACE

In August 1999 a meeting was held at Stanford University on the subject of "Topology, Geometry and Algebra: Interactions and New Directions". The goal of the conference was to bring together distinguished researchers from a variety of areas related to algebraic topology and its applications. The list of invited speakers included: Greg Arone, Sylvain Cappell, Jon Carlson, Fred Cohen, Jim Davis, Don Davis, Tom Goodwillie, Yakov Eliashberg, Tom Farrell, Mike Hopkins, Eleny Ionel, Ronnie Lee, Ib Madsen, Mark Mahowald, Bob Oliver, Peter Oszvath, John Rognes, Abigail Thompson, Ulrike Tillmann and Efim Zelmanov.

A number of topics were covered in the lectures, including homotopy theory, moduli spaces, group cohomology, manifold theory, algebraic K-theory, low dimensional topology, symplectic geometry, etc. Special emphasis was made on breadth and interactions between different areas. A large number of postdocs, graduate students and established mathematicians attended the lectures. This volume contains twelve refereed papers on a wide variety of topics reflecting the nature of the conference.

The conference was partly held to celebrate Jim Milgram's sixtieth birthday and to recognize his enormous contributions to algebraic topology over the past 35 years. His powerful mathematics, deep insight and breadth of knowledge have been an inspiration to many of us, and it is a great pleasure for us to dedicate this volume to him. The first paper in this volume is an outline of Jim Milgram's main mathematical contributions.

Acknowledgements: We would like to thank the National Science Foundation as well as Stanford University for their generous financial support. We would like to thank Willene Perez for her valuable help in organizing the conference, and Diane Reppert for her excellent technical assistance in assembling this volume.

Alejandro Adem
Gunnar Carlsson
Ralph Cohen

Contemporary Mathematics
Volume **279**, 2001

On Jim Milgram's Mathematical Work

Gunnar Carlsson

Jim Milgram has been one of the leading figures in algebraic topology over the last 35 years. He has contributed to all parts of the subject, and his work has been characterized by tremendous depth and an unsurpassed calculational power. His influence will continue to be felt throughout algebraic topology for the foreseeable future. Personally, I was one of Jim's earlier Ph.D. students. He was a superb advisor, combining the right amount of help in crucial situations with an insistence on independence. In particular, he communicated his mathematical taste, which includes an emphasis on concrete problems and computations, very effectively. I am very grateful to him for all his work with me during my student days, as well as for his friendship since them. In this note I will attempt to summarize the high points of Jim's work. The description I will give will not be exhaustive, and it is colored by my own biases and interests.

I. Iterated Loop Spaces

In the mid 1950's, I. M. James [J] had constructed the combinatorial model for the loop space of a suspension which carries his name. Many mathematicians (including Kudo-Araki [K-A], Browder [B1], and Dyer-Lashof [D-L]) had extended this work by studying the homology of iterated loop spaces of iterated suspensions. All this work was aimed at homology computations, though, and did not carry through a full extension of James' work to produce a geometric model. In his dissertation, written under the supervision of E. Calabi, Jim had carried out a detailed study of the homology of symmetric products. Motivated in part by this dissertation work, Jim set himself the task of constructing full geometric analogues for James' models in the case of iterated loop spaces. He succeeded completely, using extremely clever combinatorial constructions, and as a consequence succeeded in describing $H_*(\Omega^k\Sigma^k X, Z/pZ)$ as a functor of $H_*(X, Z/pZ)$. This piece of work was a fundamental advance in homotopy theory, and it is of a great deal of use to this day.

This work supported in part by a grant from the National Science Foundation.

II. Spherical Fibration Theory

It has been well know since the work of Steenrod that vector bundles over a fixed base space can be *classified* by maps to a classifying space BO(n), which is constructed as a direct limit of Grassmannian manifolds of n-planes in high dimensional Euclidean spaces. This statement means that isomorphism classes of vector bundles over a base space X can be identified with $[X, BO(n)]$, the collection of homotopy classes of maps from X to $BO(n)$, and that the correspondence goes via $f \to f^*\xi_n$, where ξ_n is a certain *universal bundle* over $BO(n)$. The cohomology ring of $BO(n)$ becomes and important object in bundle classification, since homotopy classes of maps can often be distinguished by their behavior on cohomology. For this reason, cohomology classes in $BO(n)$ and other classifying spaces are referred to as *characteristic classes*, since their pullbacks to the cohomology of the base space will be characteristic invariants of the bundle.

For many purposes, though, one is interested in classifying not vector bundles, but rather *spherical fibrations.* A spherical fibration is a fibration, whose fiber has the homotopy type of a sphere. Stasheff [St] showed that by analogy with the case of vector bundles, one can construct classifying spaces BG_n which classify fibrations whose fiber has the homotopy type of the n-sphere. One can also pass to a limit over n to get a space BG which classifies *stable spherical fibrations.* Jim became interested in studying the characteristic classes for spherical fibrations, i.e. the cohomology of the space BG. He succeeded completely in obtaining a description of the mod-2 cohomology ring of all the spaces BG_n [M3]. In order to do this, he first had to study the homology algebra of the space G_n of self equivalences of the n-sphere. This he carried out using his earlier work on the structure of iterated loop spaces. This calculation was then input to an Eilenberg Moore spectral sequence, which he was able to analyze. The end result is that the cohomology ring may be described as

$$H^*(BG, Z/2Z) \cong H^*(BO, Z/2Z) \otimes \Lambda(e_i; i \in I)$$

where the left hand tensor factor denotes the cohomology ring of the union of all the classifying spaces $BO(n)$ and the right hand side denotes an exterior algebra on an infinite set of generators parametrized by a set I.

III. Manifold Classification and Bordism

In the mid 1950's R. Thom [Th] defined the bordism groups of smooth unoriented manifolds. Two smooth n-manifolds M and N are said to be *bordant* if there is a smooth $n + 1$ dimensional manifold with boundary W whose boundary is the disjoint union of M and N. This gives an equivalence relation on the collection of all smooth manifolds. Further, the equivalences form an abelian group denoted Ω_n, since one may use the disjoint union operation to add bordism classes of manifolds. Thom asked what the structure of these groups were, and provided an answer via the well-known *Pontrjagin-Thom* construction. He showed that the bordism groups could be identified with homotopy groups of *Thom complexes of vector bundles*, so that the seemingly unapproachable problem of manifold classification up to bordism was reduced to a more concrete question in homotopy theory. Thom then went on to carry out the homotopy theoretic analysis of these Thom complexes, and found an explicit answer for the unoriented bordism groups. The key ingredient was an observation that the cohomology groups of the Thom complexes behaved like free modules over the Steenrod algebra.

In the late 1950's and early 1960's, topologists began the study of bordism of smooth manifolds with additional structure. For instance, one can ask for the classification of oriented manifolds, almost complex manifolds, spin manifolds, etc., up to a bordism W carrying the same structure. These cases were carried out: Wall computed the oriented bordism groups [W1], Milnor the almost complex case [Mi2], and Anderson-Brown-Petersen the spin case [A-B-P]. There are many other possible structures, and much work has gone into their analysis. The method is always to study the homotopy type of a Thom complex of a vector bundle over a classifying space.

By the late 1960's, topologists were becoming interested in studying bordism of manifolds which weren't necessarily smooth. In the earlier work, smoothness was a key ingredient, since the Pontrjagin-Thom construction required the presence of a normal bundle to an embedding as well as a good transversality theory. So, the study of topological manifolds and piecewise linear (PL) manifolds required the development of a theory of normal bundles. This was carried out by Rourke-Sanderson ([R-S1],[R-S2], and [R-S3]) and Milnor [Mi3]. Their work permitted the reduction of topological and PL bordism groups to the homotopy theory of Thom complexes of generalized "bundles" over classifying spaces $BTop$ and BPL.

The homotopy theoretic analysis of these Thom complexes was carried out by Brumfiel-Madsen-Milgram, at the prime 2 [B-M-M]. They completed a 2-step program, the first step being the analysis of the cohomology rings $H^*(BTop, Z/2Z)$ and $H^*(BPL, Z/2Z)$, and the second being the computation of the homotopy groups from this information, using the Thom isomorphism. It turns out that the first step was in this case almost the whole story, that the computation of the homotopy groups from the cohomology was almost immediate. The difficult portion was the computation of the cohomology rings. This was carried out using the two fibrations

$$G/Top \longrightarrow BTop \longrightarrow BG$$

and

$$G/PL \longrightarrow BPL \longrightarrow BG$$

G/Top and G/PL were studied by Sullivan [Su], and BG had been studied by Milgram [M3], as we saw above. Sullivan had found that G/Top and G/PL could be described as products of simple spaces, either Eilenberg-MacLane spaces or two stage Postnikov systems. The end result was the complete description of the topological and piecewise linear unoriented bordism. Madsen and Milgram carried this work further to obtain the oriented versions of these results [MaMi1].

IV. Surgery

Surgery theory is a method developed by Browder [B2] and Novikov [N] in the simply connected case, and by Wall [W2] in the non-simply connected case, for constructing and classifying manifold structures on a topological space. It has been extremely successful in addressing a number of interesting geometric problems. Jim has done a great deal of work in this area, and we will first summarize the method.

The question we are asking about a topological space is if it has the homotopy type of a compact manifold without boundary, and if so to classify the distinct such manifolds, where manifolds with the given homotopy type are distinct if they are not homeomorphic. The first observation is that because of the Poincaré Duality Theorem, the cohomology ring of a space which has the homotopy type of a closed manifold must satisfy the Poincaré duality. Such spaces are referred to as *Poincaré duality spaces*. It turns out that although Poincaré duality spaces do not come with a tangent bundle, they do have a *Spivak normal fibration* (see [Sp]). The Spivak normal fibration is a stable class of spherical fibrations, in other words it gives a homotopy class of maps from the space in question to the classifying space BG described above. If the space is known to be a smooth manifold, then the Spivak normal fibration can be identified with the unit sphere bundle in the normal bundle to a smooth embedding of the manifold in Euclidean space. This means that for a Poincaré duality space to be the underlying space of a smooth manifold, there must be a reduction of the Spivak spherical fibration to a vector bundle. This means that in the diagram

$$\begin{array}{ccc} & & BO \\ & & \downarrow \\ X & \xrightarrow{\text{Spivak}} & BG \end{array}$$

there must be a lift of the classifying map to BO. This is a real restriction, which fails for many Poincaré duality spaces, as can be verified using Jim Milgram's computation of $H^*(BG)$. Supposing that we have such a reduction, we can construct one of the objects which surgery theory deals with, a *degree one normal map*. A degree one normal map is a map $f : M \longrightarrow X$ from a smooth manifold M to X, which induces an isomorphism on cohomology in the top non-zero dimension, together with a vector bundle ξ over X, and a stable isomorphism of vector bundles from ν_M, the stable normal bundle of M, to the pullback bundle $f^*\xi$. The question asked by surgery theory is whether a degree one normal map can be modified (using surgeries) to one which is a homotopy equivalence. Surgeries are handle addition operations performed to M, in such a way that the map f and the bundle isomorphisms may be extended over the handles. The result of a surgery is a new degree one normal map, which is *normally bordant* to the original map, in the sense that there is an $(n+1)$-dimensional manifold W with boundary, with a map to X, and with an identification of the stable normal bundle to W with the pullback of ξ, so that the boundary of W is the disjoint union of the two domain manifolds, and so that the restriction of the "bundle data" is compatible with the bundle data on the two pieces. Browder and Novikov arrived at very explicit answers concerning when a degree one normal map may be modified by surgeries to one which is a homotopy equivalence. They showed the following in the simply connected case.

1. For any degree one normal map involving manifolds and Poincaré duality spaces of odd dimension greater than or equal to 5, one can always modify the map by surgeries to obtain a homotopy equivalence.
2. If the dimension n is a multiple of 4, and greater than or equal to 5, a degree one normal map can be modified by surgeries to a homotopy equivalence if and only if the difference $\mathbf{signature}(X) - \mathbf{signature}(M)$ is equal to zero. (Poincaré duality spaces have signatures just as manifolds do, since they have a non-singular middle dimensional form)
3. If n is of the form $4k+2$, and is greater than or equal to 5, then there is a $Z/2Z$-valued obstruction, referred to as the *Kervaire invariant*, so that the degree one normal map can be modified by surgeries to a homotopy equivalence if and only if this obstruction vanishes.

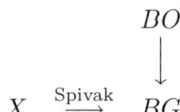

C. T. C. Wall in [W2] constructed a similar theory for non-simply connected manifolds. He showed that there are 4-periodic obstruction groups $L_n(\pi_1(X))$, depending only on the fundamental group $\pi_1(X)$, so that for a degree one normal map of dimension greater than or equal to 5, that map may be modified to a homotopy equivalence if and only if an obstruction in the appropriate obstruction group vanishes.

Here are some of Jim's contributions to this area.

1. Calculations with Brumfiel and Madsen [Ma-Mi2] which provide understanding of the homological behavior of spaces which play a key role in manifold classification. This includes the work on BG, $BTop$, and BPL, as well as the spaces G/PL and G/TOP.
2. Calculations of L-groups, joint with Hambleton [H-M1], [H-M2] and Carlsson [Ca-M].
3. Work with Hambleton, Taylor, and Williams [H-M-T-W] on the "oozing conjecture", which provides severe restrictions on which surgery obstructions can actually appear on degree one normal maps involving closed manifolds.
4. Applications of surgery and algebraic K-theory to the study of the "topological space form problem", which we will now discuss in more detail.

V. The Topological Space Form Problem

This problem is a particularly beautiful application of non-simply connected surgery theory, to which Jim has made very important contributions. This problem had its origins in the 19th century. The question that was formulated and solved then was, "which finite groups G act freely and linearly on a sphere". A linear action on a sphere is the restriction of a representation to the unit sphere in the representation space under a group invariant inner product. An equivalent problem is to ask, "which finite groups G occur as the fundamental groups of manifolds with constant positive curvature?". Representation theory allows us to give an explicit answer to this question. The following two conditions are necessary and sufficient.

1. All abelian subgroups of G are cyclic. This follows easily since non-cyclic abelian groups admit no free linear actions on spheres, as is easily verified using the character tables.
2. For any two distinct primes p and q, all subgroups of order pq are cyclic. This means that no non-trivial semidirect products of Z/pZ and Z/qZ occur. This is also relatively easy to check, using the character tables of the various non-trivial semidirect products. These conditions are referred to as the *pq-conditions.*

That these conditions are necessary is not hard to check, as we have seen, but the proof that they are also necessary requires a long argument which effectively classifies all groups satisfying these conditions.

C. T. C. Wall [W2] proposed that we remove the linearity requirement on the group action on the sphere, or equivalently that we remove any condition on curvature on the orbit manifold and instead simply require that the universal cover is topologically a sphere. Explicitly, the *topological space form problem* asks, "which finite groups G act freely on the n-sphere?".

The first thing to find out in attacking this question is how many of the necessary conditions which hold in the linear case continue to hold in the topological problem. The first condition, that all abelian subgroups must be cyclic, continues to hold. This is relatively easy to check, since a spectral sequence argument shows that for groups G which act freely on spheres, the cohomology ring $H^*(G, Z)$ must be *periodic*, i.e. that for some d, there are isomorphisms $H^\ell(G, Z) \cong H^{\ell+d}(G, Z)$ for all $\ell > 0$. Since the cohomology rings of all abelian groups are known explicitly, we can readily check that this condition doesn't hold unless the group is cyclic. R. G. Swan [Sw] showed that this condition on the abelian subgroups is sufficient to guarantee the existence of a free action of G on a finite CW-complex X having the homotopy type of a sphere. From the point of view of non-simply connected surgery theory, the orbit space X/G is now a Poincaré Duality space, to which we can apply the techniques of surgery. As for the pq-conditions, they do not hold in general, as was observed by T. Petrie [P] who produced (using surgery theory) free actions of non-trivial semidirect products of order pq on spheres. Milnor [Mi1] proved, though, that all the $2p$-conditions hold, i.e. that subgroups of order $2p$, where p is an odd prime, must be cyclic.

I. Madsen, C. B. Thomas, and C. T. C. Wall ([Ma-T-W1],[Ma-T-W2]) showed that the conditions on abelian subgroups being cyclic, together with the $2p$-conditions for all odd primes p, imply that there is a free action of G on a sphere. This theorem is one of the important achievements of surgery theory. However, it leaves open the question about which spheres G acts on. In many cases they show that G does in fact act on S^d, where d is the smallest period for the cohomology, but they are not able to resolve this question in general. Here, a more delicate analysis is required.

To describe this analysis, we first have to return to the theorem of Swan, which says that for G with only cyclic abelian subgroups, there is a finite CW-complex on which G acts. From the spectral sequence analysis and arguments about periodic cohomology, it is clear that the dimension of the sphere must be of the form $dk-1$, where d is the period in the cohomology, and where k is an arbitrary positive integer. However, the complex which Swan constructs cannot always be made d-dimensional. There is always a complex X (perhaps not finite) which has the homotopy type of S^d, so that the orbit complex has finite cohomology (and hence looks cohomologically like a finite complex), but it can't necessarily be made homotopy equivalent to a finite complex. There is a *Wall finiteness obstruction* (depending on the choice of complex X) in the algebraic K-group $K_0(Z[G])$ which must vanish if X is to have the homotopy type of a finite complex.

Jim ([M4],[Da-M]) showed that there are indeed situations in which G cannot act in the period dimension d, as a result of the fact that X cannot be chosen with vanishing finiteness obstruction. He showed that among the groups $Q(8a, b, c)$, there are values of a, b, and c for which the finiteness obstruction cannot be made to vanish, and so that these groups do not act in the period dimension. $Q(8a, b, c)$ is a group which fits into an extension

$$1 \longrightarrow C \longrightarrow Q(8a, b, c) \longrightarrow \overline{Q} \longrightarrow 1$$

where $\overline{Q}$ is a generalizes quaternion subgroup of $SU(2)$. He also showed that there are values of a, b, and c where the work of Madsen-Thomas-Wall does not guarantee an action in the period, but where an analysis of the finiteness obstruction will show that an action exists. This has shown the subtlety of the question, and Jim has shown how to perform the analysis of these delicate invariants. Jim (and Ib Madsen, independently) have also shown that even in situations where the finiteness obstruction does not preclude the existence of an action in the period dimension, the surgery obstruction may be non-zero, and so that an action in the period dimension will not exist.

VI. Atiyah Jones Conjecture

In the late 1970's, M. F. Atiyah and J. D. S. Jones [A-J] studied the moduli space solutions to the Yang-Mills equations on the space of connections on principal $SU(2)$-bundles on S^4. These solutions are named *instantons*, for reasons arising in physics. For each value of a positive integer d there is a principal $SU(2)$ bundle ξ_d with Chern class d, and we can consider the moduli space M_d of solutions to the self-duality equations on the space of connections on this bundle. We can also consider the space C_d of gauge equivalence classes of *all* connections on ξ_d, and we have the evident inclusion $i_d : M_d \longrightarrow C_d$. Atiyah and Jones found that there is an integer valued function $N(d)$ so that the induced map $H_*(i_d)$ is surjective for $* < N(d)$, and they went on to conjecture that $H_*(i_d)$ is in fact an isomorphism for $*$ less than some function $N(d)$.

Taubes [Ta] and Gravesen [G] proved a result in this direction. They observed that the spaces M_d fit together into a directed system

$$\cdots \longrightarrow M_d \longrightarrow M_{d+1} \longrightarrow M_{d+2} \longrightarrow \cdots$$

and that we therefore obtain a map on the direct limits

$$\lim M_d \longrightarrow \lim C_d.$$

They showed that this map is a homotopy equivalence. Their result is only a result about the limits, though, it does not show that this ever induces an isomorphism on homotopy groups for any particular value of d.

Jim, in collaboration with C. Boyer, J. Hurtubise, and B. Mann, was able to prove the conjecture of Atiyah-Jones, with $N(d) = \lfloor \frac{d}{2} \rfloor - 2$ [B-H-M-M]. The method used the work of Taubes-Gravesen to reduce the result to a stability argument for the maps $M_d \longrightarrow M_{d+1}$. The corresponding stability statement for the spaces C_d is immediate, since we have equivalences $C_d \cong \Omega^3_d S^3$, where the subscript denotes "degree d component", and the inclusions $C_d \longrightarrow C_{d+1}$ are given by loop sum with a degree one map. This conjecture had been the subject of intense work for a number of years, and its solution is one of the big achievements in the subject over the last 20 years.

VII. Cohomology of Simple Groups

Jim has had a large impact on the area of mathematics centered around the cohomology of finite groups. From his early days he acquired a hands-on knowledge of the cohomology of symmetric products (his thesis topic) and later the cohomology of infinite loop spaces. Along the way he developed a unique geometric insight into the cohomology of the symmetric and alternating groups, connecting Nakaoka's celebrated work with loop space techniques and invariant theory. Jim was able to use these methods to devise a method for computing the mod 2 cohomology of the finite symmetric groups. These computations and the detection arguments implemented by Jim were very influential for Quillen's subsequent ground-breaking work on group cohomology and the Adams conjecture.

Many years later (around 1988), Jim returned to his interests in finite group cohomology and he became an important catalyst for extensive ongoing interactions between algebraists and topologists. In a series of papers (many of them joint with A. Adem), Jim outlined an approach for computing the mod 2 cohomology of many of the low-rank sporadic simple groups ([A-K-M-U], [A-M-M1], [A-M-M2], [A-M1],

[A-M3], [A-M4], [A-M5], [A-M6], [Ca-M-M], [M5], [M6], [M7]) . Along the way these examples have provided testing grounds for a number of theorems and conjectures. These calculations have radically altered the nature of research in this area.

As a result of this research, Adem and Milgram published the Springer-Verlag Grundlehren text Cohomology of Finite Groups, [A-M2] which will be a standard reference in the subject for many years to come.

VIII. Mathematics Education

Over the last few years, Jim has in addition to his research work become interested in elementary and secondary mathematics education. He has been instrumental in the introduction of Content Standards in California, as well as in the development of a Framework for their implementation. In addition, he continues to play a important role as an advisor to the state of California on mathematics education, and he plays a leading role in the textbook adoption process in California. This work has had profound consequences for millions of children in California as well as throughout the country.

References

[A-K-M-U] Adem, Alejandro; Karagueuzian, Dikran; Milgram, R. James; Umland, Kristin, The cohomology of the Lyons group and double covers of alternating groups. *J. Algebra* 208 (1998), no. 2, 452–479.

[A-M-M1] Adem, Alejandro; Maginnis, John; Milgram, R. James, Symmetric invariants and cohomology of groups. *Math. Ann.* 287 (1990), no. 3, 391–411.

[A-M-M2] Adem, Alejandro; Maginnis, John; Milgram, R. James, The geometry and cohomology of the Mathieu group M_{12}. *J. Algebra* 139 (1991), no. 1, 90–133.

[A-M1] Adem, Alejandro; Milgram, R. James,Invariants and cohomology of groups. Papers in honor of José Adem (Spanish). *Bol. Soc. Mat. Mexicana* (2) 37 (1992), no. 1–2, 1–25.

[A-M2] Adem, Alejandro; Milgram, R. James, Cohomology of finite groups. Grundlehren der Mathematischen Wissenschaften vol. 309. Springer-Verlag, Berlin, 1994. viii+327 pp. ISBN: 3-540-57025-X.

[A-M3] Adem, Alejandro; Milgram, R. James, The cohomology of the Mathieu group M_{22}. *Topology* 34 (1995), no. 2, 389–410.

[A-M4] Adem, Alejandro; Milgram, R. James, The subgroup structure and mod 2 cohomology of O'Nan's sporadic simple group. *J. Algebra* 176 (1995), no. 1, 288–315.

[A-M5] Adem, Alejandro; Milgram, R. James, The mod 2 cohomology rings of rank 3 simple groups are Cohen-Macaulay. *Prospects in topology (Princeton, NJ, 1994)*, 3–12 Ann. of Math. Stud., 138 *Princeton Univ. Press, Princeton, NJ*, 1995.

[A-M6] Adem, Alejandro; Milgram, R. James, The cohomology of the McLaughlin group and some associated groups. *Math. Z.* 224 (1997), no. 4, 495–517.

[A-B-P] Anderson, D. W.; Brown, E. H., Jr.; Peterson, F. P., The structure of the Spin cobordism ring. *Ann. of Math.* (2) 86 1967 271–298.

[A-J] Atiyah, M. F.; Jones, J. D. S., Topological aspects of Yang-Mills theory. *Comm. Math. Phys.* 61 (1978), no. 2, 97–118.

[B-H-M-M] Boyer, C. P.; Hurtubise, J. C.; Mann, B. M.; Milgram, R. James, The topology of instanton moduli spaces. I. The Atiyah-Jones conjecture. *Ann. of Math.* (2) 137 (1993), no. 3, 561–609.

[B1] Browder, William, Homology operations and loop spaces. *Illinois J. Math.* 4 1960 347–357.

[B2] Browder, William, Surgery on simply-connected manifolds. Ergebnisse der Mathematik und ihrer Grenzgebiete, Band 65. Springer-Verlag, New York-Heidelberg, 1972. ix+132 pp.

[B-M-M] Brumfiel, G.; Madsen, I.; Milgram, R. James, *PL* characteristic classes and cobordism. *Ann. of Math.* (2) 97 (1973), 82–159.

[Ca-M-M] Carlson, Jon F.; Maginnis, John S.; Milgram, R. James, The cohomology of the sporadic groups J_2 and J_3. *J. Algebra* 214 (1999), no. 1, 143–173.

[Ca-M] Carlsson, G.; Milgram, R. James, The structure of odd L-groups. Algebraic topology, Waterloo, 1978 (Proc. Conf., Univ. Waterloo, Waterloo, Ont., 1978), pp. 1–72, Lecture Notes in Math., 741, Springer, Berlin, 1979.

[Da-M] Davis, J. F.; Milgram, R. James, A survey of the spherical space form problem. Mathematical Reports, 2, Part 2. Harwood Academic Publishers, Chur, 1985. xi+61 pp. ISBN: 3-7186-0250-4

[D-L] Dyer, Eldon; Lashof, R. K., Homology of iterated loop spaces. *Amer. J. Math.* 84 1962 35–88.

[G] Gravesen, Jens, On the topology of spaces of holomorphic maps. *Acta Math.* 162 (1989), no. 3-4, 247–286.

[H-M1] Hambleton, I.; Milgram, R. James, The surgery group $L_3^h(\{Z\}(G))$ for G a finite 2-group. Algebraic Topology, Waterloo, 1978 (Proc. Conf., Univ. Waterloo, Waterloo, Ont., 1978), pp. 73–89, Lecture Notes in Math., 741.

[H-M2] Hambleton, I.; Milgram, R. James, The surgery obstruction groups for finite 2-groups. *Invent. Math.* 61 (1980), no. 1, 33–52.

[H-M-T-W] Hambleton, I.; Milgram, R. J.; Taylor, L.; Williams, B., Surgery with finite fundamental group. *Proc. London Math. Soc.* (3) 56 (1988), no. 2, 349–379.

[J] James, I. M., Reduced product spaces. *Ann. of Math.* (2) 62 (1955), 170–197.

[K-A] Kudo, Tatsuji; Araki, Shôrô, Topology of H_n-spaces and H-squaring operations. *Mem. Fac. Sci. Kyushu Univ. Ser. A.* 10 (1956), 85–120.

[Ma-Mi1] Madsen, I.; Milgram, R. J., The oriented topological and *PL* cobordism rings. *Bull. Amer. Math. Soc.* 80 (1974), 855–860.

[Ma-Mi2] Madsen, I.; Milgram, R. J., The universal smooth surgery class. *Comment. Math. Helv.* 50 (1975), no. 3, 281–310.

[Ma-T-W1] Madsen, I.; Thomas, C. B.; Wall, C. T. C., The topological spherical space form problem. II. Existence of free actions. *Topology* 15 (1976), no. 4, 375–382.

[Ma-T-W2] Madsen, I.; Thomas, C. B.; Wall, C. T. C., Topological spherical space form problem. III. Dimensional bounds and smoothing. *Pacific J. Math.* 106 (1983), no. 1, 135–143.

[M1] Milgram, R. James, The homology of symmetric products. *Trans. Amer. Math. Soc.* 138 1969 251–265.

[M2] Milgram, R. James, Iterated loop spaces. *Ann. of Math.* (2) 84 1966 386–403.

[M3] Milgram, R. James, The mod 2 spherical characteristic classes. *Ann. of Math.* (2) 92 1970 238–261.

[M4] Milgram, R. James, Evaluating the Swan finiteness obstruction for periodic groups. Algebraic and geometric topology (New Brunswick, N.J.,1983), 127–158, Lecture Notes in Math., 1126, Springer, Berlin-New York, 1985.

[M5] Milgram, R. James, The 14-dimensional Kervaire invariant and the sporadic group M_{12}. *Asian J. Math.* 1 (1997), no. 2, 314–329.

[M6] Milgram, R. James, On the relation between simple groups and homotopy theory. *Group representations: cohomology, group actions and topology (Seattle, WA, 1996)*, 419–440, Proc. Sympos. Pure Math., 63, *Amer. Math. Soc., Providence, RI*, 1998.

[M7] Milgram, R. James, The cohomology of the Matheu group M_{23}. *J. Group Theory* 3 (2000), no. 1, 7–26.

[Mi1] Milnor, John, Groups which act on without fixed points. *Amer. J. Math.* 79 (1957), 623–630.

[Mi2] Milnor, J., On the cobordism ring Ω^* and a complex analogue. I. *Amer. J. Math.* 82 1960 505–521.

[Mi3] Milnor, J., Microbundles. I. *Topology* 3 1964 suppl. 1, 53–80.

[N] Novikov, S. P., Homotopically equivalent smooth manifolds. I. (Russian) *Izv. Akad. Nauk SSSR Ser. Mat.* 28 1964 365–474.

[P] Petrie, Ted, The existence of free metacyclic actions on homotopy spheres. *Bull. Amer. Math. Soc.* 76 1970 1103–1106.

[R-S1] Rourke, C. P.; Sanderson, B. J., Block bundles. I. *Ann. of Math.* (2) 87 1968 1–28.

[R-S2] Rourke, C. P.; Sanderson, B. J., Block bundles. II. Transversality. *Ann. of Math.* (2) 87 1968 256–278.

[R-S3] Rourke, C. P.; Sanderson, B. J., Block bundles. III. Homotopy theory. *Ann. of Math.* (2) 87 1968 431–483.

[Sp] Spivak, Michael, Spaces satisfying Poincaré duality. *Topology* 6 1967 77–101.

[St] Stasheff, James, A classification theorem for fibre spaces. *Topology* 2 1963 239–246.

[Su] Sullivan, D., On the Hauptvermutung for manifolds. *Bull. Amer. Math. Soc.* 73 1967 598–600.

[Sw] Swan, Richard G., Periodic resolutions for finite groups. *Ann. of Math.* (2) 72 1960 267–291.

[Ta] Taubes, Clifford Henry, The stable topology of self-dual moduli spaces. *J. Differential Geom.* 29 (1989), no. 1, 163–230.

[Th] Thom, René, *Variétés differentiubles cobornnts.* (French) *C. R. Acad. Sci. Paris* 236, (1953), 1733–1735.

[T-W] Thomas, C. B.; Wall, C. T. C., The topological spherical space form problem. I. *Compositio Math.* 23 (1971), 101–114.

[W1] Wall, C. T. C., Determination of the cobordism ring. *Ann. of Math.* (2) 72 1960 292–311.

[W2] Wall, C. T. C., Surgery on compact manifolds. Second edition. Edited and with a foreword by A. A. Ranicki. Mathematical Surveys and Monographs, 69. American Mathematical Society, Providence, RI, 1999. xvi+302 pp. ISBN: 0-8218-0942-3.

Department of Mathematics, Stanford University, Stanford, CA 94305
E-mail address: `gunnar@math.stanford.edu`

Contemporary Mathematics
Volume **279**, 2001

A renormalized Riemann-Roch formula and the Thom isomorphism for the free loop space

Matthew Ando and Jack Morava

ABSTRACT. Let E be a circle-equivariant complex-orientable cohomology theory. We show that the fixed-point formula applied to the free loop space of a manifold X can be understood as a Riemann-Roch formula for the quotient of the formal group of E by a free cyclic subgroup. The quotient is not representable, but (locally at p) its p-torsion subgroup is, by a p-divisible group of height one greater than the formal group of E.

I believe in the fundamental interconnectedness of all things.
—Dirk Gently [**Ada88**]

1. Introduction

Let $\mathbb{T}$ denote the circle group, and, if X is a compact smooth manifold, let $\mathcal{L}X \stackrel{\text{def}}{=} C^\infty(\mathbb{T}, X)$ denote its free loop space. The group $\mathbb{T}$ acts on $\mathcal{L}X$, and the fixed point manifold is again X, considered as the subspace of constant loops. In the 1980's, Witten showed that the fixed-point formula in ordinary equivariant cohomology, applied to the free loop space $\mathcal{L}X$ of a spin manifold X, yields the index of the Dirac operator (i.e. the $\hat{A}$-genus) of X—a fundamentally K-theoretic quantity [**Ati85**]. He also applied the fixed-point theorem in equivariant K-theory to a Dirac-like operator on $\mathcal{L}X$ to obtain the elliptic genus and "Witten genus" of X [**Wit88**]—quantities associated with elliptic cohomology.

Among homotopy theorists, these developments generated considerable excitement. The chromatic program organizes the structure of finite stable homotopy types, locally at a prime p, into layers indexed by nonnegative integers. The nth layer is detected by a family of cohomology theories $\mathcal{E}_n$; rational cohomology, K-theory, and elliptic cohomology are detecting theories for the first three layers [**Mor85, DHS88, HS98**].

2000 *Mathematics Subject Classification.* Primary 57R91,55N20; Secondary 14L05, 19L10, 55P92.

Key words and phrases. Free loop space, fixed-point formula, quotients of formal groups, Riemann-Roch, equivariant Thom isomorphism, prospectra.

Both authors were supported by the NSF.

The geometry and analysis related to rational cohomology and K-theory are reasonably well-understood, but for $n \geq 2$ and for elliptic cohomology in particular, very little is known. Witten's work provides a major suggestion: for $n = 1$ and $n = 2$ his analysis gives a correspondence

$$\begin{matrix}\text{analysis underlying } \mathcal{E}_n \\ \text{applied to } X\end{matrix} \leftrightarrow \begin{matrix}\text{analysis underlying } \mathcal{E}_{n-1} \\ \text{applied to } \mathcal{L}X.\end{matrix} \tag{1.1}$$

This paper represents our attempt to understand why Witten's procedure appears to connect the chromatic layers in the manner of (1.1). To do this we consider very generally the fixed-point formula attached to a complex-oriented theory E with formal group law F. We recall that for $n > 0$, such a theory detects chromatic layer n if the formal group law F has height n.

Our first result is that the fixed-point formula of a suitable equivariant extension of E (Borel cohomology is fine, as is the usual equivariant K-theory) applied to the free loop space yields a formula which is *identical* to the Riemann-Roch formula for the quotient $F/(\hat{q})$ of the formal group law F by a free cyclic subgroup $(\hat{q})$ (compare formulae (3.4) and (4.5)).

The quotient $F/(\hat{q})$ is not a formal group, so to understand its structure, we work p-locally and study its p-torsion subgroup $F/(\hat{q})[p^\infty]$. We construct a group $\mathrm{Tate}(F)$ with a canonical map

$$\mathrm{Tate}(F) \to F/(\hat{q}),$$

which induces an isomorphism of torsion subgroups in a suitable setting. Our second result is that the group $\mathrm{Tate}(F)[p^\infty]$ is a p-divisible group, fitting into an extension

$$F[p^\infty] \to \mathrm{Tate}(F) \to \mathbb{Q}_p/\mathbb{Z}_p$$

of p-divisible groups. If the height of F is n, then the height of $\mathrm{Tate}(F)[p^\infty]$ is $n+1$, but its étale quotient has height 1. In a sense we make precise in §5.3, it is the universal such extension.

Thus the fixed-point formula on the free loop space interpolates between the chromatic layers in the same way that p-divisible groups of height $n+1$ with étale quotient of height 1 interpolate between formal groups of height n and formal groups of height $n+1$. This is discussed in more detail, from the homotopy-theoretic point of view, in our earlier paper [**AMS98**] with Hal Sadofsky; this paper is a kind of continuation, concerned with analytic aspects of these phenomena. We show that Witten's construction in rational cohomology produces K-theoretic genera because of the exponential exact sequence

$$0 \to \mathbb{Z} \to \mathbb{C} \to \mathbb{C}^\times \to 1 \tag{1.2}$$

expressing the multiplicative group (K-theory) as the quotient of the additive group (ordinary cohomology) by a free cyclic subgroup; while his work in K-theory produces elliptic genera because of the exact sequence

$$0 \to q^{\mathbb{Z}} \to \mathbb{C}^\times \to \mathbb{C}^\times/q^{\mathbb{Z}} \to 1 \tag{1.3}$$

(where q is a complex number with $|q| < 1$), expressing the Tate elliptic curve $\mathbb{C}^\times/q^{\mathbb{Z}}$ as the quotient of the multiplicative group by a free cyclic subgroup.

These analytic quotients have already been put to good use in equivariant topology. Grojnowski constructs from equivariant ordinary cohomology a complex $\mathbb{T}$-equivariant elliptic cohomology using the elliptic curve $\mathbb{C}/\Lambda$ which is the quotient of the complex plane by a lattice; and Rosu uses Grojnowski's functor to give a striking conceptual proof of the rigidity of the elliptic genus. Grojnowski's ideas applied to the multiplicative sequence (1.3) give a construction of complex $\mathbb{T}$-equivariant elliptic cohomology based on equivariant K-theory; details will appear elsewhere. Completing this circle, Rosu has used the quotient (1.2) to give a construction of complex equivariant K-theory [**Gro94, Ros99, RK99**].

Several of the formulae in this paper involve formal infinite products; see for example (3.4) and (4.5). On the fixed-point formula side, the source of these is the Euler class of the normal bundle ν of X in $\mathcal{L}X$ (3.2). From this point of view, the problem is that the bundle ν is infinite-dimensional, so it does not have a Thom spectrum in the usual sense. However, ν has a highly nontrivial circle action, which defines a locally finite-dimensional filtration by eigenspaces. Following the program sketched in [**CJS95**], we construct from this filtration a Thom *pro-spectrum*, whose Thom class is the infinite product.

In the particular cases of the additive and multiplicative formal groups ($n = 1, 2$ above), one can also control the infinite products by replacing them with products which converge to holomorphic functions on $\mathbb{C}$; this construction of elliptic functions goes back to Eisenstein. We are grateful to Kapranov for pointing out to us that Eistenstein considered the the analogous problem for $n > 2$. In [**Eis44**] he described the difficulty of interpreting such infinite products. He went on to hint that he perceived a useful approach, and concluded the following.

> Die Functionen, zu welchen man auf diesen Wege geführt wird, scheinen sehr merkwürdige Eigenschaften zu besitzen; sie eröffnen ein Feld, auf dem sich Stoff zu den reichhaltigsten Untersuchungen darbietet, und welches der eigentliche Grund und Boden zu sein scheint, auf welchem die schwierigsten Theile der Analysis und Zahlentheorie ineinander greifen.

1.1. Formal group schemes. In this paper (especially in section 5) we shall consider formal schemes in the sense of [**Str99, Dem72**]. A *formal scheme* is a filtered colimit of affine schemes. For example the "formal line"

$$\hat{\mathbb{A}}^1 \stackrel{\text{def}}{=} \operatorname*{colim}_n \operatorname{spec} \mathbb{Z}[x]/x^n$$

is a formal scheme. Note that an affine scheme is a formal scheme in a trivial way. An important feature of this category which we shall use is that it has finite products. For example,

$$\hat{\mathbb{A}}^1 \times \hat{\mathbb{A}}^1 = \operatorname{colim} \operatorname{spec}\big(\mathbb{Z}[x]/(x^n) \otimes \mathbb{Z}[y]/(y^m)\big).$$

In particular a *formal group scheme* means an abelian group in the category of formal schemes. A formal group scheme whose underlying formal scheme is isomorphic to the formal scheme $\hat{\mathbb{A}}^1$ is called a commutative one-dimensional formal Lie group. We shall simply call it a *formal group*.

The first reason for considering formal schemes is that formal groups are not quite groups in the category of affine schemes, because a group law

$$F(s,t) = s + t + \cdots \in R[\![s,t]\!]$$

over a ring R gives a diagonal

$$R[\![s]\!] \to R[\![s,t]\!] \cong R[\![s]\!]\hat{\otimes} R[\![t]\!]$$

only to the completed tensor product.

The second reason for considering formal schemes is that, if G is an affine group scheme, then its torsion subgroup G_{tors} is a formal scheme (the colimit of the affine schemes $G[N]$ of torsion of order N), but not in general a scheme.

If X is a formal scheme over R, and S is an R-algebra, then X_S will denote the resulting formal scheme over S.

2. The umkehr homomorphism and an ungraded analogue

2.1. Let E be a complex-oriented multiplicative cohomology theory with formal group law F, and let $h\colon X \to Y$ be a proper complex-oriented map of smooth finite-dimensional connected manifolds, of fiber dimension $d = \dim X - \dim Y$. The Pontrjagin-Thom collapse associates to these data an "umkehr" homomorphism [**Qui71**]

$$h_! : E^*(X) \to E^{*-d}(Y).$$

We will be concerned with similar homomorphisms in certain infinite-dimensional contexts. In order to do so, we systematically eliminate the shift of $-d$ in the degree by restricting our attention to *even periodic* cohomology theories E. The examples show (3.3) that this amounts to measuring quantities relative to the vacuum.

2.2. Even periodic ring theories. Let E be a cohomology theory. If X is a space, then $E^*(X)$ will denote its *unreduced* cohomology; if A is a spectrum, then $E^*(A)$ will denote its cohomology in the usual sense. These notations are related by the isomorphism $E^*(X) \cong E^*(\Sigma^\infty X_+)$, where X_+ denotes the union of X and a disjoint basepoint. The reduced cohomology of X will be denoted $\widetilde{E}(X)$. Let $*$ denote the one-point space.

A cohomology theory E with commutative multiplication is *even* if $E^{\mathrm{odd}}(*) = 0$. It is *periodic* if $E^2(*)$ contains a unit of $E^*(*)$. If E is an even periodic theory, then we write $E(X)$ for $E^0(X)$ and E for $E^0(*)$. We sometimes write $X_E = \operatorname{spec} E(X)$ for the spectrum, in the sense of commutative algebra, of the commutative ring $E(X)$.

A space X is *even* if $H_*(X)$ is a free abelian group, concentrated in even degrees. In that case the natural map

$$\operatorname{colim} F_E \to X_E, \tag{2.1}$$

where F is the filtered system of maps of finite CW complexes to X, is an isomorphism. This gives X_E the structure of a formal scheme. The functor $X \mapsto X_E$ from even spaces to formal schemes over E preserves finite products and coproducts: if X and Y are two even spaces, then

$$(X \times Y)_E \cong X_E \times Y_E \cong \operatorname{spec} E(X)\hat{\otimes} E(Y).$$

Here $\hat{\otimes}$ refers to the completion of the tensor product with respect to the topology defined by the filtrations of $E(X)$ and $E(Y)$.

2.3. Orientations and coordinates. Let $P \stackrel{\text{def}}{=} \mathbb{C}P^\infty$ be the classifying space for complex line bundles. Let $m\colon P \times P \to P$ be the map classifying the tensor product of line bundles. It induces a map

$$P_E \times P_E \xrightarrow{m_E} P_E,$$

which makes P_E a formal group scheme over E. Of course it is a formal group: let $i\colon S^2 \to P$ be the map classifying the Hopf bundle. A choice of element $x \in \widetilde{E}(P)$ such that $v = i^*x \in \widetilde{E}(S^2) \cong E^{-2}(*)$ is a unit is called a *coordinate* on P_E. There is then an isomorphism

$$E(P) \cong E[\![x]\!],$$

which determines a formal group *law* F over E by the formula

$$F(x, y) = m^*x \in E(P \times P) \cong E[\![x, y]\!].$$

Any even-periodic cohomology theory E is complex-orientable. An *orientation* on E is a multiplicative natural transformation

$$MU \to E.$$

These correspond bijectively with elements $u \in \widetilde{E}^2(P)$ such that

$$i^*u = \Sigma^2(1), \tag{2.2}$$

where Σ is the suspension isomorphism [**Ada74**]. A coordinate x thus determines an orientation $u = v^{-1}x$.

DEFINITION 2.3. We shall use the notation (E, x, F) to denote an even periodic cohomology theory E with coordinate x and group law F. We shall call such a triple a *parametrized* theory.

2.4. Thom isomorphism. An orientation $u \in \widetilde{E}^2(P)$ gives the usual Thom classes and characteristic classes for complex vector bundles. If k is an integer, let $\underline{k}$ denote the trivial complex vector bundle of rank k. If X is a connected space and V is a complex vector bundle of rank d over X, then we write

$$X^V \stackrel{\text{def}}{=} \Sigma^\infty(\mathbb{P}(V \oplus \underline{1})/\mathbb{P}(V))$$

for the suspension spectrum of its Thom space, with bottom cell in degree $2d$. We write α^V_{usual} for the Thom isomorphism

$$\alpha^V_{\text{usual}}\colon E^*(X) \cong E^{*+2d}(X^V).$$

In the same way, a coordinate $x \in \widetilde{E}(P)$ gives rise to a Thom isomorphism

$$\alpha^V\colon E(X) \cong E(X^V).$$

If $v = i^*x \in \widetilde{E}(S^2)$ is the associated orientation, the isomorphisms α_{usual} and α are related by the formula

$$\alpha^V = v^{\text{rank } V} \alpha^V_{\text{usual}}.$$

REMARK 2.4. One effect of condition (2.2) is that $\alpha^{\underline{d}}_{\mathrm{usual}}$ coincides with the suspension isomorphism

$$\alpha^{\underline{d}}_{\mathrm{usual}} = \Sigma^{2d} \colon E^*(X) \cong E^{*+2d}(X^{\underline{d}}).$$

The Thom isomorphism α defined by a coordinate chooses $v \in \widetilde{E}(S^2) \cong E(*^{\underline{1}})$ as $\alpha^{\underline{1}}$. Thus α_{usual} may be viewed as a composition of Thom isomorphisms

$$\alpha^V_{\mathrm{usual}} \colon E(X^{\underline{d}}) \xrightarrow{(\alpha^{\underline{d}})^{-1}} E(X) \xrightarrow{\alpha^V} E(X^V).$$

If $\zeta \colon \Sigma^\infty X_+ \to X^V$ denotes the zero section, then we write

$$e_{\mathrm{usual}}(V) \overset{\mathrm{def}}{=} \zeta^* \alpha^V_{\mathrm{usual}}(1) \in E^{2d}(X)$$
$$e(V) \overset{\mathrm{def}}{=} \zeta^* \alpha^V(1) \in E(X)$$

for the usual and degree-zero Euler classes of V; these are related by the formula

$$e(V) = v^{\mathrm{rank}\, V} e_{\mathrm{usual}}(V).$$

If $U(n)$ denotes the unitary group and T is its maximal torus of diagonal matrices, then the map

$$E(BU(n)) \to E(BT) \cong E((B\mathbb{T})^n)$$

is the inclusion of the ring of invariants under the action of the Weyl group W. The coordinate gives an isomorphism

$$E(BT) \cong E((B\mathbb{T})^n) \cong E[\![x_1, ..., x_n]\!],$$

with W acting as the permutation group Σ_n on the x_i's. Thus we can define degree-zero Chern classes c_i in $E(BU(n))$ by the formula

$$\sum_{i=0}^{n} c_i z^{n-i} = \prod_{i=0}^{n} (z + x_i). \tag{2.5}$$

If F is the group law resulting from the coordinate x, then we call the c_i the "F–Chern classes".

Returning to the map

$$X \xrightarrow{h} Y,$$

as in (2.1), we can now define an umkehr map

$$E(X) \xrightarrow{h_F} E(Y),$$

using the degree-zero Thom isomorphism α. We write F to indicate the dependence on the coordinate.

DEFINITION 2.6. If X is any manifold, we denote by p^X the map

$$X \xrightarrow{p^X} *.$$

If E is an even periodic theory with group law F, then its F*–genus* is the element $p^X_F(1)$ of E.

2.5. The Riemann-Roch formula. The Riemann-Roch formula compares the umkehr homomorphisms h_F and h_G of two coordinates with formal group laws F and G, related by an isomorphism

$$\theta \colon F \to G.$$

The book of Dyer [**Dye69**] is a standard reference.

PROPOSITION 2.7. *If $h\colon X \to Y$ is a proper complex-oriented map of fiber dimension $2d$, then*

$$h_G(u) = h_F\left[u \cdot \prod_{j=1}^{d} \frac{x_j}{\theta(x_j)}\right], \tag{2.8}$$

where the x_i are the terms in the factorization

$$z^d + c_1 z^{d-1} + \ldots + c_d = \prod_{j=1}^{d} (z + x_i)$$

of the total F–Chern class of the formal inverse of the normal bundle of h. □

REMARK 2.9. Changing the coordinate by a unit $u \in E$ multiplies the umkehr homomorphism by u^d; by such a renormalization, we can always assume that θ is a *strict* isomorphism.

2.6. The fixed-point formula.

Notations for circle actions. Let $\mathbb{T}$ denote the circle group $\mathbb{R}/\mathbb{Z}$. If X is a $\mathbb{T}$-space then we write

$$X_{\mathbb{T}} \stackrel{\text{def}}{=} E\mathbb{T} \underset{\mathbb{T}}{\times} X$$

for the Borel construction and $X^{\mathbb{T}}$ for the fixed-point set. Let $\mathbb{T}^* = \operatorname{Hom}[\mathbb{T}, \mathbb{C}^\times]$ be the character group of $\mathbb{T}$; we will also write $\hat{\mathbb{T}} = \mathbb{T}^* - \{1\}$ for the set of nontrivial irreducible representations. For $k \in \mathbb{T}^*$, let $\mathbb{C}(k)$ be the associated one-dimensional complex representation. There is then an associated complex line bundle $\mathbb{C}(k)_{\mathbb{T}}$ over $B\mathbb{T}$.

It is convenient to choose an an isomorphism $\mathbb{T}^* \cong \mathbb{Z}$; this determines, in particular, an isomorphism $B\mathbb{T} \cong \mathbb{C}P^\infty$. For $k \in \mathbb{Z}$ we have $\mathbb{C}(k) = \mathbb{C}(1)^{\otimes k}$ and $\mathbb{C}(k)_{\mathbb{T}} = \mathbb{C}(1)_{\mathbb{T}}^{\otimes k}$. If $\hat{q} \in E(B\mathbb{T})$ is the Euler class of $\mathbb{C}(1)_{\mathbb{T}}$, then the Euler class of $\mathbb{C}(k)_{\mathbb{T}}$ is $[k](\hat{q})$.

Equivariant cohomology.

DEFINITION 2.10. Let (E, x, F) be a parametrized theory. A $\mathbb{T}$-equivariant cohomology theory $E_{\mathbb{T}}$ is an *extension* of (E, x, F) if

(1) There is a natural transformation

$$E(X/\mathbb{T}) \to E_{\mathbb{T}}(X),$$

which is an isomorphism if $\mathbb{T}$ acts freely on X. In particular the coefficient ring $E_{\mathbb{T}}(*)$ is an algebra over $E(*)$, and so it is 2-periodic.

(2) There is a natural forgetful transformation

$$E_{\mathbb{T}}(X) \to E(X).$$

If X is a trivial $\mathbb{T}$-space then the composition

$$E(X) \to E_{\mathbb{T}}(X) \to E(X)$$

is the identity.

(3) $E_{\mathbb{T}}$ has Thom classes and so Euler classes for complex $\mathbb{T}$-vector bundles, which are multiplicative and natural under pull-back. If V/X is such a bundle, then we write $e_{\mathbb{T}}(V) \in E_{\mathbb{T}}(X)$ for its (degree-zero) Euler class. These are compatible with the Thom isomorphism in E in the sense that, if the $\mathbb{T}$-action on V/X is trivial, then

$$e_{\mathbb{T}}(V) = e(V).$$

(4) If L_1 and L_2 are complex $\mathbb{T}$-line bundles, then

$$e_{\mathbb{T}}(L_1 \otimes L_2) = e_{\mathbb{T}}(L_1) +_F e_{\mathbb{T}}(L_2).$$

DEFINITION 2.11. If $E_{\mathbb{T}}$ is equivariantly complex oriented as above, a homomorphism $E_{\mathbb{T}} \to \hat{E}_{\mathbb{T}}$ of multiplicative $\mathbb{T}$-equivariant cohomology theories is a *suitable* localization if

(1) $\hat{E}_{\mathbb{T}}(*)$ is flat over $E_{\mathbb{T}}(*)$,
1. When $k \neq 0$, $e_{\mathbb{T}}(\mathbb{C}(k))$ maps to a unit of $\hat{E}_{\mathbb{T}}(*)$, and
(2) The fixed-point formula (2.13) holds for $\hat{E}_{\mathbb{T}}$.

In order to state the fixed-point formula, we need the following observation of [**AS68**].

LEMMA 2.12. *Let $E_{\mathbb{T}}$ be a suitable theory. Let S be a compact manifold with trivial $\mathbb{T}$-action, and let V be a complex $\mathbb{T}$-vector bundle over S. If the fixed-point bundle $V^{\mathbb{T}}$ is zero, then $e_{\mathbb{T}}(V)$ is a unit of $E_{\mathbb{T}}(S)$.*

PROOF. Recall [**Seg68**] that the natural map

$$\bigoplus_{k \in \hat{\mathbb{T}}} V(k) \otimes \mathbb{C}(k) \to V$$

is an isomorphism, where $V(k) \overset{\mathrm{def}}{=} \mathrm{Hom}[\mathbb{C}(k), V]$ is the evident vector bundle over S with trivial $\mathbb{T}$-action. By applying the ordinary splitting principle to $V(k)$, we are reduced to the case that $V = L \otimes \mathbb{C}(k)$, where L is a complex line bundle over S with trivial $\mathbb{T}$-action. If $E_{\mathbb{T}}$ is suitable then the Euler class of V is

$$e_{\mathbb{T}}(L \otimes \mathbb{C}(k)) = e(L) +_F e_{\mathbb{T}}(\mathbb{C}(k)).$$

Since S is a compact manifold, $e(L)$ is nilpotent in $E(S)$, so $e_{\mathbb{T}}(V)$ is a unit of $E_{\mathbb{T}}(S)$ because $e_{\mathbb{T}}(\mathbb{C}(k))$ is a unit of $E_{\mathbb{T}}(*)$. □

Now suppose that M is a compact almost-complex manifold with a compatible $\mathbb{T}$-action. Let

$$j \colon S \to M$$

denote the inclusion of the fixed-point set; it is a complex-oriented equivariant map, with $\mathbb{T}$–equivariant normal bundle ν. The fixed-point formula which we require in Definition 2.11 is the equation

$$p_F^M(u) = p_F^S\left(\frac{j^*u}{e_{\mathbb{T}}(\nu)}\right) . \tag{2.13}$$

By Lemma 2.12 $e_{\mathbb{T}}(\nu)$ is a unit of $E_{\mathbb{T}}(S)$, so this localization theorem is a corollary to the projection formula

$$j_F j^*(x) = x \cdot e_{\mathbb{T}}(\nu)$$

for the umkehr of the inclusion of the fixed-point set.

EXAMPLE 2.14. The Borel extension

$$E_{\mathrm{Borel}}(X) \stackrel{\mathrm{def}}{=} E(X_{\mathbb{T}})$$

of an even periodic ring theory has Thom classes $e_{\mathbb{T}}(V) = e(V_{\mathbb{T}})$ for complex $\mathbb{T}$-vector bundles, and the localization defined by inverting the multiplicative subset generated by $e_{\mathbb{T}}(\mathbb{C}(k)), k \neq 0$ will be suitable.

EXAMPLE 2.15. Let $K_{\mathbb{T}}$ denote the usual equivariant K-theory. Then $K_{\mathbb{T}} = \mathbb{Z}[q, q^{-1}]$, where q is the representation $\mathbb{C}(1)$, considered as a vector bundle over a point. The Euler class of a line bundle is $e_{\mathbb{T}}(L) = 1 - L$, so $\hat{q} = 1 - q$. The group law is multiplicative:

$$\mathbb{G}_m(x, y) = x + y - xy. \tag{2.16}$$

We have $[k](\hat{q}) = 1 - q^k$, and consequently

$$\hat{K}(X) \stackrel{\mathrm{def}}{=} K_{\mathbb{T}}(X) \otimes_{K_{\mathbb{T}}} \mathbb{Z}((q))$$

is suitable.

EXAMPLE 2.17. If

$$H^*_{\mathbb{T}}(X) \stackrel{\mathrm{def}}{=} H^*(X_{\mathbb{T}}; \mathbb{Q}[v, v^{-1}])$$

is Borel cohomology with two-periodic rational coefficients, and $\hat{q} = e(\mathbb{C}(1)_{\mathbb{T}})$, then $H_{\mathbb{T}}(*) \cong \mathbb{Q}[\![\hat{q}]\!]$, and $e(\mathbb{C}(k)_{\mathbb{T}}) = k\hat{q}$. The rational Tate cohomology

$$\hat{H}^*(X) \stackrel{\mathrm{def}}{=} H^*_{\mathbb{T}}(X)[\hat{q}^{-1}]$$

is suitable.

EXAMPLE 2.18. More generally, two–periodic $\hat{K}(n)_{\mathbb{T}}$ (with n finite positive) is suitable: if $[p]_{K(n)}(X) = X^{p^n}$ and $k = k_0 p^s$ with $(k_0, p) = 1$ then

$$[k](\hat{q}) = [k_0](\hat{q}^{p^{ns}}) = k_0 \hat{q}^{p^{ns}} + \cdots \in \mathbb{F}_p((\hat{q}))$$

has invertible leading term. Integral lifts of $K(n)$ behave similarly; the Cohen ring [**AMS98**] of $\mathbb{F}_p((\hat{q}))$ defines a completion of the Borel-Tate localization.

These coefficient rings have natural topologies, which are relevant to the convergence of infinite products in Corollary 6.6.

3. Application to the free loop space

Let $\hat{E}_{\mathbb{T}}$ be a suitable localization (2.11) of an equivariantly complex oriented cohomology theory, let X be a compact complex-oriented manifold, and let $\mathcal{L}X$ be its free loop space. Since $\mathcal{L}X$ is not finite-dimensional, the existence of an umkehr homomorphism $p_F^{\mathcal{L}X}$ is not clear. However, $\mathbb{T}$ acts on $\mathcal{L}X$ by rotations with fixed set X of constant loops, and Witten discovered that the fixed-point formula (2.13) for the F-genus $p_F^{\mathcal{L}X}(1)$ of $\mathcal{L}X$ continues to yield interesting formulae. In this section we review his calculation.

3.1. The normal bundle to the constant loops and its Euler class. One approximates the space $C^\infty(S^1, \mathbb{C})$ by the sub-vector space of Laurent polynomials

$$\mathbb{C}[\mathbb{T}^*] \cong \bigoplus_{k\in\mathbb{T}^*} \mathbb{C}(k) \hookrightarrow C^\infty(S^1, \mathbb{C}).$$

The tangent space of $\mathcal{L}X$ is $C^\infty(S^1, TX)$. If $p \in X$ is considered as a constant loop, then the tangent space to $\mathcal{L}X$ at p is the $\mathbb{T}$–space $T\mathcal{L}X_p \cong C^\infty(S^1, TX_p)$. It is a $\mathbb{T}$-bundle with a Laurent polynomial approximation

$$TX_p \otimes \mathbb{C}[\mathbb{T}^*]$$

Thus the normal bundle ν of the inclusion of X in $\mathcal{L}X$ has approximation

$$\nu \simeq \bigoplus_{k\in\hat{\mathbb{T}}} TX \otimes \mathbb{C}(k). \tag{3.1}$$

If

$$z^d + c_1 z^{d-1} + \ldots + c_d = \prod_{j=1}^{d} (z + x_i)$$

is the formal factorization of the total F-Chern class of TX, then

$$e_{\mathbb{T}}(\nu) = \prod_{j=1}^{d} \prod_{k\neq 0} (x_j +_F [k]_F(\hat{q})), \tag{3.2}$$

where $\hat{q} = e_{\mathbb{T}}(\mathbb{C}(1))$.

3.2. The fixed point formula. Applying (2.13) to the inclusion

$$X \to \mathcal{L}X$$

yields the formula

$$p_F^{\mathcal{L}X}(1) = p_F^X \left[\prod_{j=1}^{d} \prod_{k\neq 0} \frac{1}{x_j +_F [k]_F(\hat{q})} \right]. \tag{3.3}$$

Equation (3.3) requires some interpretive legerdemain. For example, the leading coefficient of

$$\prod_{k\neq 0} (x +_F [k]_F(\hat{q}))$$

is the objectionable expression $\prod_{k\neq 0}(k\hat{q})$; but, as physicists say, this quantity is not directly 'observable'. For this reason, we consider the the *renormalized* formal product

$$\Theta_F(x;\hat{q}) \stackrel{\text{def}}{=} x\prod_{k\neq 0}\frac{(x+_F[k](\hat{q}))}{[k](\hat{q})}.$$

In section 6 below we provide a natural setting for such formal products.

The fixed point formula suggests that we define the equivariant F-genus of $\mathcal{L}X$ to be

$$\tilde{p}_F^{\mathcal{L}X}(1) \stackrel{\text{def}}{=} p_F^X\left[\prod_{j=1}^{d}\frac{x_j}{\Theta_F(x_j;\hat{q})}\right]. \tag{3.4}$$

3.3. Examples.

The additive group law. When F is the additive group law Θ_F becomes

$$\begin{aligned}\Theta_{\mathbb{G}_a}(x,\hat{q}) &= x\prod_{k\neq 0}(\frac{x}{k\hat{q}}+1)\\ &= x\prod_{k>0}(1-\frac{x^2}{k^2\hat{q}^2}).\end{aligned}$$

This is the Weierstrass product for $\pi^{-1}\hat{q}\sin\hat{q}^{-1}\pi x$, so for the theory $\hat{H}$ of (2.17), formula (3.4) gives

$$\tilde{p}_F^{\mathcal{L}X}(1) = (2\pi i/\hat{q})^d\left[\prod_{j=1}^{d}\frac{x_j/2}{\sinh(x_j/2)}\right][X]\,.$$

This is just the $\hat{A}$-genus of X, up to a normalization depending on the dimension of X. In [**Ati85**], Atiyah rewrites the formal product

$$x\prod_{k\neq 0}(x+k\hat{q})$$

as

$$x\left(\prod_{k>0}k\right)^2\prod_{k>0}\left(\frac{x^2}{k^2}-\hat{q}^2\right)$$

and invokes zeta-function renormalization [**Den92**] to replace $(\prod_{k>0}k)^2$ with 2π, yielding

$$2\pi x\prod_{k>0}(\frac{x^2}{k^2}-\hat{q}^2)\ ;$$

specializing $\hat{q}$ to i then gives the classical expression. From our point of view it's natural to think of the Chern class $\hat{q}$ of $\mathbb{C}(1)$ as the holomorphic one-form $z^{-1}dz$ on the complex projective line, and thus to identify $\hat{q}$ with its period $2\pi i$ with respect to the equator of $\mathbb{C}P_1$ as in §2.1 of [**Del89**]: the 'Betti realization' of the Tate motive $\mathbb{Z}(n)$ is $(2\pi i)^n\mathbb{Z}\subset\mathbb{C}$.

The multiplicative group law. In the case of the equivariant K-theory $\hat{K}$ of example (2.15), the Euler class of a line bundle L is $e_{\mathbb{T}}(L) = 1 - L$. Writing q for the generator $\mathbb{C}(1)$ of $\mathbb{T}^*$, the Euler class of $\mathbb{C}(1)$ is

$$\hat{q} = e_{\mathbb{T}}(\mathbb{C}(1)) = 1 - q.$$

The multiplicative group law (2.16) gives

$$e_{\mathbb{T}}(L) +_{\mathbb{G}_m} [k](\hat{q}) = 1 - q^k L,$$

and so the formal product $\Theta_{\mathbb{G}_m}(x, \hat{q})$ becomes

$$\begin{aligned} \Theta_{\mathbb{G}_m}(x,\hat{q}) &= (1-L)\prod_{k>0}\frac{(1-Lq^k)(1-Lq^{-k})}{(1-q^k)(1-q^{-k})} \\ &= \left(\prod_{k>0} L\right)(1-L)\prod_{k>0}\frac{(1-Lq^k)(1-L^{-1}q^k)}{(1-q^k)^2}. \end{aligned} \tag{3.5}$$

Aside for the powers of L, this is essentially the product expansion for the Weierstrass σ function

$$\sigma(L,q) = (1-L)\prod_{k>0}\frac{(1-q^kL)(1-q^kL^{-1})}{(1-q^k)^2} \in \mathbb{Z}[L,L^{-1}][[q]]$$

(see for example [**MT91**] or p. 412 of [**Sil94**]). The infinite factor is objectionable: in the product (3.4) defining the hypothetical $\hat{K}$-genus of $\mathcal{L}X$ this factor contributes an infinite power of $\Lambda^{\mathrm{top}}TX$, but if $c_1X = 0$ (*e.g.* if X is Calabi-Yau) and we are careful with the product, we can replace $\Theta_{\mathbb{G}_m}$ with θ. The resulting invariant is the Witten genus [**Wit88, AHS98**]. Segal [**Seg88**] replaces the formal product

$$\prod_{k\neq 0}(1-q^kL)(1-q^{-k}L)$$

which arises in the multiplicative case with

$$(\prod_{k>0} q^{-k}L)\prod_{k>0}(1-q^kL)(1-q^kL^{-1}).$$

He eliminates the infinite product of L's by assuming $\Lambda^{\mathrm{top}}TX$ trivial and he uses zeta-function renormalization to replace $q^{-\sum k}$ with $q^{-1/12}$.

4. A Riemann-Roch formula for the quotient of a formal group by a free subgroup

The starting point for this paper was the discovery that the formal products (3.3) and (3.4) which arise in applying the fixed point formula to study the F-genus of the free loop space are *precisely* the same as those obtained from the Riemann-Roch theorem for the quotient of F by a free cyclic subgroup. We explain this in section 4.2, after briefly reviewing finite quotients of formal groups, following [**Lub67, And95**].

4.1. The quotient of a formal group by finite subgroup. In this section, we assume that F is a formal group law over a complete local domain R of characteristic 0 and residue characteristic $p > 0$. If A is a complete local R-algebra, the group law F defines a new abelian group structure on the maximal ideal $\mathbf{m}_A$ of A. We will refer to $(\mathbf{m}_A, +_F)$ as the group $F(A)$ of A-valued points of F.

If H is a *finite* subgroup of $F(R)$, then Lubin shows that there is a formal group law F/H over R, determined by the requirement that the power series

$$f_H(x) \overset{\text{def}}{=} \prod_{h \in H} (x +_F h) \in R[\![x]\!] \tag{4.1}$$

is a homomorphism of group laws $F \xrightarrow{f_H} F/H$; in other words there is an equation

$$F/H(f_H(x), f_H(y)) = f_H(F(x, y)).$$

The main point is that the power series f_H is constructed so the kernel of f_H applied to $F(R)$ is the subgroup H.

The coefficient $f'_H(0) = \prod_{h \neq 0} h$ of the linear term of $f_H(x)$ is not a unit of R, and so f_H is an isomorphism of formal group laws only over $R[f'_H(0)^{-1}]$. Over this ring, we might as well replace f_H with the strict isomorphism

$$g_H(x) \overset{\text{def}}{=} x \prod_{h \neq 0} \frac{x +_F h}{h} = \frac{f_H(x)}{f'_H(0)},$$

and define G to be the formal group law

$$G(x, y) = g_H\left(F(g_H^{-1}(x), g_H^{-1}(y))\right)$$

over $R[f'_H(0)^{-1}]$; then F/H and G are related by the isomorphism

$$t(x) = f'_H(0)x.$$

If F is the group law and R the ring of coefficients of a parametrized theory, then the Riemann-Roch formula (2.8) for a compact complex-oriented manifold X is the equation

$$p_G^X(u) = p_F^X \left[u \prod_{j=1}^{d} \frac{x_j}{g_H(x_j)} \right] \tag{4.2}$$

over $R[f'_H(0)^{-1}]$.

4.2. The case of a free cyclic subgroup. Now suppose that (E, x, F) is a parametrized theory. Let R be the even periodic theory defined by the formula

$$R(X) \overset{\text{def}}{=} E(B\mathbb{T} \times X).$$

The projection $B\mathbb{T} \to *$ gives a natural transformation $E(X) \to R(X)$. In particular the coordinate $x \in E(P)$ gives a coordinate $x \in R(P)$. The group law $\mathbf{F}$ is just the group law F, considered over the E-algebra R.

What extra structure is available over R? A character $\lambda \in \mathbb{T}^*$ gives a map $B\mathbb{T} \to P$, and so an R-valued point $u(\lambda)$ of P_E. As λ varies through the group of characters, these points assemble into a homomorphism of groups

$$u \colon \mathbb{T}^* \to \mathbf{F}(R). \tag{4.3}$$

It is easy to see that this is the inclusion of a subgroup. As usual we write $\hat{q} = u(1)$ for the generator of $\mathbb{T}^*$ using the isomorphism $\mathbb{T}^* \cong \mathbb{Z}$. We write $(\hat{q})$ for the subgroup $u(\mathbb{T}^*)$ generated by $\hat{q}$.

The analogue of Lubin's formula (4.1) is

$$f_{(\hat{q})}(x) = \prod_{k \in \mathbb{Z}} (x +_F [k](\hat{q})), \tag{4.4}$$

and the resulting Riemann-Roch formula

$$p^X_{\mathbf{F}/(\hat{q})}(u) = p^X_F \left[u \prod_{j=1}^{d} \prod_{0 \neq k} \frac{1}{x_j +_F [k]_F(\hat{q})} \right]$$

has right-hand side identical to the formal product (3.3) arising from the fixed-point formula.

Of course this does not avoid the problems of applying the fixed-point formula to the free loop space. From this point of view, the trouble is that the quotient object $\mathbf{F}/(\hat{q})$ isn't a formal group; for example the coefficient of x is the product (4.4) is $\prod_{k \neq 0} (k\hat{q})$ This particular problem is fixed by using

$$\begin{aligned} g_{(\hat{q})}(x) &= x \prod_{0 \neq k \in \mathbb{T}^*} \frac{x +_F [k]_F(\hat{q})}{[k]_F(\hat{q})} \\ &= \Theta_F(x, \hat{q}). \end{aligned}$$

The Riemann-Roch formula in this case is

$$p^X_G(u) = p^X_F \left[u \prod_{j=1}^{d} \frac{x_j}{\Theta_F(x_j, \hat{q})} \right] \tag{4.5}$$

with right-hand-side identical to the renormalized genus (3.4).

However, expanding such Weierstrass products as formal power series is still highly nontrivial, as the examples above have shown.

5. The structure of $\mathbf{F}/(\hat{q})$

Suppose now that F is a formal group law over a complete local ring E with residue field k of characteristic $p > 0$, and that $\mathbf{F}$ is its pullback over the power series algebra $R = E[\![\hat{q}]\!]$. The coordinate defines a homomorphism

$$\begin{aligned} u \colon \mathbb{Z} &\to \mathbf{F}(R) \\ u(n) &= [n]_{\mathbf{F}}(\hat{q}); \end{aligned} \tag{5.1}$$

but there is no reason to expect the cokernel of this homomorphism to be a formal group. However, $\mathbf{F}/(\hat{q})$ certainly makes sense as a group-valued functor of complete local R-algebras.

In section 5.1, we show by construction that the torsion subgroup of $\mathbf{F}/(\hat{q})$ has a natural approximation by a representable functor. We construct a formal group scheme $\mathrm{Tate}(F)$ over R together with a natural transformation

$$\mathrm{Tate}(F)_{\mathrm{tors}} \to \mathbf{F}/(\hat{q})_{\mathrm{tors}}$$

of group-valued functors, which is an isomorphism if $\hat{q}$ has infinite order in F. The formal group scheme $\mathrm{Tate}(F)_{\mathrm{tors}}$ is our model for $\mathbf{F}/(\hat{q})$.

Because E has finite residue characteristic, we work p-locally: the formal group scheme $\mathrm{Tate}(F)[p^\infty]$ is a p-divisible group in the sense of [**Tat67, Dem72**]. The p-torsion subgroup $G[p^\infty]$ of a formal group G of finite height is a *connected* p-divisible group, but $\mathrm{Tate}(F)[p^\infty]$ is not; its maximal étale quotient is a constant height-one p-divisible group $\mathbb{Q}_p/\mathbb{Z}_p$, and its connected component is just the p-divisible group $F[p^\infty]$ of F. In other words, there is an extension

$$\mathbf{F}[p^\infty] \longrightarrow \mathrm{Tate}(F)[p^\infty] \longrightarrow \mathbb{Q}_p/\mathbb{Z}_p$$

of p-divisible groups over R. We will see in (5.7) that it is in fact the *universal* example of such an extension, and it follows that if E_n is the ring which classifies lifts of a formal group G of height n over an algebraically closed field k [**LT66**], then $R = E_n[\![\hat{q}]\!]$ represents the functor which classifies lifts of a p-divisible group of height $n+1$ with connected component $G[p^\infty]$. Thus $\hat{q}$ may be viewed as a "Serre-Tate parameter" in the sense of [**Kat81**].

5.1. A model for the torsion subgroup of $F/(\hat{q})$. One difficulty in representing the quotient $\mathbf{F}/(\hat{q})$ over $E[\![\hat{q}]\!]$ is that the subgroup $(\hat{q})$ does not act universally freely on $\mathbf{F}$: consider, for example, the specialization $\hat{q} = 0$. It turns out that, as long as one restricts to the torsion subgroup of $\mathbf{F}/(\hat{q})$, this is the only obstruction. By freeing up the action, we are able to construct a representable functor whose torsion subgroup coincides with that of $F/(\hat{q})$ whenever $\hat{q}$ is of infinite order.

Following [**KM85**, §8.7], let $\mathrm{Tate}(F)$ be the scheme over $R = E[\![\hat{q}]\!]$ defined by the disjoint union

$$\mathrm{Tate}(F) = \bigcup_{a \in \mathbb{Q}\cap[0,1)} F_R \times \{a\}.$$

If A is a complete local R-algebra, then

$$\mathrm{Tate}(F)(A) = \{\text{pairs } (g,a) \text{ with } g \in F(A) \text{ and } a \in \mathbb{Q} \cap [0,1).\}$$

This has a group structure given by

$$(g,a)\cdot(h,b) = \begin{cases} (g +_F h, a+b) & \text{if } a+b<1 \\ (g +_F h -_F \hat{q}, a+b-1) & \text{if } a+b \geq 1. \end{cases}$$

By construction, $\mathrm{Tate}(F)(A)$ is the quotient in the exact sequence

$$\begin{array}{ccccc} 0 \to & \mathbb{Z} \to & F(A)\times\mathbb{Q} & \to \mathrm{Tate}(F)(A) \to 0 \\ & n \mapsto & ([n](\hat{q}), n) & \\ & & (x,a) & \mapsto (x -_F [\flat(a)](\hat{q}), \sharp(a)), \end{array} \tag{5.2}$$

where $\flat(a)$ and $\sharp(a)$ are the integral and fractional parts of the rational number a. Equivalently, $\mathrm{Tate}(F)$ is the pushout of $\mathbb{Z} \to \mathbb{Q} \to \mathbb{Q}/\mathbb{Z}$ along the homomorphism u:

$$\begin{array}{ccccc} \mathbb{Z} & \longrightarrow & \mathbb{Q} & \longrightarrow & \mathbb{Q}/\mathbb{Z} \\ {\scriptstyle u}\downarrow & & \downarrow & & \downarrow{\scriptstyle =} \\ F_R & \longrightarrow & \mathrm{Tate}(F) & \longrightarrow & \mathbb{Q}/\mathbb{Z}. \end{array}$$

Thus $\mathrm{Tate}(F)$ is a kind of homotopy quotient of $\mathbf{F}$ by $\mathbb{Z}$.

PROPOSITION 5.3. *Projection onto the first factor in the construction above defines a natural transformation*

$$\mathrm{Tate}(F)_{\mathrm{tors}} \to (\mathbf{F}/(\hat{q}))_{\mathrm{tors}},$$

which is an isomorphism if $\hat{q}$ is not in $F(A)_{\mathrm{tors}}$.

PROOF. Let A be a complete local R algebra, and suppose that $\hat{q}$ is not torsion in $F(A)$. We see from (5.2) that there is an isomorphism

$$\mathrm{Tate}(F)(A) \cong \mathbf{F}/(\hat{q})(A) \times \mathbb{Q}$$

which is compatible with the projection and clearly induces an isomorphism

$$\mathrm{Tate}(F)_{\mathrm{tors}}(A) \cong (\mathbf{F}/(\hat{q}))_{\mathrm{tors}}(A).$$

□

REMARK 5.4. The Proposition holds in any context in which $\hat{q}$ is not torsion. Another source of examples is p-adic fields. Let $S = E[\frac{1}{p}]((\hat{q}))$. Let L be a complete nonarchimedean field with norm $\| - \|$. Given a continuous map $S \to L$, the group law $\mathbf{F}$ defines a group structure on the set $\{v \in L | \|v\| < 1\}$; and we write $\mathbf{F}(L)$ for this group. The subgroup generated by $\hat{q}$ is necessarily free, and the argument shows that there is an isomorphism of groups

$$\mathrm{Tate}(F)_{\mathrm{tors}}(L) \cong (\mathbf{F}/(\hat{q}))_{\mathrm{tors}}(L).$$

REMARK 5.5. If the ring E is the ring of coefficients in an even periodic ring theory, then the preceding construction could be carried out with $R = E(B\mathbb{T})$ and the homomorphism u (4.3). The result is an extension

$$P_E \to \mathrm{Tate}(P_E) \to (\mathbb{T}^* \otimes \mathbb{Q})/\mathbb{T}^*$$

of group schemes over R. Proposition 5.3 becomes an isomorphism

$$\mathrm{Tate}(P_E)_{\mathrm{tors}} \cong ((P_E)_R/\mathbb{T}^*)_{\mathrm{tors}}.$$

5.2. Notation. If A is an abelian group, then $A_T = \operatorname{spec} T^A$ is the resulting constant formal group scheme over T. The category of *test rings* is the category of Artin local E-algebras with residue field k.

DEFINITION 5.6. If G is a formal group scheme over a ring T, and A is an abelian group, let $\mathrm{Hom}[A, G]$ be the group of homomorphisms

$$A_T \to G$$

of formal group schemes over T. Similarly, let $\mathrm{Ext}[A, G]$ be the set of isomorphisms classes of extensions of formal group schemes

$$G \to X \to A_T.$$

If G is a formal group scheme over E, let $\underline{\mathrm{Hom}}[A, G]$ and $\underline{\mathrm{Ext}}[A, G]$ be the functors from test rings to groups such that

$$\underline{\mathrm{Hom}}[A, G](T) \stackrel{\mathrm{def}}{=} \mathrm{Hom}[A_T, G_T]$$

and

$$\underline{\mathrm{Ext}}[A, G](T) \stackrel{\mathrm{def}}{=} \mathrm{Ext}[A_T, G_T].$$

Now if G is a formal group over E, pulling back over G defines a natural point

$$\Delta \colon G \to G \times G = G_G$$

and hence a homomorphism

$$u \colon \mathbb{Z}_G \to G_G.$$

It is clear that this gives an isomorphism $G \cong \underline{\mathrm{Hom}}[\mathbb{Z}, G]$.

Equivalently, if F is a formal group law over E, then $R = E[\![\hat{q}]\!]$ pro-represents the functor $\underline{\mathrm{Hom}}[\mathbb{Z}, F]$ on the category of test rings, with universal example $u(n) = [n]_F(\hat{q})$ (5.1). Similarly, if E is the ring of coefficients of an even periodic ring theory, then $R = E(B\mathbb{T})$ pro-represents the functor $\underline{\mathrm{Hom}}[\mathbb{T}^*, P_E]$, with the homomorphism u of (4.3) as the universal example.

5.3. Universal properties.

A universal extension. A continuous homomorphism of E-algebras from R to a test algebra T defines an extension

$$F_T \to (\mathrm{Tate}(F))_T \to (\mathbb{Q}/\mathbb{Z})_T$$

and hence an extension

$$F[p^\infty]_T \to (\mathrm{Tate}(F)[p^\infty])_T \to (\mathbb{Z}[\tfrac{1}{p}]/\mathbb{Z})_T \ .$$

of torsion subgroups.

LEMMA 5.7. *The ring R pro-represents the functor* $\underline{\mathrm{Ext}}[\mathbb{Q}_p/\mathbb{Z}_p, F]$, *with* $\mathrm{Tate}(F)$ *as the universal example.*

PROOF. In the exact sequence

$$\underline{\mathrm{Hom}}[\mathbb{Q}, F] \to \underline{\mathrm{Hom}}[\mathbb{Z}, F] \to \underline{\mathrm{Ext}}[\mathbb{Q}/\mathbb{Z}, F] \to \underline{\mathrm{Ext}}[\mathbb{Q}, F],$$

the first and last terms are zero because p acts nilpotently on F and as an isomorphism on $\mathbb{Q}$. □

A universal p-divisible group. Any p-divisible group Γ over a field k is naturally an extension

$$\Gamma^0 \to \Gamma \to \Gamma_{\mathrm{et}} \tag{5.8}$$

of a connected group by an étale group. If the residue field k is algebraically closed, then the sequence (5.8) has a canonical splitting [**Dem72**, p. 34]. We will be interested in the case when Γ_{et} has height one, which is to say that it is isomorphic to the constant group scheme $(\mathbb{Q}_p/\mathbb{Z}_p)_k$.

Tate showed [**Tat67**] that the functor $G \mapsto G[p^\infty]$ is an equivalence between the categories of formal groups of finite height and connected p-divisible groups. Let's fix a one-dimensional formal group Γ^0 of height n over the algebraically closed field k and define Γ to be the product extension

$$\Gamma^0 \to \Gamma \to (\mathbb{T}^* \otimes \mathbb{Q}_p/\mathbb{Z}_p)_k \ .$$

DEFINITION 5.9. If G is a p-divisible group over k, and if T is a test ring, then a *lift* of G to R is a pair (F, δ) consisting of a p-divisible group F over R and an isomorphism

$$F_k \underset{\cong}{\xrightarrow{\delta}} G$$

of p-divisible groups over k. An equivalence of lifts (F, δ) and (F', δ') is an isomorphism $f \colon F \to F'$ such that

$$\delta = \delta' f_k.$$

The set of isomorphism classes of lifts of G to R will be denoted $\mathrm{Lifts}_G(R)$. As R varies, $\mathrm{Lifts}_G(R)$ defines a functor from test rings to sets.

Lubin and Tate construct a formal power series algebra E_n over the Witt ring of k which pro-represents the functor $\mathrm{Lifts}_{\Gamma^0}$. There is an even periodic cohomology theory with E_n as ring of coefficients, and the universal lift F of Γ^0 as formal group.

THEOREM 5.10. *The ring* $R = E_n(B\mathbb{T})$ *pro-represents* Lifts_Γ, *with universal example* $\mathrm{Tate}(F)[p^\infty]$.

PROOF. Let T be a test ring. Suppose that (H, ϵ) is a lift of Γ. Then (H^0, ϵ^0) is a lift of Γ^0. According to [**LT66**], there is a unique pair (f, a) consisting of a map $f \colon E_n \to T$ and an isomorphism

$$a \colon (F_T, \epsilon_T^{\mathrm{univ}}) \cong (H^0, \epsilon^0)$$

of lifts of Γ^0. On the other hand, ϵ_{et} induces an isomorphism

$$H_{\mathrm{et}} = (H_{\mathrm{et}}(k))_T \cong (\Gamma_{\mathrm{et}}(k))_T = (\mathbb{T}^* \otimes \mathbb{Q}_p/\mathbb{Z}_p)_T.$$

Assembling these gives an extension

$$F_T \to H \to (\mathbb{T}^* \otimes \mathbb{Q}_p/\mathbb{Z}_p)_T$$

which defines an isomorphism

$$\mathrm{Lifts}_\Gamma \cong \underline{\mathrm{Ext}}[F, \mathbb{T}^* \otimes \mathbb{Q}_p/\mathbb{Z}_p] .$$

R pro-represents the right-hand side by Lemma 5.7. □

REMARK 5.11. The analogous result for the ordinary multiplicative group $\mathbb{G}_m$ is described in [**KM85**, §8.8]. Closely related examples occur in [**AMS98**], which are motivated by purely homotopy-theoretic questions about Mahowald's root invariant.

5.4. Examples. If $L \in A((q))$ is a unit, the formal product defining the Weierstrass function of3.3 defines an element $\sigma(L, q) \in A((q))$. The functional equation $\sigma(qL, q) = (-L)^{-1}\sigma(L, q)$ then implies that the modified product

$$\sigma[L, r] \stackrel{\mathrm{def}}{=} q^{-\flat(r)(\flat(r)+1)/2}(-L)^{\flat(r)}\sigma(L, q)$$

(where $\flat(r)$ is the integral part of $r \in \mathbb{Q}$) satisfies the identity $\sigma[qL, r+1] = \sigma[L, r]$ and can thus be regarded as a function from a localization of $\mathrm{Tate}(\mathbb{G}_m)$ to the $\mathbb{Z}((q))$-line. Since $\sigma(L, 0) = 1 - L$ we can think of the modified function as a deformation of the usual coordinate at the identity of the multiplicative group.

Similarly, if we regard $\hat{q}$ as an element of the locally compact field $\mathbb{C}$, the modified sine function

$$s[x,r] \stackrel{\text{def}}{=} \pi^{-1}\hat{q}\sin\hat{q}^{-1}(\pi x - r)$$

satisfies the identity $s[x+\hat{q}, r+1] = s[x,r]$ and so can be interpreted as a function from $\mathrm{Tate}(\mathbb{G}_a)(A)$ to A. It is also a deformation of the usual coordinate on the additive group, in that $s[x,0] \to 0$ as $\hat{q} \to \infty$.

In general, however, there seems to be no reason to expect that a coordinate on F will extend to a coordinate on $\mathrm{Tate}(F)$: our construction yields a group object, but not a group *law*. In the two examples above, we do have (something like) coordinates, which define interesting genera: ordinary cohomology leads to the $\hat{A}$-genus, suitably normalized, if $r = 0$; but if $r \neq 0 \in \mathbb{Q}/\mathbb{Z}$, Cauchy's theorem (applied to a small circle $\mathbf{C}$ around the origin) yields

$$\frac{1}{2\pi i}\int_{\mathbf{C}} p^X_{\mathbb{G}_a}\left[\prod_{j=1}^{j=d}\frac{\hat{q}^{-1}\pi x_j z}{\sin\hat{q}^{-1}\pi(x_j z + r\hat{q})}\right] z^{-d-1}\,dz = (\frac{\pi}{\hat{q}\sin\pi r})^d\chi(X)\,.$$

The K-theory genus extends similarly, but the resulting function differs from the standard elliptic genus by multiplication by $(-1)^d q^{-d\flat(r)(\flat(r)+1)/2}$.

Any multiplicative cohomology theory E can be described as taking values in a category of sheaves over $\mathrm{spec}\,E(*)$, and the Borel extension of such a theory takes values in sheaves over $\mathrm{spec}(E(*)[\![\hat{q}]\!])$. The construction in 5.1 of the Tate group as a disjoint union of copies of such affines implies that a theory E with formal group F has a natural extension to an equivariant theory taking values in a category of sheaves over the group object $\mathrm{Tate}(F)$. Similarly, a suitable localization of the Borel extension defines an equivariant theory taking values in sheaves over a suitable localization of $\mathrm{Tate}(F)$. This resembles (but is easier than) the constructions of [**GKV, Gro94, RK99**], for here we're only patching together Borel extensions.]

6. Prospectra and equivariant Thom complexes

Cohen, Jones, and Segal show that pro-objects in the category of spectra are the appropriate context in which to study Thom complexes, and so umkehr maps, for semi-infinite vector bundles. In this section we observe that the ideas of this paper fit very naturally into their framework.

6.1. Thom prospectra. If V and W are complex vector bundles over a space X, we extend the notation for Thom isomorphisms in 2.4 by writing

$$\alpha^V_W \stackrel{\text{def}}{=} \alpha^V \circ (\alpha^W)^{-1} \colon E(X^W) \to E(X^V)\,.$$

'Desuspending' by $V \oplus W$ then gives a homomorphism

$$\alpha^{-W}_{-V} : E(X^{-V}) \to E(X^{-W})\,.$$

The inclusions of a filtered vector bundle

$$\mathcal{V} : 0 = V_0 \subset V_1 \subset \cdots$$

define maps

$$i_{V_n} : X^{V_{n-1}} \to X^{V_n}$$

of Thom spectra, which desuspend to define a pro-object

$$X^{-\mathcal{V}} \stackrel{\text{def}}{=} \{\cdots \to X^{-V_n} \xrightarrow{i^{V_n}} X^{-V_{n-1}} \to \dots\}$$

in the category of spectra. The E-cohomology of $X^{-\mathcal{V}}$ is the colimit

$$E(X^{-\mathcal{V}}) \stackrel{\text{def}}{=} \operatorname*{colim}_n E(X^{-V_n}),$$

as in the appendix to [**CJS95**].

LEMMA 6.1. *In E-cohomology, the homomorphism induced by i^{V_n} is*

$$E(i^{V_n}) = e(V_n/V_{n-1})\alpha^{-V_n}_{-V_{n-1}}.$$

□

EXAMPLE 6.2. For any integer n, let

$$I_n = \bigoplus_{n \geq |k| > 0} \mathbb{C}(k);$$

thus $\mathcal{I} = \operatorname{colim}_n I_n$ is a filtered $\mathbb{T}$-vector bundle over a point. More generally, if V is a complex vector bundle over a space X, let

$$\mathcal{I}V \stackrel{\text{def}}{=} V \otimes \mathcal{I}$$

be the corresponding filtered $\mathbb{T}$-vector bundle. In this notation the Laurent approximation (3.1) to the formal normal bundle of the constant loops in $\mathcal{L}X$ is $\mathcal{I}TX$.

We write j_n for the map

$$j_n = i^{I_n V} \colon X^{-I_n V} \to X^{-I_{n-1} V}.$$

EXAMPLE 6.3. If V_n is the sum of n copies of $\mathbb{C}(1)$, considered as a $\mathbb{T}$-vector bundle over a point, then the Borel construction on $\mathcal{V}$ is the Thom prospectrum for $\mathbb{C}P^\infty_\infty$ constructed in [**CJS95**].

6.2. Equivariant cohomology. Suppose now that $\hat{E}_{\mathbb{T}}$ is a suitable extension of an equivariantly oriented theory $E_{\mathbb{T}}$, and let X be a finite complex with trivial $\mathbb{T}$-action, as above. As n varies, the Thom isomorphisms

$$\alpha^{I_n V} \colon E_{\mathbb{T}}(X) \xrightarrow[\cong]{E_{\mathbb{T}}} (X^{-I_n V})$$

are not compatible with the maps j_n; but this can be cured over $\hat{E}_{\mathbb{T}}$ by a suitable renormalization. By (2.12) the class

$$u_n(V) \stackrel{\text{def}}{=} e(V \otimes I_n) = \prod_{0<|k|\leq n} e_{\mathbb{T}}(V \otimes \mathbb{C}(k)).$$

is a unit of $\hat{E}_{\mathbb{T}}(X)$, and the homomorphism

$$\omega_n(V) \colon \hat{E}_{\mathbb{T}}(X) \to \hat{E}_{\mathbb{T}}(X^{-I_n V})$$

defined by the formula $\omega_n(V) = u_n(V)\alpha^{-I_n V}$ is an isomorphism.

THEOREM 6.4. *The diagram*

$$\begin{array}{ccc} \hat{E}_{\mathbb{T}}(X) & = & \hat{E}_{\mathbb{T}}(X) \\ \downarrow{\scriptstyle \omega_{n-1}(V)} & & \downarrow{\scriptstyle \omega_n(V)} \\ \hat{E}_{\mathbb{T}}(X^{-I_{n-1}V}) & \xrightarrow{j_n} & \hat{E}_{\mathbb{T}}(X^{-I_n V}) \end{array}$$

commutes; in particular, the maps $\omega_n(V)$ assemble into a "Thom isomorphism"

$$\omega^V : \hat{E}_{\mathbb{T}}(X) \xrightarrow{\cong} \hat{E}_{\mathbb{T}}(X^{-\mathcal{I}V}) .$$

□

If V and W are two complex vector bundles over X, then (as with the usual Thom isomorphism) we define

$$\omega_W^V : \hat{E}_{\mathbb{T}}(X^{-\mathcal{I}W}) \xrightarrow{\cong} \hat{E}_{\mathbb{T}}(X^{-\mathcal{I}V}).$$

COROLLARY 6.5. *If the vector bundle V has rank d, then the relative isomorphism*

$$\omega_d^V : \hat{E}_{\mathbb{T}}(X^{-\mathcal{I}\underline{d}}) \to \hat{E}_{\mathbb{T}}(X^{-\mathcal{I}V})$$

is given by the formula

$$\omega_d^V = \left(\prod_{j=1}^{d} \prod_{k \neq 0} \frac{x_j +_F [k](\hat{q})}{[k](\hat{q})} \right) \alpha_{-\mathcal{I}\underline{d}}^{-\mathcal{I}V},$$

where the x_j are the terms in the formal factorization (2.5) *of the total F-Chern class of V.* □

The terms in this product are well-defined at any finite stage, if not in the limit. If $E_{\mathbb{T}}$ is $\hat{H}$ or $\hat{K}$ (and $c_1(V) = 0$), then the infinite products converge and we have the

COROLLARY 6.6. *If the vector bundle V has rank d, then the diagram*

$$\begin{array}{ccc} \hat{E}_{\mathbb{T}}(X^{\underline{d}}) & \xrightarrow{\alpha_{\underline{d}}^{-\mathcal{I}\underline{d}}} & \hat{E}_{\mathbb{T}}(X^{-\mathcal{I}\underline{d}}) \\ {\scriptstyle \hat{\alpha}_{\underline{d}}^V} \downarrow & & \downarrow {\scriptstyle \omega_{\underline{d}}^V} \\ \hat{E}_{\mathbb{T}}(X^V) & \xrightarrow{\alpha_V^{-\mathcal{I}V}} & \hat{E}_{\mathbb{T}}(X^{-\mathcal{I}V}) . \end{array}$$

commutes; where

$$\hat{\alpha}_{\underline{d}}^V \stackrel{def}{=} \frac{\Theta_F(V)}{c_d(V)} \alpha_{\underline{d}}^V .$$

More precisely, the assertion is that these examples are naturally topologized, and that at any finite stage the diagram commutes modulo error terms which converge in this topology to zero as n grows. □

In other words, in a suitable theory with group law F the isomorphism $\omega_{\underline{d}}^V$ is very much like a Thom isomorphism $\hat{\alpha}_{\underline{d}}^V$ for a theory with F replaced by its extension Tate(F). From this point of view, the fact that Witten's formula for the α-genus of $\mathcal{L}X$ equals the $\hat{\alpha}$-genus of X can be interpreted as saying that the inclusion $j : X \to \mathcal{L}X$ behaves as if there is a cohomological analogue

$$j_* : \hat{E}_{\mathbb{T}}(X^{-\mathcal{I}TX}) \to \hat{E}_{\mathbb{T}}(\mathcal{L}X)$$

of the Thom collapse map, with an associated umkehr $j_F = j_* \omega^{TX}$ satisfying the projection formula

$$j_F j^*(x) = x \cdot e_{\mathbb{T}}(\nu) .$$

Perhaps the intuition underlying the physicists' interest in elliptic cohomology is that a reasonable equivariant theory applied to the geometrical object X^{-TTX} captures more information than that theory does, when applied directly to X. Thus the equivariant K-theory of this formal neighborhood of X is the Tate elliptic cohomology of X. This seems to be related to the recent construction (by Kontsevich and others) of new invariants for singular complex algebraic varieties, by considering them as varieties over the Laurent series field $\mathbb{C}((q))$.

6.3. Polarizations. In the preceding account, the role of the rational parameter in the construction of Tate(F) is geometrically unmotivated, because we have ignored some issues connected with the polarization [**CJS95**] of the loopspace.

Such a structure is an equivalence class of splittings of the tangent bundle of $\mathcal{L}X$ into a sum of positive- and negative-frequency components: if X is complex-oriented (*e.g.* symplectic, with a choice of compatible almost-complex structure), then the composition

$$\mathcal{L}X \to \mathcal{L}(BU) = U \times BU \to U/SO$$

of the map induced by the classifying map for the tangent bundle of X with the projection to a classifying space for such splittings defines a canonical polarization.

The universal cover of the free loopspace. If X is simply-connected, the fundamental group of its free loopspace will be isomorphic to $H_2(X, \mathbb{Z})$, which will be nontrivial in general. There is thus good reason to consider the simply-connected cover $\widetilde{\mathcal{L}X}$ of the loopspace: this can be defined as the space of smooth maps of a two-disk to X, modulo the relation which identifies two maps if their restrictions to the boundary circle agree, and if furthermore their difference, regarded as an element of the deck-transformation group $\pi_2(X) = H_2(X, \mathbb{Z})$, is null-homotopic. The circle acts on $\widetilde{\mathcal{L}X}$ by rotating loops, as does the fundamental group of $\mathcal{L}X$, and in general the fixed-point set

$$\widetilde{\mathcal{L}X}^{\mathbb{T}} \cong H_2(X, \mathbb{Z}) \times X$$

will have many components. The choice of a basepoint defines a lift of the canonical polarization to a map

$$\widetilde{\mathcal{L}X} \to \mathbb{R} \times SU/SO$$

which restricts to a locally constant map

$$H_2(X, \mathbb{Z}) \times X \to \mathbb{R} \times SU/SO \to \mathbb{R}$$

defined by evaluating $\alpha \in H_2(X, \mathbb{Z})$ on the first Chern class of X. We think of the polarization as defining the zero-frequency modes in the Fourier decomposition of small loops near a fixed-point component, so that shifting by $\beta \in H_2(X, \mathbb{Z})$ gives an isomorphism

$$T\widetilde{\mathcal{L}X}^{\mathbb{T}}_{x,\alpha} \cong T\widetilde{\mathcal{L}X}^{\mathbb{T}}_{x,\alpha+\beta} \otimes \mathbb{C}(\langle c_1(V), \beta\rangle) \,.$$

$\mathbb{T}$-equivariant Picard groups and orientations. We can regard the cohomology of a space Y as a sheaf of rings over the zero-dimensional scheme

$$(\pi_0 Y)_{\mathbb{Z}} \stackrel{\text{def}}{=} \operatorname{spec} H^0(Y, \mathbb{Z}) \,.$$

Similarly, the set of equivalence classes of $\mathbb{T}$-line bundles over a $\mathbb{T}$-space X is naturally isomorphic to $H^2(Y_{\mathbb{T}}, \mathbb{Z})$, which can be interpreted as a sheaf of groups over $(\pi_0 Y)_{\mathbb{Z}}$, the fiber above component Y_i being the constant group scheme

$$\operatorname{coker}\,[H^2_{\mathbb{T}}(*, \mathbb{Z}) \to H^2_{\mathbb{T}}(Y_i)]\ ;$$

when the circle action on X is trivial, this is just a complicated way of indexing the summands of $H^2(Y, \mathbb{Z})$.

An orientation on a complex-oriented cohomology theory E with formal group F defines a natural homomorphism from the Picard group of line bundles over Y to the group of $E(Y)$-valued points of F. This suggests that a complex orientation for the Tate extension of an equivariantly complex-oriented theory should be defined as a natural transformation from the Picard group of $\mathbb{T}$-line bundles over X to $\mathrm{Tate}(F)(E(Y))$, both regarded as schemes over $(\pi_0 Y)_{\mathbb{Z}}$; by restriction such a transformation defines a map

$$\pi_0 Y \to \mathbb{Q}/\mathbb{Z}\ .$$

In other words, such a generalized orientation assigns to a component of Y and a $\mathbb{T}$-line bundle over it, a characteristic class in the cohomology of the component, together with an $r \in \mathbb{Q}$ depending on the component, which shifts by an integer when the bundle is twisted by a character.

When Y is $\widetilde{\mathcal{L}X}^{\mathbb{T}}$, the polarization defines a very natural map of this sort, which sends α to $\langle c_1(V), \alpha\rangle$. In more general situations, *e.g.* in Givental's work [**Vo99**] on the quantum cohomology of a symplectic manifold (X, ω) the evaluation map $\alpha \mapsto \langle \omega, \alpha \rangle$ plays a similar role; but in this generalization, ω no longer needs to be an integral class.

7. Concluding remarks

The parts of this paper which deal principally with the fixed point formula on the free loop space are formulated in terms of a general equivariant cohomology theory $E_{\mathbb{T}}$, but those which relate to the quotient $\mathbf{F}/(\hat{q})$ use only the group law $\mathbf{F}$ of the theory $R(X) = E(B\mathbb{T} \times X)$, and so essentially use the Borel theory E_{Borel}. We do this because in that case we can be more specific about our constructions. In good cases one can hope to do better.

Specifically, suppose that the multiplication homomorphism $\mathbb{T} \times \mathbb{T} \to \mathbb{T}$ induces a map

$$E_{\mathbb{T}} \to E_{\mathbb{T} \times \mathbb{T}}$$

and that one has an isomorphism $E_{\mathbb{T}\times\mathbb{T}}(*) \cong E_{\mathbb{T}}(*) \otimes_E E_{\mathbb{T}}(*)$. The upshot is a group scheme $G = \operatorname{spec} E_{\mathbb{T}}(*)$ over E, such that the formal group P_E associated to $E(P)$ is the completion of G. Indeed [**GKV, Gro94**], one expects in general that there is an abelian group scheme G which is the more fundamental object in $\mathbb{T}$-equivariant E-theory, with $E_{\mathbb{T}}(*)$ as structure sheaf.

In any case, as remarked in §5.2, over $G_G = G \times G$ there is a natural map

$$\mathbb{Z}_G \to G_G, \tag{7.1}$$

and the natural map $G_G \to \mathrm{Hom}[\mathbb{Z}, G]$ is an isomorphism. One could consider the group Tate(G) over G, fitting into a short exact sequence

$$G_G \to \mathrm{Tate}(G) \to \mathbb{Q}/Z.$$

The group Tate(P_E) considered in the main text would then arise from Tate(G) by completing in both copies of G.

This is the situation in K-theory. Over the ring $K_{\mathbb{T}} = \mathbb{Z}[q, q^{-1}]$ one has a homomorphism

$$\begin{aligned} u \colon \mathbb{Z} &\to \mathbb{G}_{m\,K_{\mathbb{T}}} \\ n &\mapsto q^n \end{aligned}$$

and this gives rise to a group Tate$(\mathbb{G}_m)$ over $K_{\mathbb{T}}$. Over the completion $K_{\mathbb{T}} \to K(B\mathbb{T}) = \mathbb{Z}[\![\hat{q}]\!]$ one recovers the group Tate$(\mathbb{G}_m)$ considered in the main text. On the other hand, the group Tate$(\mathbb{G}_m)_{\mathrm{tors}}$ becomes isomorphic to the torsion subgroup of the classical Tate curve Tate(q) already over the suitable localization $\hat{K} = \mathbb{Z}((q))$ of (2.15), where u is the inclusion of a sub-groupscheme. See [**KM85**, §8.8] for details.

Interestingly enough, in the case of K-theory this *is* the solution to the problem of the infinite products. We have purposefully written the equivariant Euler class of $L \otimes \mathbb{C}(k)$ as $e(L) +_F [k]_F(\hat{q})$, as in any finite situation the formula

$$1 - L_1L_2 = (1 - L_1) + (1 - L_2) - (1 - L_1)(1 - L_2)$$

makes it possible to calculate the Euler class of a vector bundle using the formal multiplicative group, i.e. the right-hand side. However, in order to calculate the infinite product $\Theta_{\mathbb{G}_m}(x, \hat{q})$ (3.5), one is *forced* to use the left-hand side. Thus from this point of view the renormalization is handled by knowing about the global multiplicative group, instead of merely its formal completion.

It seems reasonable to hope that the rich theory of $\mathbb{T}$-equivariant ring spectra will provide additional examples of such $E_{\mathbb{T}}$. If $E_{\mathbb{T}}$ is a complex-oriented $\mathbb{T}$-equivariant ring spectrum, then the cohomology $E_{\mathbb{T}}(\mathbb{CP})$ (where $\mathbb{CP}$ is the space of lines in the ambient $\mathbb{T}$-universe) carries the structure of a $\mathbb{T}$-equivariant formal group law (in the sense of [**CGK**]). This is, among other things, a formal group law F and a homomorphism

$$v \colon \mathbb{T}^* \to F(E_{\mathbb{T}}(\mathbb{CP})).$$

In the situation we are considering, F is the completion of the group $G = \mathrm{spec}\, E_{\mathbb{T}}$, and the homomorphism v is obtained from the homomorphism u by completion.

References

[Ada88] Douglas Adams. *Dirk Gently's holistic detective agency.* Pocket Books, 1988.

[Ada74] J. Frank Adams. *Stable homotopy and generalised homology.* Univ. of Chicago Press, 1974.

[And95] Matthew Ando. Isogenies of formal group laws and power operations in the cohomology theories E_n. *Duke Math.* J. 79:423–485, 1995

[AHS98] Matthew Ando, Michael J. Hopkins, and Neil P. Strickland. Elliptic spectra, the Witten genus, and the theorem of the cube, 1998. Preprint available at `hopf.math.purdue.edu`.

[AMS98] Matthew Ando, Jack Morava, Hal Sadofsky. Completions of Tate $\mathbb{Z}/(p)$-cohomology of periodic spectra. *Geometry and Topology* 2:145–174, 1998.

[Ati85] Michael F. Atiyah. Circular symmetry and stationary phase approximation. *Asterisque*, 132:43–60, 1985.

[AS68] M. F. Atiyah and G. B. Segal. The index of elliptic operators. II. *Annals of Mathematics*, 87:531–545, 1968.

[CJS95] R. L. Cohen, J. D. S. Jones, and G. B. Segal. Floer's infinite-dimensional Morse theory and homotopy theory. In *The Floer memorial volume*, pages 297–325. Birkhäuser, Basel, 1995.

[CGK] Michael Cole, Igor Kriz, John Greenlees. Equivariant formal groups. Preprint available at `hopf.math.purdue.edu`

[Del89] Pierre Deligne. Le groupe fondamental de la droite projective moins trois points. In *Galois groups over* **Q**. ed. Y. Ihara et al (Berkeley, CA, 1987), 79–297, Springer, New York, 1989.

[Dem72] Michel Demazure. *Lectures on p-divisible groups*, Lecture Notes in Mathematics 302. Springer, 1972.

[Den92] Christopher Deninger. Local L-factors of motives and regularized determinants. *Inventiones Math¿* 107:135-150, 1992.

[DHS88] Ethan S. Devinatz, Michael J. Hopkins, and Jeffrey H. Smith. Nilpotence and stable homotopy theory I. *Annals of Mathematics*, 128:207–241, 1988.

[Dye69] Eldon Dyer. *Cohomology theories*. W. A. Benjamin, Inc., New York-Amsterdam, 1969. Mathematics Lecture Note Series.

[Eis44] Gotthold Eisenstein. Bemerkungen zu den elliptischen und Abelschen Transcendenten. *Crelle's Journal*, 27:185–191, 1844.

[GKV] Viktor Ginzburg, Mikhail Kapranov, Eric Vasserot. Elliptic algebras and equivariant elliptic cohomology Preprint available at `xxx.lanl.gov`

[Gro94] Ian Grojnowski. Delocalized equivariant elliptic cohomology. Unpublished manuscript, 1994.

[HS98] Michael J. Hopkins and Jeffrey H. Smith. Nilpotence and stable homotopy theory II. *Annals of Mathematics*, 148:1–49, 1998

[Kat81] Nicholas M. Katz. Serre-Tate local moduli. In *Surfaces algébriques: séminaire de géométrie algébrique d'Orsay 1976–1978*, Lecture Notes in Mathematics 868 Springer, 1981.

[KM85] Nicholas M. Katz and Barry Mazur. *Arithmetic moduli of elliptic curves*, Annals of Mathematics Studies 108. Princeton University Press, 1985.

[LT66] Jonathan Lubin and John Tate. Formal moduli for one-parameter formal lie groups. *Bull. Soc. math. France*, 94:49–60, 1966.

[Lub67] Jonathan Lubin. Finite subgroups and isogenies of one-parameter formal Lie groups. *Annals of Math.*, 85:296–302, 1967.

[Mor85] Jack Morava. Noetherian localizations of categories of cobordism comodules. *Annals of Mathematics*, 121:1–39, 1985.

[MT91] B. Mazur and J. Tate. The p-adic sigma function. *Duke Math. J.*, 62(3):663–688, 1991.

[Qui71] Daniel M. Quillen. Elementary proofs of some results of cobordism theory using Steenrod operations. *Advances in Mathematics*, 7:29–56, 1971.

[Ros99] Ioanid Rosu. Equivariant elliptic cohomology and rigidity. To appear in *American J. Math.*

[RK99] Ioanid Rosu and Allen Knutson. Equivariant K-theory and equivariant cohomology. Preprint available at `hopf.math.purdue.edu`

[Seg68] Graeme Segal. Equivariant K-theory. *Inst. Hautes Études Sci. Publ. Math. No.*, 34:129–151, 1968.

[Seg88] Graeme Segal. Elliptic cohomology. *Seminaire Bourbaki*, (695), 1988.

[Sil94] Joseph Silverman. *Advanced topics in the arithmetic of elliptic curves*, Graduate Texts in Mathematics 151. Springer, 1994.

[Str99] Neil P. Strickland. Formal schemes and formal groups. In J. P. Meyer, J. Morava, and W. S. Wilson, editors, *Homotopy-invariant algebraic structures: in honor of J. M. Boardman*, Contemporary mathematics 239 American Mathematical Society, 1999.

[Tat67] John Tate. p-divisible groups. In T.A. Springer, editor, *Proceedings of a Conference on Local Fields (Driebergen)*. Springer, 1967.

[Vo99] Claire Voisin *Mirror symmetry*. Translated from the 1996 French original by Roger Cooke, American Mathematical Society, Providence, RI, 1999.

[Wit88] Edward Witten. The index of the Dirac operator in loop space. In Peter S. Landweber, editor, *Elliptic curves and modular forms in algebraic topology*, Lecture Notes in Mathematics 1326 Springer, 1988

THE UNIVERSITY OF ILLINOIS AT URBANA-CHAMPAIGN

E-mail address: mando@math.uiuc.edu

THE JOHNS HOPKINS UNIVERSITY

E-mail address: jack@math.jhu.edu

Contemporary Mathematics
Volume **279**, 2001

The 1-line of the K-theory Bousfield-Kan spectral sequence for $Spin(2n+1)$

Martin Bendersky and Donald M. Davis

ABSTRACT. For X a simply-connected finite H-space, there is a Bousfield-Kan spectral sequence which converges to the homotopy groups of its K-completion. When $X = Spin(2n+1)$, we expect that these homotopy groups equal the v_1-periodic homotopy groups of X in dimension greater than n^2. In this paper, we accomplish two things. (1) We prove that for any X, the 1-line of this spectral sequence is determined in an explicit way from the K-theory and its Adams operations. (2) For $X = Spin(2n+1)$, we make an explicit calculation of this 1-line.

1. Statement of results

The p-primary v_1-periodic homotopy groups, $v_1^{-1}\pi_*(X;p)$, of a topological space X, as defined in [**13**], are a localization of the portion of the actual homotopy groups of X detected by K-theory. In [**10, 4, 6**], $v_1^{-1}\pi_*(X;p)$ was calculated for classical groups X and primes p in all cases except $(SO(n),2)$. In this paper we make a first step toward the calculation of $v_1^{-1}\pi_*(SO(n);2)$, which is of course isomorphic to $v_1^{-1}\pi_*(Spin(n);2)$.

In [**7**], a Bousfield-Kan-type spectral sequence $E_r^{s,t}(X)$ based on periodic K-theory was introduced. It converges to the homotopy groups of the K-completion $X_K^\wedge$ of the space X. In this paper, we will deal exclusively with the localization of this spectral sequence at the prime 2. We expect to prove in a subsequent paper that if $X = Spin(n)$ then $\pi_i(X_K^\wedge) \approx v_1^{-1}\pi_i(X;2)$ for sufficiently large values of i. Thus, since v_1-periodic homotopy groups are periodic, a computation of this spectral sequence for $X = Spin(n)$ and $t-s$ large would yield a complete computation of $v_1^{-1}\pi_*(Spin(n))$.

If X is an H-space such that $K_*(X)$ is a free commutative algebra on odd-dimensional classes (e.g. $X = Spin(n)$), then ([**7**, 4.9]) $E_2(X)$ is the homology of an unstable cobar complex determined by these classes and their K_*K-coaction. The main result of this paper is an explicit computation of the 1-line $E_2^{1,t}(Spin(2n+1))$ of this spectral sequence when $p=2$ and $X = Spin(2n+1)$, .

The first step toward this computation is the following general result, which was inspired by the results of Bousfield ([**9**]) at the odd primes. This result will

2000 *Mathematics Subject Classification.* 55T15,55Q52.
Key words and phrases. homotopy groups, Adams operations, Spinor groups.

be proved in Section 2. In it, $(-)^{\#}$ denotes the Pontryagin dual, and $PK^1(X)$ the primitives in $K^1(X)$.

THEOREM 1.1. *If X is a simply-connected finite H-space with K_*X an exterior algebra on odd-dimensional classes of degree $\le 2M+1$, then $E_2^{1,2m}(X)=0$ for all m, and, if $m>M$, then*

$$E_2^{1,2m+1}(X)\approx(PK^1(X)/\operatorname{im}(\psi^r-r^m:r=2,3,-1))^{\#}.$$

For simplicity of notation, we define

$$\widetilde{v}^{2m}(X):=PK^1(X)/\operatorname{im}(\psi^r-r^m:r=2,3,-1)$$

and the closely-related functor

$$v^{2m}(X):=PK^1(X)/\operatorname{im}(\psi^2,\psi^3-3^m,\psi^{-1}-(-1)^m).$$

The advantage of $\widetilde{v}^*(-)$ is that it is more closely related to $E_2^{1,*+1}(-)$, while the advantage of $v^*(-)$ is that it is periodic and has simpler formulas. They are related by the following result, which will be proved at the end of this section.

PROPOSITION 1.2. *If X is a simply-connected finite H-space with K_*X an exterior algebra on classes of degrees $2d_1+1,\dots,2d_r+1$, then $v^{2m}(X)\approx\widetilde{v}^{2m}(X)$ if $m>\sum d_i$.*

The second purpose of this note is to compute $v^{2m}(Spin(2n+1))$, which, by Theorem 1.1 and Proposition 1.2, is isomorphic to $E_2^{1,2m+1}(Spin(2n+1))$ if $m>n^2$. Here we use the rational equivalence of $Spin(2n+1)$ with $\prod_{i=1}^n S^{4i-1}$, and that the sum of the odd integers up to $2n-1$ is n^2.

We will show in Proposition 3.1 that there is an injection of Adams-modules

$$PK^1(Sp(n))\to PK^1(Spin(2n+1)),$$

and hence a morphism

$$v^{2m}(Sp(n))\to v^{2m}(Spin(2n+1)),\tag{1.3}$$

which is dual to a morphism of $E_2^{1,2m+1}(-)$. The following result was proved in [**5**]. Here and throughout, $\nu(-)$ denotes the exponent of 2 in a number.

THEOREM 1.4. *([**5**]) If m is odd and $m\ge 2n$, then*

$$E_2^{1,2m+1}(Sp(n))\approx \mathbf{Z}/2^{eSp(m,n)},$$

where

$$eSp(m,n):=\min(\nu(S_{m,j}):j>2n),$$

with $S_{m,j}=\sum(-1)^k\binom{j}{k}k^m$. If m is even, then $E_2^{1,2m+1}(Sp(n))=\mathbf{Z}/2$.

Actually, the result in [**5**] was proved for the BP-based unstable Novikov spectral sequence, but one might use the change of rings theorem in [**7**], or mimic the calculation in [**5**] to yield the result for the K-based spectral sequence.

Our second main result is as follows. In light of 1.1, it gives $E_2^{1,2m+1}(Spin(2n+1))$ when $m>n^2$.

THEOREM 1.5. *If $n=3$ or $n\ge 5$, and m is odd, then*

$$v^{2m}(Spin(2n+1))\approx \mathbf{Z}/2^{\min(eSp(m,n),\nu(R_1(m,n)),\nu(R_2(m,n)))}\oplus \mathbf{Z}/2^{\min(2+\nu(m+1),n)},$$

where

$$R_1(m,n) = \sum_{\text{odd } k\geq 1} k^m \left(\sum_{i=0}^{n-k} \binom{2n+1}{i} - 2\sum_{t\geq 0} \binom{2n+2}{n-1-k-4t} \right)$$

and $R_2(m,n) = T/2^{\min(2+\nu(m+1),n)}$, *where*

$$T = (2^{2n+1}-3^{m+1}+1) \sum_{\text{odd } k\geq 1} k^m \sum_{t\geq 0} \binom{2n+2}{n-1-k-4t} - 3\cdot 2^{2n} \sum_{\text{odd } k\geq 1} k^m \sum_{t\geq 0} \binom{2n+1}{n-1-k-3t}.$$

If m *is even, then* $v^{2m}(Spin(2n+1)) \approx \mathbf{Z}/2 \oplus \mathbf{Z}/2$.

The sums over odd k could include even values as well, as long as $m > n^2$, since this is greater than the largest possible exponent of $v^*(Spin(2n+1))$. We will discuss the slight anomaly when $n = 4$ in Proposition 4.21. Since $Spin(5) \approx Sp(2)$, the case $n = 2$ has been covered in [**6**] (and is quite different).

The morphism (1.3) sends a generator to $g_1 + ag_2$, for some integer a. We could certainly determine a, but shall not bother to do so here.

We illustrate with concrete calculations when $n = 5$ and 6. These results are obtained by computer calculation, although we shall show after (4.20) how to obtain some of them by hand.

PROPOSITION 1.6. *For* $n = 5$ *and 6,* $v^{2m}(Spin(2n+1)) \approx \mathbf{Z}^{e_1(2n+1)} \oplus \mathbf{Z}^{e_2(2n+1)}$, *where* $e_1(2n+1)$ *and* $e_2(2n+1)$ *are as in the following table, which also presents* $eSp(m,n)$ *for comparison.*

n	m	$eSp(m,n)$	$e_1(2n+1)$	$e_2(2n+1)$
5	3 mod 8	8	5	4
5	7 mod 8	$\min(11, \nu(m-7)+6)$	$\min(11, \nu(m-103)+2)$	5
5	1 mod 4	$\min(14, \nu(m-73)+6)$	$\min(12, \nu(m-73)+4)$	3
6	15 mod 16	11	6	6
6	7 mod 16	11	$\min(9, \nu(m-23)+4)$	5
6	3 mod 8	$\min(15, \nu(m-75)+9)$	$\min(15, \nu(m-523)+5)$	4
6	1 mod 4	$\min(14, \nu(m-9)+8)$	$\min(13, \nu(m-201)+5)$	3

PROOF OF PROPOSITION 1.2. There is a rational equivalence $X \simeq \prod S^{2d_i+1}$. Thus there are elements $x_i \in PK^1(X)$ for which $\psi^k(x_i) = k^{d_i}x_i$ for all k. Similarly to the procedure in [**11**], we can find a basis $\{w_1, \ldots, w_r\}$ of $PK^1(X)_{(2)}$ such that each w_i is a rational linear combination of $\{x_j : j \geq i\}$, and hence the matrix of each ψ^k is triangular with $\{k^{d_1}, \ldots, k^{d_r}\}$ along the diagonal. Thus the Adams module $PK^1(X)$ can be built from $PK^1(S^{2d_1+1}), \ldots, PK^1(S^{2d_r+1})$ by short exact sequences, and by the Snake Lemma this implies the same of $v^{2m}(X)$ (built from $v^{2m}(S^{2d_i+1})$). Since $2^{d_i}v^{2m}(S^{2d_i+1}) = 0$ by [**7**], we deduce $2^{\sum d_i}v^{2m}(X) = 0$ (a very conservative estimate). If $m \geq \sum d_i$, then $\cdot 2^m$ is 0 in $v^{2m}(X)$. Since the only difference between $v^{2m}(X)$ and $\widetilde{v}^{2m}(X)$ is multiples of 2^m, we conclude that they are equal. ■

2. Proof of Theorem 1.1

In this section, we prove Theorem 1.1. We begin by observing that the exact sequence in E_2 induced by the coefficient sequence

$$0 \to \mathbf{Z} \to \mathbf{Q} \to \mathbf{Q}/\mathbf{Z} \to 0$$

induces isomorphisms $E_2^{s,t}(X;\mathbf{Q}/\mathbf{Z}) \to E_2^{s+1,t}(X)$ except when $s = 0$, $t = 2d_i + 1$. Hence there is an isomorphism when $s = 0$, $t > 2d_r + 1$. Everything is localized at 2. We have

$$E_2^{0,2m+1}(X;\mathbf{Q}/\mathbf{Z}) \approx \ker\big(\bar{d} : K_{2m+1}(X) \otimes \mathbf{Q}/\mathbf{Z} \to (K_*K \otimes K_*X \otimes \mathbf{Q})/U(X)\big), \tag{2.1}$$

where $\bar{d}$ is induced from

$$d : (K_*(X) \otimes \mathbf{Q}, K_*(X)) \to (K_*K \otimes K_*X \otimes \mathbf{Q}, U(X)),$$

the boundary in the unstable cobar complex described in [**7**, §4]. The following result, whose proof is deferred until the end of this section, is crucial. Here $QK_*(-)$ denotes the indecomposables.

PROPOSITION 2.2. *Suppose $QK_*(X)$ has basis $\{y_1, \dots, y_r\}$ as a K_*-module with $y_j \in K_{2d_j+1}(X)$, and $x \in K_*(X) \otimes \mathbf{Q}$ has $d(x) = \sum s_j \otimes y_j$, with $s_j \in K_*K \otimes \mathbf{Q}$. Then $x \in \ker(\bar{d})$ (of (2.1)) if and only if for all positive integers k and all subscripts j*

$$k^{d_j} \langle \psi^k, s_j \rangle \in K_{*(2)}. \tag{2.3}$$

The condition (2.3) says that each $k^{d_j}\langle \psi^k, s_j \rangle$ is a 2-local integer times the generator of $K_{|s_j|}$.

Now we prove Theorem 1.1. Suppose $\{y_1, \dots, y_r\}$ is as in Proposition 2.2 and

$$\psi : K_*X \to K_*K \otimes K_*X$$

satisfies

$$\psi(y_j) = \sum_i s_{i,j} \otimes y_i. \tag{2.4}$$

Let $u \in K_2$ denote the Bott class, so that $K_* = \mathbf{Z}_{(2)}[u]$, and let $v = \eta_R(u)$. Then

$$d(u^e y_j) = (v^e - u^e) \otimes y_j - \sum_{i \neq j} u^e s_{i,j} \otimes y_i.$$

Since, by [**8**, p.676], $\langle \psi^k, v \rangle = ku$ and $\langle \psi^k, u \rangle = u$, the expression whose coefficients correspond to $\langle \psi^k, s_j \rangle$ in (2.3) equals

$$k^e u^e y_j - \sum_{\text{all } i} \langle \psi^k, s_{i,j} \rangle u^e y_i.$$

A basis for $QK_{2m+1}(X)$ is given by $\{u^{m-d_1}y_1, \dots, u^{m-d_r}y_r\}$, and, by Proposition 2.2, $E_2^{0,2m+1}(X;\mathbf{Q}/\mathbf{Z})$ is the intersection, over all integers $k > 1$, of the kernel of the morphism

$$\delta_k : K_{2m+1}(X) \otimes \mathbf{Q}/\mathbf{Z} \to K_{2m+1}(X) \otimes \mathbf{Q}/\mathbf{Z}$$

defined by

$$\delta_k(u^{m-d_j} y_j) = -k^{m-d_j} u^{m-d_j} k^{d_j} y_j + u^{m-d_j} \sum_i \langle \psi^k, s_{i,j} \rangle k^{d_i} y_i.$$

Here we have negated the differential d for later convenience. Define $\alpha_{i,j,k} \in \mathbf{Z}$ by $\langle \psi^k, s_{i,j} \rangle k^{d_i} = \alpha_{i,j,k} u^{d_j - d_i}$, and define an $r \times r$-matrix $A_k = (\alpha_{i,j,k})$. Then δ_k is a linear transformation of free $\mathbf{Q}/\mathbf{Z}$-modules with matrix $A_k - k^m I$.

Now let $w_i \in K^{2d_i+1}(X)$ be dual to y_i, and let $z_i = u^{d_i}w_i \in K^1(X)$. By [**8**, 11.19] applied to (2.4), we have

$$\psi^k(w_i) = \sum_j \langle \psi^k, s_{i,j}\rangle w_j.$$

Thus

$$\psi^k(z_i) = k^{d_i}u^{d_i}\sum_j \langle \psi^k, s_{i,j}\rangle u^{-d_j}z_j = \sum_j \alpha_{i,j,k}z_j.$$

Thus $A_k - k^m I$ is the matrix whose ith row gives $(\psi^k - k^m)(z_i)$ expressed in terms of the basis $\{z_1, \dots, z_r\}$. Then $\widetilde{v}^{2m}(X)$ is the abelian group presented by a matrix obtained by stacking the matrices $A_k - k^m I$ for all k. Thus we see that the same matrix A, defined to be all matrices $A_k - k^m I$ stacked, yields $E_2^{0,2m+1}(X; \mathbf{Q}/\mathbf{Z})$ if we take the kernel of the linear transformation of free $\mathbf{Q}/\mathbf{Z}$-modules whose matrix is A, and it yields $\widetilde{v}^{2m}(X)$ if we think of it as the presentation of an abelian group. Theorem 1.1 follows now from the following proposition.

PROPOSITION 2.5. *If A is an integer matrix, let $K(A)$ denote the kernel of the linear transformation of free $\mathbf{Q}/\mathbf{Z}$-modules whose matrix is A, and let $G(A)$ denote the abelian group presented by A. Then there is an isomorphism of abelian groups $K(A) \approx (G(A))^{\#}$. The same is true if we are localized at a prime p.*

PROOF. The matrix A can be brought to diagonal form (with rows and columns of 0's adjoined) by integer row and column operations. These operations consist of interchanging, multiplying by -1, and adding a multiple of one to another. For the first interpretation of the matrix, they correspond to change-of-basis in the domain and range, and hence do not affect the kernel (up to isomorphism), while in the second interpretation they correspond to invertible change of generating set and to simplification of relations, and hence do not affect the isomorphism class of the group presented. Let D be a diagonal matrix with entries $d_i > 1$, m 1's, and n 0's on the diagonal, and possibly many additional rows of 0's. Then $K(D) \approx \bigoplus \mathbf{Z}/d_i \oplus (\mathbf{Q}/\mathbf{Z})^n$, while $G(D) \approx \bigoplus \mathbf{Z}/d_i \oplus (\mathbf{Z})^n$. Since $(\mathbf{Z})^{\#} \approx \mathbf{Q}/\mathbf{Z}$ and $(\mathbf{Z}/d)^{\#} \approx \mathbf{Z}/d$, the proposition follows. ∎

PROOF OF PROPOSITION 2.2. We must show that if $d(x) = \sum s_j \otimes y_j$, then $\bar{d}(x) \in U(X)$ if and only if $k^{d_j}\langle \psi^k, s_j\rangle$ is 2-integral for all j.

The following analysis of the unstable condition (i.e. the condition for being in $U(X)$) is based on [**7**, §7], which derived from [**3**]. Let $\mathbf{K}_{2n}$ denote the $2n$th space in the Ω-spectrum for K. There is a homotopy equivalence $\mathbf{K}_{2n} \to BU$, and the structure maps $\Sigma^2\mathbf{K}_{2n} \to \mathbf{K}_{2n+2}$ are the Bott maps B. These combine to a tower

$$K_{2e}(BU) \xrightarrow{B_*} K_{2e+2}(BU) \xrightarrow{B_*} \cdots \xrightarrow{B_*} K_{2e+2d_j}(BU) \xrightarrow{B_*} \cdots \to K_{2e}(K),$$

with $K_{2e}(K)$ the direct limit of the tower. There is a similar tower after tensoring with $\mathbf{Q}$. Suppose $s_j \in K_{2e}(K) \otimes \mathbf{Q}$. Then $s_j \otimes y_j \in U(X)$ if and only if s_j pulls back to an element of $K_{2e+2d_j}(BU)$. This follows from the description of unstable comodules in the fourth paragraph after [**7**, 4.3].

For purposes of calculation, it is convenient to write the above tower in the equivalent form

$$K_0(BU) \xrightarrow{B_*} K_0(BU) \xrightarrow{B_*} \cdots \xrightarrow{B_*} K_0(BU) \xrightarrow{B_*} \cdots \to K_0(K).$$

In [**3**, 1.2-1.4], it is observed that $K_0(BU) \subset \mathbf{Q}[w]$ is a polynomial algebra generated by $\binom{w}{n} = w(w-1)\cdots(w-n+1)/n! \in \mathbf{Q}[w]$ for $n \geq 1$. Let A denote the free abelian group on the rational polynomials $\binom{w}{n}$. Then A consists of all rational polynomials $f(w)$ such that $f(n)$ is an integer for every integer n, so-called *numerical polynomials*. Then, still following [**3**], $B_*(f) - wf$ is decomposable for all $f \in A$, and B_* annihilates decomposables, from which it is deduced that $K_0(K) \approx w^{-1}A$, with an element f in the dth factor corresponding to $w^{-d}f$ in the limit.

It is also noted in [**3**, p.390] that $w = u^{-1}v \in K_0(K)$. This implies that $\langle \psi^k, w^n \rangle = k^n$. Thus $\langle \psi^k, f(w) \rangle = f(k)$, and so A consists of those polynomials $f(w)$ for which $\langle \psi^k, f(w) \rangle$ is integral for all k.

Now $s_j \in K_0(K) \otimes \mathbf{Q}$ is represented by a finite Laurent series $s_j(w)$ and pulls back to an element in the d_jth copy of $K_0(BU)$ if and only if $w^{d_j}s_j(w)$ is a numerical polynomial if and only if $\langle \psi^k, w^{d_j}s_j(w) \rangle$ is integral for all k. Since $\langle \psi^k, w^{d_j}s_j \rangle = k^{d_j}\langle \psi^k, s_j \rangle$, our desired conclusion follows. ■

3. Proof of Theorem 1.5

We begin with a result culled from [**16**].

PROPOSITION 3.1. *There are morphisms of Adams modules*

$$\begin{array}{ccccc} PK^1(SU(2n+1)) & \xrightarrow{j^*} & PK^1(Sp(n)) & \xrightarrow{\theta} & PK^1(Spin(2n+1)) \\ \langle B_1, \ldots, B_{2n} \rangle & \longrightarrow & \langle B'_1, \ldots, B'_n \rangle & \longrightarrow & \langle B''_1, \ldots, B''_{n-1}, D \rangle, \end{array}$$

where $\langle -, \ldots, - \rangle$ denotes the free abelian group on the indicated classes, $\theta(B'_i) = B''_i$ if $i < n$, $\theta(B'_1 + \cdots + B'_n) = 2^{n+1}D$, and

$$j^*(B_i) = \begin{cases} B'_i & i \leq n \\ B'_{2n+1-i} & i > n. \end{cases}$$

PROOF. We use Hodgkin's result ([**14**]) that $PK^1(G) \approx \langle \beta(\rho_1), \ldots, \beta(\rho_l) \rangle$, where $\rho_1, \ldots, \rho_l$ are the fundamental representations of G, and $\beta(\rho)$ is the virtual vector bundle over ΣG associated to $\rho - \dim(\rho)$. Naylor ([**16**, p.151]) notes that the maps

$$Sp(n) \xrightarrow{k} SU(2n) \xrightarrow{i} SU(2n+1)$$

induce morphisms of representation rings

$$R(SU(2n+1)) \xrightarrow{i^*} R(SU(2n)) \xrightarrow{k^*} R(Sp(n))$$

satisfying

$$k^*(\mu'_i) = \begin{cases} \eta_i & i \leq n \\ \eta_{2n-i} & i > n \end{cases}$$

and $i^*(\mu_i) = \mu'_i + \mu'_{i-1}$, where η_i, $1 \leq i \leq n$, μ'_i, $1 \leq i \leq 2n-1$, and μ_i, $1 \leq i \leq 2n$, are the representations given by exterior power operations on a canonical representation. We have $\mu'_0 = \mu'_{2n} = 1$. Letting $B'_i = \beta(\eta_i) + \beta(\eta_{i-1})$ and $B_i = \beta(\mu_i)$ yields the desired result about j^*, which is the composite $k^* \circ i^*$.

Naylor also considers the composite

$$j' : Spin(2n+1) \to SO(2n+1) \to SU(2n+1).$$

We let λ_i denote the exterior powers of the canonical unitary representation of $Spin(2n+1)$, Δ the Spin representation, $B_i'' = \beta(\lambda_i)$, and $D = \beta(\Delta)$. Naylor shows that

$$j'^*(B_i) = \begin{cases} B_i'' & i \leq n \\ B_{2n+1-i}'' & i > n \end{cases}$$

and $2^{n+1}D = B_1'' + \cdots + B_n''$. Since $\ker(j^*) = \ker(j'^*)$, there is a homomorphism of abelian groups

$$PK^1(Sp(n)) \xrightarrow{\theta} PK^1(Spin(2n+1))$$

defined by $\theta(B_i') = B_i''$ for $i \leq n$, and it will be an Adams-module homomorphism since j^* and j'^* are Adams-module homomorphisms with j^* surjective. ∎

Next we relate the above basis of $PK^1(SU(n))$ with one used by Bousfield ([**9**]).

PROPOSITION 3.2. *There is an isomorphism of Adams modules*

$$PK^1(SU(n)) \approx \widetilde{K}^0(CP^{n-1}). \tag{3.3}$$

Let ξ_k denote the element of $PK^1(SU(n))$ which corresponds to $\xi^k - 1$ under this isomorphism. Here ξ denotes the Hopf bundle over CP^{n-1}. Then

$$B_j = \sum (-1)^{k+1} \tbinom{n}{j-k} \xi_k.$$

PROOF. If I denotes the augmentation ideal of $R(G)$, then there is an isomorphism $I/I^2 \approx PK^1(G)$ under which $(-1)^{k+1}\lambda^k$ corresponds to ψ^k. (See, e.g., [**11**, 2.1ff].) We first claim that under this isomorphism and (3.3) (which is well-known) $\{\theta_n - n\} \in I/I^2$ corresponds to $\{\xi - 1\} \in \widetilde{K}^0(CP^{n-1})$. To see this, we use a result in [**15**, p.206] which states that there is a commutative diagram

$$\begin{array}{ccc} CP^{n-1} & \xrightarrow{f} & BU \\ {\scriptstyle j}\downarrow & & \downarrow{\scriptstyle B} \\ \Omega SU(n) & \xrightarrow{\Omega i} & \Omega SU \end{array}$$

in which j is a canonical map, i is the inclusion, B the Bott map, and f satisfies $f^*(c_i) = 0$ unless $i = 1$, and $f^*(c_1)$ is the usual generator of $H^2(CP^{n-1})$. This implies that f classifies ξ. The isomorphism (3.3) is defined using the maps j and B. Hence ξ corresponds under this isomorphism to the map i, which is the canonical representation θ_n.

Since $\psi^k(\xi) = \xi^k$, we obtain that

$$\xi_k = (-1)^{k+1}\lambda^k(\theta_n - n) = (-1)^{k+1}\sum_{j=0}^{k} \tbinom{-n}{k-j}\lambda^j(\theta_n).$$

But $\lambda^j(\theta_n) = B_j$. To invert this equation, we take a formal sum of the equations for all values of k, using powers of an indeterminate x. We have

$$\sum (-1)^{k+1}\xi_k x^k = \sum x^k \sum_{j=0}^{k} \tbinom{-n}{k-j} B_j = (1+x)^{-n} \sum x^j B_j$$

and hence

$$\sum x^j B_j = \sum_i \tbinom{n}{i} x^i \sum_k (-1)^{k+1}\xi_k x^k = \sum x^j \sum_k \tbinom{n}{j-k}(-1)^{k+1}\xi_k,$$

as desired. ■

The following corollary is immediate from the description of j^* in Proposition 3.1 along with Proposition 3.2.

COROLLARY 3.4. *$PK^1(Sp(n)) \approx \langle \xi_1, \xi_2, \dots : R_{n+1}, \dots, R_{2n}, S_j, j > 2n \rangle$, where*

$$R_j = \sum (-1)^{k+1} \tbinom{2n+1}{j-k} \xi_k - \sum (-1)^{k+1} \tbinom{2n+1}{2n+1-j-k} \xi_k,$$

$S_j = \sum (-1)^k \binom{j}{k} \xi_k$, and $\psi^t \xi_k = \xi_{kt}$.

Note that the relations allow one to express each ξ_i with $i > n$ in terms of $\xi_1, \dots, \xi_n$. From Corollary 3.4, we easily deduce the following result.

COROLLARY 3.5. *$v^{2m}(Sp(n)) \approx \mathbf{Z}/2^{eSp'(m,n)}$, where*

$$eSp'(m,n) := \min(\nu(R_{m,n+1}), \dots, \nu(R_{m,2n}), \nu(S'_{m,j}), j > 2n),$$

with $R_{m,j}$ and $S'_{m,j}$ defined by

$$R_{m,j} = \sum_{k \text{ odd}} \tbinom{2n+1}{j-k} k^m - \sum_{k \text{ odd}} \tbinom{2n+1}{2n+1-j-k} k^m,$$

$$S'_{m,j} = \sum_{k \text{ odd}} \tbinom{j}{k} k^m.$$

PROOF. The sums are taken over all odd values of k for which the binomial coefficients are nonzero. The generator of the group is ξ_1, and the relations are obtained from those in 3.4 by $\xi_k = \psi^2(\xi_{k/2}) \equiv 0$ if k is even, while if k is odd, $\xi_k = \psi^k(\xi_1) \equiv k^m \xi_1$. Here we are using Corollary 3.9 to imply that $\psi^k \equiv k^m$ in $v^{2m}(-)$ for all odd k. ■

Because of Theorems 1.1 and 1.4 and Corollary 3.5, we have

PROPOSITION 3.6. *$eSp'(m,n) = eSp(m,n)$; i.e.,*

$$\min(\nu(R_{m,n+1}), \dots, \nu(R_{m,2n}), \nu(S_{m,j}), j > 2n) = \min(\nu(S_{m,j}), j > 2n),$$

and so the relations $R_{m,j}$ are superfluous in $eSp'(m,n)$

We should remark on the minor difference between $S_{m,j}$ in 1.4 and $S'_{m,j}$ in 3.5, in that one involves a sum over all values of k, while the other involves a sum only over odd values of k. As remarked in Section 1, we only consider values of m large enough that summands which are multiples of 2^m would not affect divisibility by 2 of the sum. Thus $\nu(S_{m,j}) = \nu(S'_{m,j})$ for such values of m.

Next we explain why the relations $\psi^3 - 3^m$ and $\psi^{-1} - (-1)^m$ imply relations $\psi^k - k^m$ for all odd integers k. We contrast with the situation when we are localized at an odd prime. The difference is due to the following result, which is stated in [**1**, 2.9].

PROPOSITION 3.7. *If p is an odd prime, then the multiplicative group of units $(\mathbf{Z}/p^n)^\times$ is cyclic of order $(p-1)p^{n-1}$. If r generates $(\mathbf{Z}/p^2)^\times$, then any integer congruent to r mod p^2 generates $(\mathbf{Z}/p^n)^\times$. The group $(\mathbf{Z}/2^n)^\times$ is the direct product of the subgroup consisting of ± 1 and the subgroup of classes congruent to 1 mod 4; the latter subgroup is cyclic of order 2^{n-2} generated by any integer congruent to 5 mod 8.*

This implies that, for any n, any odd integer is congruent mod p^n to $\pm 3^e$ for some e. We also need the following result of Adams.

PROPOSITION 3.8. ([**2**, 5.1]) *If X is a finite complex, there exists an integer e so that for all k and m, $\psi^k \equiv \psi^{k+m^e}$ in $K(X)/mK(X)$.*

We will use the following consequence of these results.

COROLLARY 3.9. *Let X be a finite complex, and if p is odd let r generate $(\mathbf{Z}/p^2)^\times$. Let*

$$Q_m = \begin{cases} PK^1(X;\mathbf{Z}_{(p)})/(\mathrm{im}(\psi^r - r^m)) & \text{if } p \text{ is odd} \\ PK^1(X;\mathbf{Z}_{(2)})/(\mathrm{im}(\psi^3 - 3^m, \psi^{-1} - (-1)^m)) & \text{if } p = 2. \end{cases}$$

Then $\psi^s = s^m$ in Q_m for all $s \not\equiv 0$ mod p.

PROOF. We consider just the case $p = 2$. The case when p is odd is similar and slightly easier. We show $\psi^s \equiv s^m$ mod 2^n for all positive integers n. This implies that they are equal in the finitely generated $\mathbf{Z}_{(2)}$-module.

We first show that Proposition 3.8 is valid for $K^1(-)$ provided k and m are relatively prime. Adams' result dealt with $K^0(-)$. The validity in $K^{-1}(-)$ is immediate since $K^{-1}(X) = K^0(\Sigma X)$. Since ψ^k in $K^{-1}(-)$ corresponds to $k\psi^k$ in $K^1(-)$, the periodicity in $K^1(-)$ follows, as long as k is a unit mod m.

Choose an integer e that works for X in Proposition 3.8. Use Proposition 3.7 to find an integer ℓ and $\epsilon = 0$ or 1 so that $s \equiv (-1)^\epsilon 3^\ell$ mod 2^{ne}. Then, for some k, we have, in Q_m mod 2^n,

$$\psi^s = \psi^{(-1)^\epsilon 3^\ell + k2^{ne}} \equiv \psi^{(-1)^\epsilon}\psi^{3^\ell} \equiv ((-1)^\epsilon)^m (3^\ell)^m \equiv s^m.$$

■

The following result is immediate from Propositions 3.1 and 3.2.

PROPOSITION 3.10. *The group $PK^1(Spin(2n+1))$ has a subgroup isomorphic to $PK^1(Sp(n))$, as described in Corollary 3.4, and an additional generator D satisfying*

$$2^{n+1}D = \sum_{j=1}^{n}\sum_{k=1}^{j}(-1)^{k+1}\binom{2n+1}{j-k}\xi_k. \tag{3.11}$$

The Adams operations are determined by $\psi^t\xi_k = \xi_{kt}$.

The next two theorems, 3.15 and 3.18, determine $v^{2m}(Spin(2n+1))$. Two of the relations, $\mathcal{R}_2$ and $\mathcal{R}_3$, are defined in terms of the following algorithms. The algorithms will be justified in the proof of 3.15 and computed in 3.18. We use the relations R_j, $n+1 \le j \le 2n$, and S_j, $2n < j \le 3n$, of Corollary 3.4. Note that each equals $\pm\xi_j$ mod terms with smaller subscripts of ξ.

For $\mathcal{R}_2$, begin with

$$\sum_{j=1}^{n}\sum_{k=1}^{j}(-1)^{k+1}\binom{2n+1}{j-k}\xi_{2k}, \tag{3.12}$$

and subtract multiples of $R_{2n}, \dots, R_{n+1}$ to eliminate $\xi_{2n}, \dots, \xi_{n+1}$. This results in a linear expression E in $\xi_1, \dots, \xi_n$ which, we will show in (4.2), equals $(-1)^{n+1}2^n\xi_n$ plus lower terms. Write (3.11) in the form

$$2^{n+1}D + (-1)^n\xi_n + \sum_{k=1}^{n-1}(-1)^k\xi_k \sum_{i=0}^{n-k}\tbinom{2n+1}{i}. \tag{3.13}$$

Add 2^n times (3.13) to E to eliminate ξ_n. The resulting linear expression in $\xi_1, \dots, \xi_{n-1}$, and D is divisible by 2^{n+1}. Divide it by 2^{n+1}, and then replace ξ_k by k^m (resp. 0) if k is odd (resp. even) to get $\mathcal{R}_2$.

For $\mathcal{R}_3$, begin with

$$\sum_{j=1}^{n}\sum_{k=1}^{j}(-1)^{k+1}\tbinom{2n+1}{j-k}\xi_{3k}, \tag{3.14}$$

and subtract multiples of $S_{3n}, \dots, S_{2n+1}, R_{2n}, \dots, R_{n+1}$, and of (3.13) to eliminate $\xi_{3n}, \dots, \xi_{n+1}, \xi_n$. This results in a linear expression in $\xi_1, \dots, \xi_{n-1}$, and D which is divisible by 2^{n+1}. Divide it by 2^{n+1} and replace ξ_k by k^m or 0 as before to get an expression E'. The relation $\mathcal{R}_3$ is $3^mD - E'$.

THEOREM 3.15. *The abelian group $v^{2m}(Spin(2n+1))$ has generators ξ_1 and D with relations*

$$2^{n+1}D = (\sum_{j=1}^{n}\sum_{odd\ k}\tbinom{2n+1}{j-k}k^m)\xi_1, \tag{3.16}$$

$2^{eSp(m,n)}\xi_1$, *$\mathcal{R}_2$ and $\mathcal{R}_3$.*

PROOF. The second relation is the one obtained in Propositions 3.5 and 3.6. This relation due to Adams operations on ξ_i-classes is still present in $v^{2m}(Spin(2n+1))$. The first relation is the definition of D, together with $\xi_k \sim \psi^k\xi_1 \sim k^m\xi_1$ if k is odd, and $\xi_k \sim \psi^k\xi_1 \sim 0$ if k is even.

The relations $\mathcal{R}_2$ and $\mathcal{R}_3$ are consequences of $\psi^2D \sim 0$ and $\psi^3D \sim 3^mD$, respectively. We evaluate $\psi^2(2^{n+1}D)$ and $\psi^3(2^{n+1}D)$ using (3.11) and $\psi^\ell\xi_k \sim \xi_{k\ell}$. The relations are reduced using R_j and S_j. Since ψ^kD must be an integral combination of the generators $\xi_1, \dots, \xi_{n-1}$, and D, it follows that $\psi^2(2^{n+1}D)$ and $\psi^3(2^{n+1}D)$, when expressed in terms of these classes, must be divisible by 2^{n+1}. Performing the division yields the desired relations for $\psi^2(D)$ and $\psi^3(D)$.

By Corollary 3.9, we need only to append the relation $\psi^{-1}(D) \sim (-1)^mD$ in order to achieve all relations $\psi^k(D) \sim k^mD$ for all odd k. However, by Proposition 3.17, this relation adds no information if m is odd, while if m is even, which will be treated at the end of this section, it implies $2D \sim 0$. ■

PROPOSITION 3.17. *$\psi^{-1} = -1$ in $PK^1(Spin(2n+1))$.*

PROOF. First note that $\psi^{-1} = (-1)^m$ in $PK^1(S^{2m+1})$. We use this to prove, by induction on n, that $\psi^{-1} = -1$ in $PK^1(Sp(n))$.

Indeed, consider the short exact sequence

$$0 \to PK^1(S^{4n-1}) \xrightarrow{p^*} PK^1(Sp(n)) \xrightarrow{i^*} PK^1(Sp(n-1)) \to 0,$$

and assume $\psi^{-1} = -1$ in $PK^1(Sp(n-1))$. Let y generate $\mathrm{im}(p^*)$. Note $\psi^{-1}(y) = -y$. Let $x \in PK^1(Sp(n))$. By naturality of ψ^{-1} and the induction hypothesis, we have $i^*(\psi^{-1}(x)+x)=0$, and hence $\psi^{-1}(x)+x=\alpha y$ for some integer α. Apply ψ^{-1} and use $\psi^{-1}\psi^{-1}=1$ to obtain $\alpha y=-\alpha y$, and hence $\alpha=0$.

Now we use the morphism $PK^1(Sp(n))\xrightarrow{\theta}PK^1(Spin(2n+1))$ considered in Proposition 3.1 to deduce the proposition. Since $2^{n+1}D\in\mathrm{im}(\theta)$, we obtain

$$\psi^{-1}(2^{n+1}D)=-2^{n+1}D,$$

and hence $\psi^{-1}(D)=-D$ since there is no torsion in $PK^1(Spin(2n+1))$. All other elements are in the image from $PK^1(Sp(n))$). ■

The algorithm distills down to the following closed form for the relations in Theorem 3.15, which was observed from extensive computer calculations by the second author and then proved (see Section 4) utilizing a number of elaborate combinatorial arguments.

THEOREM 3.18. *$v^{2m}(Spin(2n+1))$ is the abelian group with generators ξ_1 and D and four relations as below:*

$$2^{eSp(m,n)}\xi_1,$$

$$\Big(\sum_{\text{odd }k}k^m\sum_{i=0}^{n-k}\tbinom{2n+1}{i}\Big)\xi_1-2^{n+1}D, \tag{3.19}$$

$$\Big(\sum_{\text{odd }k}k^m\sum_{t\ge0}\tbinom{2n+2}{n-1-k-4t}\Big)\xi_1-2^nD, \tag{3.20}$$

and

$$\Big(2^n\sum_{\text{odd }k}k^m\sum_{t\ge0}\tbinom{2n+1}{n-1-k-3t}\Big)\xi_1-\tfrac13(2^{2n+1}+1-3^{m+1})D. \tag{3.21}$$

If m is odd and $n\ne4$, the coefficients of ξ_1 in (3.19) and (3.20) are divisible by 2^n.

From this, we easily deduce Theorem 1.5. Subtract 2 times (3.20) from (3.19) to obtain the relation $R_1(m,n)$ of 1.5 on the generator ξ_1. Note that the exponent of 2 in the coefficient of D in (3.21) is $\min(2n+1,\nu(m+1)+2)$ if m is odd. If $n\le\nu(m+1)+2$, divide (3.20) by 2^n to split off $\mathbf{Z}/2^n$, the second summand in 1.5, and then add an appropriate multiple of (3.20) to (3.21) to eliminate D, yielding the relation $R_2(m,n)$ of 1.5 on ξ_1. The last part of Theorem 3.18 is necessary in order to know that we can split off $\mathbf{Z}/2^n$.

Similarly, if $\nu(m+1)+2<n$, use (3.21) to split off $\mathbf{Z}/2^{\nu(m+1)+2}$, and then subtract an appropriate multiple of (3.21) from (3.20) to obtain $R_2(m,n)$. This completes the proof of Theorem 1.5 when m is odd.

To prove that $v^{2m}(Spin(2n+1))\approx\mathbf{Z}/2\oplus\mathbf{Z}/2$ when m is even, we begin by showing that $eSp'(m,n)=1$ in Corollary 3.5 when m is even. It suffices to show that $R_{m,2n}\equiv2\bmod 4$ and that $R_{m,j}$ and $S'_{m,j}$ are even for all relevant j. Since $k^m\equiv1\bmod 4$ if k is odd and m even, we have, mod 4,

$$S'_{m,j}\equiv\sum_{k\text{ odd}}\tbinom{j}{k}=2^{j-1},$$

$$R_{m,2n}\equiv\sum_{k\text{ odd}}\tbinom{2n+1}{k}-2\tbinom{2n+1}{2n+1}=2^{2n}-2,$$

and other $R_{m,j}$ are congruent to 2^{2n} minus 2 times a sum of binomial coefficients.

As observed in the proof of 3.15, ψ^{-1} implies $2D \sim 0$ when m is even. Thus it remains to show that the coefficients of relations in Theorem 3.18 are even when m is even, which reduces to showing

$$\sum_{k \text{ odd}} \sum_{i=0}^{n-k} \binom{2n+1}{i} \equiv 0 \mod 2 \tag{3.22}$$

and

$$\sum_{k \text{ odd}} \sum_{t \geq 0} \binom{2n+2}{n-k-1-4t} \equiv 0 \mod 2. \tag{3.23}$$

The sum in (3.22) is $\sum_{i=0}^{n-1} \binom{2n+1}{i} \left[\frac{n-i+1}{2}\right]$. If n is even, terms $2j$ and $2j+1$ in this sum will have the same parity. If $n \equiv 3 \mod 4$, terms $2j+1$ and $2j+2$ will have the same parity. If $n \equiv 1 \mod 4$, the only odd terms will occur in pairs $i = 8j$ and $i = 8j+3$.

The sum in (3.23) is $\sum \binom{2n+2}{n-2j} \left[\frac{j+1}{2}\right]$. If n is odd, all coefficients $\binom{2n+2}{n-2j}$ are even, while if $n = 2N$, this becomes $\sum \binom{2N+1}{N-j} \left[\frac{j+1}{2}\right]$ mod 2, which equals the first sum, and hence is even. This completes the proof of Theorem 1.5.

4. Proof of Theorem 3.18

In this section, we perform the algorithm in Theorem 3.15 and obtain Theorem 3.18. First note that (3.19) is an elementary manipulation of (3.16).

We begin by deriving (3.20). Let M be a matrix whose rows present the relations $R_{2n}, \ldots, R_{n+1}$ of 3.4 expressed in terms of $\xi_{2n}, \ldots, \xi_1$. Then

$$M_{i,j} = (-1)^{j-i}\binom{2n+1}{j-i} + (-1)^{i+j}\binom{2n+1}{j+i-2n-1}.$$

Let M_L (resp. M_R) denote the left (resp. right) half of M.

Let $\sigma_j := \sum_{k=0}^{j} \binom{2n+1}{k}$. Let P denote a row vector of length $2n$ whose $(2j+1)$st entry is $(-1)^{n+1+j}\sigma_j$ for $0 \leq j < n$, with other entries 0. This represents (3.12). Write $P = (P_L|P_R)$, with P_L and P_R of length n. We wish to add multiples of the rows of M to P to annihilate P_L. This is facilitated by first replacing M by the equivalent set of relations $M_L^{-1}M = [I|M_L^{-1}M_R]$. The linear expression E in the algorithm for $\mathcal{R}_2$ described before Theorem 3.15 corresponds to $P_R - P_L M_L^{-1} M_R$. We will show that this equals

$$(-1)^{n+1}2^n \sum_{j=0}^{n-1} (-1)^j \binom{2n+1}{j} V_j, \tag{4.1}$$

where V_j is a vector of length n which begins with j 0's followed by

$$1, 1, -1, -1, 1, 1, -1, \ldots.$$

The expression corresponding to (4.1) is

$$2^n \Big(\sum_{k=1}^{n} (-1)^{k+1} \xi_k \Big(\sum_{j=0}^{n-k} \binom{2n+1}{j} - 2 \sum_{t \geq 0} \binom{2n+2}{n-1-k-4t} \Big)\Big). \tag{4.2}$$

We have used

$$\binom{2n+2}{a} = \binom{2n+1}{a} + \binom{2n+1}{a-1} \tag{4.3}$$

to show that this corresponds to (4.1). Adding 2^n times (3.13) to (4.2) and replacing ξ_k by k^m or 0 yields -2^{n+1} times (3.20), as desired.

To complete the proof of (3.20), it remains to prove (4.1). Let $c_i = \binom{2n+1}{i}$ and $d_i = \binom{2n+i}{i}$. Consideration of generating functions implies that for any positive integer m

$$\sum_{j=0}^{m}(-1)^j d_j c_{m-j} = 0 \tag{4.4}$$

and that the (i,j)th entry of M_L^{-1} is d_{j-i}. Using these facts, one can easily prove the following lemma.

LEMMA 4.5. *$M_L^{-1}M_R = \sum_{j=1}^{n} D_jC_j$, where D_j is a column vector of length n with t^{th} entry $d_{n+1-t-j} - d_{n+j-t}$, and C_j is a row vector of length n with t^{th} entry $(-1)^{t-j}c_{t-j}$.*

To simplify notation, we shall complete the proof of (4.1) when n is odd. The proof when n is even is virtually the same. Let $n = 2e+1$. We shall prove the following lemma.

LEMMA 4.6. *For $1 \le j \le n$,*

$$-P_LD_j = (-1)^{[(j-1)/2]}2^n + \sum_{u=1}^{[j/2]}(-1)^{e+u+1}\sigma_{e+u}d_{j-2u}.$$

Then the entry in the tth position of $-P_LM_L^{-1}M_R$ is

$$(-P_L\sum_j D_jC_j)_t = A_{1,t} + A_{2,t},$$

where

$$A_{1,t} = 2^n\sum_{j=1}^{t}(-1)^{[(j-1)/2]}(-1)^{t-j}c_{t-j},$$

which equals the tth entry of (4.1), and

$$A_{2,t} = \sum_{j=1}^{t}(-1)^{t-j}c_{t-j}\sum_{u=1}^{[j/2]}(-1)^{e+u+1}\sigma_{e+u}d_{j-2u}.$$

Using (4.4), one can check that

$$A_{2,t} = \begin{cases} 0 & t \text{ odd} \\ (-1)^{e+T+1}\sigma_{e+T} & t = 2T. \end{cases}$$

Thus $P_R + A_2 = 0$, and hence $P_R - P_LM_L^{-1}M_R$ equals (4.1), as claimed. Once we have given the proof of Lemma 4.6, we will be done with the proof of the $\mathcal{R}_2$-part of Theorem 3.18.

PROOF OF LEMMA 4.6. We use generating functions. The coefficients of x_0 through x_{n-1} in $p(x) := (1-x^2)^{2n+1}/(1+x^2)$ give the entries of P_L, while $q(x) := (1-x)^{-(2n+1)} = \sum d_ix^i$. Note that $p(x)q(x) = (1+x)^{2n+1}/(1+x^2)$. We obtain

$$\begin{aligned} -P_LD_j &= \sum_{t=1}^{n}\operatorname{coef}(x^{t-1}, p(x))\cdot\big(\operatorname{coef}(x^{n+j-t}, q(x)) - \operatorname{coef}(x^{n-j+1-t}, q(x)\big) \\ &= \operatorname{coef}(x^{n+j-1}, p(x)q(x)) - \operatorname{coef}(x^{n-j}, p(x)q(x)) \qquad (4.7) \\ &\quad -\sum_{t>n}\operatorname{coef}(x^{t-1}, p(x))\cdot d_{n+j-t}. \end{aligned}$$

The sum subtracted at the end is due to the truncation of D_j; one easily checks that this equals the u-sum in Lemma 4.6. Finally we observe that, if we employ the symmetry property of binomial coefficients, then the expression in line (4.7) equals the following expression if $j \equiv 1$ or 2 mod 4, and equals the negative of this expression otherwise.

$$\binom{2n+1}{n} - \binom{2n+1}{n-1} - \binom{2n+1}{n-2} + \binom{2n+1}{n-3} + \binom{2n+1}{n-4} - \binom{2n+1}{n-5} - \cdots$$

This expression equals 2^n, as can be seen by expanding $(1-i)^{2n+1}$ or by induction on n. ■

The derivation of (3.21) is similar to that of (3.20) just performed, although somewhat more complicated. Replacing the matrix M used in (3.20) is a matrix N whose rows present the relations $S_{3n}, \dots, S_{2n+1}, R_{2n}, \dots, R_{n+1}$ of 3.4 in terms of $\xi_{3n}, \dots, \xi_1$. Then

$$N_{i,j} = \begin{cases} (-1)^{j-i}\binom{3n+1-i}{j-i} & 1 \le i \le n \\ (-1)^{j-i}\binom{2n+1}{j-i} + (-1)^{i+j}\binom{2n+1}{j+i-4n-1} & n < i \le 2n. \end{cases}$$

Divide N into $n \times n$ submatrices which we name as follows:

$$N = \begin{pmatrix} T_1 & U_1 & U_2 \\ 0 & T_2 & U_3 \end{pmatrix}.$$

We perform row operations on N to reduce it to

$$N' = \begin{pmatrix} I & 0 & B_2 - B_1 W \\ 0 & I & W \end{pmatrix},$$

where $B_i = T_1^{-1}U_i$, and $W = T_2^{-1}U_3 = \sum_{j=1}^n D_j C_j$, exactly as in Lemma 4.5.

Let σ_j, c_i, d_i, C_j, and D_j be as in the preceding proof, with $\sigma_j = 0$ if j is not an integer. Let Q denote a row vector of length $3n$ whose $(3j+1)$st entry is $(-1)^{n+1+j}\sigma_j$ for $0 \le j < n$, with other entries 0. This represents (3.14). Write $Q = (Q_0|Q_1|Q_2)$ with each Q_i of length n. We use N' to reduce Q to $(0|0|Q_2')$, where

$$Q_2' = Q_2 - Q_0(B_2 - B_1 W) - Q_1 W. \tag{4.8}$$

We will show this equals

$$(-1)^n\Big(\tfrac{1}{3}(2^{2n+1}+1)(-\sigma_0, \sigma_1, -\sigma_2, \sigma_3, \dots, \sigma_{n-1}) \\ + 2^{2n+1}(0, -c_0, c_1, -c_2, c_3 + c_0, -(c_4 + c_1), \dots)\Big). \tag{4.9}$$

Thus the vector (4.9) expresses $\psi^3(2^{n+1}D)$ in terms of $\xi_n, \dots, \xi_1$ as

$$\tfrac{1}{3}(2^{2n+1}+1)\sum_{k=1}^{n}(-1)^{k+1}\xi_k \sum_{i=0}^{n-k}\binom{2n+1}{i} + 2^{2n+1}\sum_{k=1}^{n-1}(-1)^k \xi_k \sum_{t\ge 0}\binom{2n+1}{n-1-k-3t}. \tag{4.10}$$

We add $\frac{1}{3}(2^{2n+1}+1)$ times (3.13) to eliminate ξ_n and get

$$\tfrac{1}{3}(2^{2n+1}+1)2^{n+1}D + 2^{2n+1}\sum_{k=1}^{n-1}(-1)^k \xi_k \sum_{t\ge 0}\binom{2n+1}{n-1-k-3t}.$$

Divide by 2^{n+1}, replace ξ_k by k^m or 0, and subtract from $3^m D$; this yields exactly (3.21).

It remains to show that Q_2' equals (4.9). We show this when $n \equiv 0 \bmod 6$; other congruences can be handled similarly. Let $r = n/3$. It is not difficult to prove that

$$(B_1)_{i,j} = (-1)^j \binom{n+j-i-1}{n-i}\binom{3n+1-i}{n+j-i} \text{ and } (B_2)_{i,j} = (-1)^j \binom{2n+j-i-1}{n-i}\binom{3n+1-i}{2n+j-i}.$$

We obtain that, for $0 \le t < n$, the $(t+1)$st entry of Q_2' is

$$(-1)^{t+1}\sigma_{2r+t/3} - \sum_{u=0}^{r-1}(-1)^{u+t}\sigma_u\binom{2n+t-3u-1}{n+t}\binom{3n-3u}{n-t} \tag{4.11}$$

$$+ \sum_{s=0}^{n-1}\sum_{u=0}^{r-1}(-1)^{u+s}\sigma_u\binom{n+s-3u-1}{s}\binom{3n-3u}{2n-s} \tag{4.12}$$

$$\cdot \sum_{j=0}^{t}(d_{n-s+j-t-1} - d_{n+t-j-s})(-1)^j c_j$$

$$+ \sum_{u=0}^{r-1}(-1)^u\sigma_{r+u}\sum_{j=0}^{t}(d_{n-3u-1+j-t} - d_{n+t-j-3u})(-1)^j c_j. \tag{4.13}$$

It follows from (4.4) that the j-sum in (4.12) is 0 for $s \ge n$, except that it is -1 when $s = n+t$. Hence (4.12) can be made into a sum over all $s \ge 0$ with the sum-part of (4.11) incorporated into it.

Next we note that if $u = r + \delta$ with $\delta \ge 0$ in (4.13), then, again using (4.4), the j-sum in (4.13) is nonzero only when $\delta = t/3$, in which case it is -1. Thus (4.13) can be extended to $u \ge 0$ if the initial term of (4.11) is included. Similarly we note that if $u = r + \Delta$ with $\Delta \ge 0$ in (4.12), then $\binom{n+s-3u-1}{s}\binom{3n-3u}{2n-s}$ is nonzero only when $s = 3\Delta$, in which case it is $(-1)^\Delta$. Thus the modified (4.13) can be incorporated into (4.12) by extending the u-sum of (4.12) to $u \ge 0$. Now the $(t+1)$st entry of Q_2' has been simplified to

$$\sum_{s\ge 0}\sum_{u\ge 0}(-1)^{u+s}\sigma_u\binom{n+s-3u-1}{s}\binom{3n-3u}{2n-s}\sum_{j=0}^{t}(d_{n-s+j-t-1} - d_{n+t-j-s})(-1)^j c_j. \tag{4.14}$$

We will prove the following lemma at the end of this section.

LEMMA 4.15. *If $n \ge 0$, and v and w are arbitrary integers, then*

$$\sum_{s\ge 0}(-1)^s\binom{n+s-1-w}{s}\binom{3n-w}{2n-s}\binom{3n-s+v}{2n} = \binom{4n-w+v}{2n}.$$

Using this with $w = 3u$, (4.14) reduces to

$$\sum_{j=0}^{t}(-1)^j c_j \sum_{u\ge 0}(-1)^u\sigma_u\left(\binom{4n-3u+j-t-1}{2n} - \binom{4n-3u+t-j}{2n}\right). \tag{4.16}$$

Viewing $\binom{4n-3u+\epsilon}{2n}$ as $(-1)^{u+\epsilon}\operatorname{coef}(x^{2n-3u+\epsilon}, (1+x)^{-(2n+1)})$, and σ_u as

$$\operatorname{coef}(x^{3u}, (1+x^3)^{2n+1}/(1-x^3)),$$

we obtain a simplification of (4.16) to

$$(-1)^{t+1}\sum_{j=0}^{t} c_j\big(\mathrm{coef}(x^{2n+j-t-1}, \tfrac{(1-x+x^2)^{2n+1}}{1-x^3}) + \mathrm{coef}(x^{2n+t-j}, \tfrac{(1-x+x^2)^{2n+1}}{1-x^3})\big). \tag{4.17}$$

We will prove the following lemma at the end of this section.

LEMMA 4.18. *Let* $f_m(x) = (1-x+x^2)^{m+1}/(1-x^3)$. *Then for* $m \geq 0$ *and* $j \geq 0$,

$$\mathrm{coef}(x^{m-j-1}, f_m(x)) + \mathrm{coef}(x^{m+j}, f_m(x)) = \begin{cases} (1+2(-2)^m)/3 & j \equiv 0, 2 \mod 3 \\ (1+2(-2)^{m+1})/3 & j = 1 \mod 3. \end{cases}$$

We obtain that (4.17) equals

$$(-1)^{t+1}\left(\tfrac{1}{3}(2^{2n+1}+1)\sum_{j=0}^{t} c_j - 2^{2n+1}\sum_{\substack{0\leq j\leq t \\ t-j\equiv 1\ (3)}} c_j\right),$$

which equals the component of (4.9) in position $t+1$, as desired, completing the proof of (3.21), as described surrounding (4.9).

Next we prove the portion of Theorem 3.18 which states that

$$\sum_{\text{odd } k} k^m \sum_{t\geq 0} \binom{2n+2}{n-1-k-4t} \tag{4.19}$$

is divisible by 2^n if m is odd and $n \neq 4$. This occurs in (3.20). We accomplish this by showing later in this section that, if $m = 2a+1$, then (4.19) equals

$$\binom{n-1}{1}2^{2n-4} + \sum_{j=2}^{n/2}\binom{n-j}{j}2^{2n-4j}\sum_{i\geq j-1} 8^i\binom{a}{i}\sum_{t=0}^{j-2}(-1)^t\binom{2j-1}{t}(2j-2t-1)\binom{j-t}{2}^i. \tag{4.20}$$

Then we note that the first term of (4.20) is divisible by 2^n provided $n \geq 3$. For each value of $j \geq 2$, the j-summand is nonzero for $n \geq 2j$ and is clearly divisible by $2^{2n-4j+3(j-1)}$, which is $\geq 2^n$ provided $n \geq j+3$. The case $n=4$ and $j=2$ is the only time that this gives a nonzero term which might not be divisible by 2^n; it is what causes the restriction $n \neq 4$ in the last part of Theorem 3.18.

The expression (4.20), while not as elegant as (4.19), is often more useful in performing specific calculations of $\nu(-)$, such as in Proposition 1.6. Before proving (4.20), we handle the case $n=4$, which was excluded from Theorem 1.5.

PROPOSITION 4.21.

$$v^{2m}(Spin(9)) \approx \begin{cases} \mathbf{Z}/2 \oplus \mathbf{Z}/2 & m \text{ even} \\ \mathbf{Z}/8 \oplus \mathbf{Z}/2^{e(m)} & m \text{ odd,} \end{cases}$$

where

$$e(m) = \begin{cases} \min(\nu(m-5)+2, 6) & \text{if } m \equiv 1 \text{ mod } 4 \\ \min(\nu(m-7)+2, 8) & \text{if } m \equiv 3 \text{ mod } 4. \end{cases}$$

PROOF. The numbers $eSp(m,n)$ are as in [**6**, 5.1] (suitably interpreted), and large enough as to not matter here. Let $P = 3^m - 3$. The other three relations of 3.18 are $(160+10P)\xi_1 - 32D$, $(48+P)\xi_1 - 16D$, and $16(39+P)\xi_1 + (P-168)D$. If $m \equiv 3$ mod 4, then $P \equiv 8$ mod 16, so we get a $\mathbf{Z}/8$ due to $(48+P)\xi_1$. (This is the anomaly.) If $m \equiv 1$ mod 4, $P \equiv 0$ mod 16, and we get a $\mathbf{Z}/8$ due to $(P-168)D$.

If $m \equiv 1 \bmod 4$, let $Q = 3(3^{m-5} - 1)$, so that $\nu(Q) = \nu(m-5) + 2$. Replace P by $3^4Q + 3^5 - 3$. Subtract multiples of the third relation from the other two so as to eliminate the D-term from each. The relations obtained are, up to odd multiples, $2^6 + 2Q + Q^2/4$ and $2^8 + Q + Q^2/8$. For $\nu(Q) \geq 4$, the minimum of the exponent of 2 of these two expressions is $\min(6, \nu(Q))$, as claimed. The case $m \equiv 3 \bmod 4$ is handled similarly. ■

Now we establish the equivalence of (4.19) and (4.20). A key ingredient is the following lemma.

LEMMA 4.22. *If $n \geq 0$ and A is an integer, then*

$$\sum_{j=0}^{n/2} (-1)^j \binom{n-j}{j} 4^{n-2j} \binom{2j-1}{j-A} = (-1)^A \sum_{t \geq 0} \binom{2n+2}{n-2A-4t}.$$

PROOF. Thinking of either expression as a function $f(n, A)$, we establish that each satisfies

$$f(0, A) = \begin{cases} 0 & A > 0 \\ 1 & A \leq 0, \text{ even} \\ -1 & A < 0, \text{ odd,} \end{cases}$$

$f(-1, A) = 0$, and the recursive formula

$$f(n, A) = 4f(n-1, A) - f(n-2, A-1) - 2f(n-2, A) - f(n-2, A+1).$$

Thus they are equal. The recursive formula for the LHS of 4.22 is proved using several applications of (4.3), while for the RHS one proves the closely related formula

$$\binom{2n+2}{m} = 4\binom{2n}{m-1} + \binom{2n-2}{m} - 2\binom{2n-2}{m-2} + \binom{2n-2}{m-4}$$

by multiplying both sides by x^m and summing over all values of m. ■

Replace k in (4.19) by $2s+1$, and m by $2a+1$. Using 4.22 with $A = s+1$, (4.19) becomes

$$\begin{aligned} &\sum_{s \geq 0} (2s+1)^{2a+1} (-1)^{s+1} \sum_{j=0}^{n/2} (-1)^j \binom{n-j}{j} 4^{n-2j} \binom{2j-1}{j-s-1} \\ = &\sum_{j=0}^{n/2} (-1)^j \binom{n-j}{j} 4^{n-2j} \sum_{s \geq 0} (-1)^{s+1} (2s+1) \binom{2j-1}{j-s-1} \sum_{i \geq 0} \binom{a}{i} (4s^2 + 4s)^i. \end{aligned}$$

Replacing s by $j - t - 1$ yields a formula which can easily be manipulated to (4.20), except that the i-sum is unrestricted. The restriction to $i \geq j-1$ follows from the following lemma.

LEMMA 4.23. *If $1 \leq i \leq j-2$, then*

$$\sum_{t=0}^{j-2} (-1)^t \binom{2j-1}{t} (2j - 2t - 1) \binom{j-t}{2}^i = 0.$$

PROOF. We solve the system of equations

$$\begin{array}{lcl} a_1\binom{3}{2} + a_2\binom{4}{2} + \cdots + a_{j-2}\binom{j}{2} & = & 1 \\ a_1\binom{3}{2}^2 + a_2\binom{4}{2}^2 + \cdots + a_{j-2}\binom{j}{2}^2 & = & 1 \\ & \vdots & \\ a_1\binom{3}{2}^{j-2} + a_2\binom{4}{2}^{j-2} + \cdots + a_{j-2}\binom{j}{2}^{j-2} & = & 1 \end{array}$$

by Cramer's rule and the Vandermonde evaluation of determinants, obtaining

$$\begin{aligned} a_{s-2} &= \frac{1}{\binom{s}{2}}\frac{\prod_{t>s}(\binom{t}{2}-1)\prod_{t<s}(1-\binom{t}{2})}{\prod_{t>s}(\binom{t}{2}-\binom{s}{2})\prod_{t<s}(\binom{s}{2}-\binom{t}{2})} \\ &= \frac{2}{s(s-1)}\frac{\prod_{t\neq s}(t(t-1)-2)}{\prod_{t\neq s}(t(t-1)-s(s-1))} \\ &= \frac{2}{s(s-1)}\frac{\prod_{t\neq s}(t-2)(t+1)}{\prod_{t\neq s}(t-s)(t+s-1)} \\ &= (-1)^{s+1}\frac{2s-1}{3}\frac{\binom{2j-1}{j-s}}{\binom{2j-1}{j-2}}. \end{aligned}$$

The values of t in the products always run from 3 to j. Substituting the solutions into the equations yields equations which reduce to those that were to be proved. ■

This completes the proof that the coefficient of ξ_1 in (3.20) is divisible by 2^n if $n \neq 4$. The proof that the coefficient of ξ_1 in (3.19) is also divisible by 2^n is extremely similar. Using virtually identical methods, we show it equals

$$\binom{n+1}{1}2^{2n-3}$$
$$+\sum_{j\geq 2}2^{2n+1-4j}\left(\binom{n+2-j}{j}-\binom{n-j}{j-2}\right)\sum_{i\geq j-1}8^i\binom{a}{i}\sum_{t=0}^{j-2}(-1)^t\binom{2j-1}{t}(2j-2t-1)\binom{j-t}{2}^i.$$

Each term of this is divisible by 2^{n+1} except when $n=4$ and $j=2$, in which case it is divisible by 2^n.

We complete the paper by giving deferred proofs of two lemmas.

PROOF OF LEMMA 4.15. This lemma can probably be proved by some sort of generating function argument. However, we present a proof along the lines by which it was discovered, utilizing `Maple` and the method of [**17**].

As described in [**17**], one can often prove $f(n) := \sum_s F(n,s)$ is constant (independent of n) by finding a function G such that

$$F(n+1,s)-F(n,s) = G(n,s+1)-G(n,s). \tag{4.24}$$

If such a function G can be found, then summing this equation over all values of s yields the conclusion $f(n+1)-f(n)=0$, and hence f is constant.

In our application, we run the software associated with the book [**17**] on the function

$$F(n,s) := (-1)^s\binom{n+s-1-w}{s}\binom{3n-w}{2n-s}\binom{3n-s+v}{2n}\Big/\binom{4n-w+v}{2n}. \tag{4.25}$$

Here n, w, and v are all free parameters, but we tell `Maple` that n is the induction parameter by running `zeil(F(n,s),s,n,N)`, where F is as in (4.25). The second argument of `zeil` is the summation variable, the third is the parameter to be used for the recursion, and the fourth is the symbol to use for the operator which increases the value of the third parameter by 1. The output in this case is `N-1,` followed by $G(n,s)$, an explicit polynomial, involving v and w in addition to n and s, for which (4.24) holds. In this case, the polynomial G involves approximately 500 monomials. The significance of the `N-1` is that this operator acting on F equals $G(n, s+1) - G(n,s)$. This says exactly (4.24). So we know that the ratio of the two expressions involved in the lemma is constant. To know that this ratio is 1, we note that when $n = 0$ each equals 1. ∎

PROOF OF LEMMA 4.18. The proof is by induction on m. The lemma is readily verified if $m = 0$. Let $s_{m,j}$ denote the sum of the two coefficients with which the lemma deals. Since $f_{m+1} = (1 - x + x^2)f_m$, we obtain

$$s_{m+1,j} = \begin{cases} s_{m,1} & \text{if } j = 0 \\ s_{m,j-1} - s_{m,j} + s_{m,j+1} & \text{if } j > 0. \end{cases}$$

The induction step follows easily. ∎

References

[1] J. F. Adams, *On the groups $J(X)$-II*, Topology **3** (1965) 137-171.

[2] ———, *On the groups $J(X)$-III*, Topology **3** (1965) 193-222.

[3] A. Baker, F. Clarke, N. Ray, and L. Schwartz, *On the Kummer congruences and the stable homotopy of BU*, Trans Amer Math Soc **316** (1989) 385-432.

[4] M. Bendersky and D. M. Davis, *2-primary v_1-periodic homotopy groups of $SU(n)$*, Amer Jour Math **114** (1991) 529-544.

[5] ———, *The unstable Novikov spectral sequence for $Sp(n)$, and the power series* $\sinh^{-1}(x)$, Proc Adams Symp, London Math Soc Lecture Note Series, **176** (1992) 55-72.

[6] M. Bendersky, D. M. Davis, and M. Mahowald, *v_1-periodic homotopy groups of $Sp(n)$*, Pac Jour Math **170** (1995) 319-378.

[7] M. Bendersky and R. D. Thompson, *The Bousfield-Kan spectral sequence for periodic homology theories*, to appear in Amer Jour Math.

[8] J. M. Boardman, *Stable operations in generalized cohomology*, Handbook of Algebraic Topology, (1995) 585-686, Elsevier.

[9] A. K. Bousfield, *The K-theory localization and v_1-periodic homotopy groups of H-spaces*, Topology **38** (1999) 1239-1264.

[10] D. M. Davis, *v_1-periodic homotopy groups of $SU(n)$ at odd primes*, Proc London Math Soc **43** (1991) 529-544.

[11] ———, *From representation theory to homotopy groups*, to appear.

[12] ———, *Elements of large order in $\pi_*(SU(n))$*, Topology **37** (1998) 293-327.

[13] D. M. Davis and M. Mahowald, *Some remarks on v_1-periodic homotopy groups*, London Math Society Lecture Notes, **176** (1992) 55-72.

[14] L. Hodgkin, *On the K-theory of Lie groups*, Topology **6** (1967) 1-36.

[15] M. Mimura and H. Toda, *Topology of Lie groups*, Translations of Math Monographs, Amer Math Soc **91** (1991).

[16] C. M. Naylor, *Cohomology operations in the K-theory of the classical groups*, Port Math **38** (1979) 145-153.

[17] M. Petkovsek, H.S.Wilf, and D. Zeilberger, *A=B*, (1996) A.K. Peters.

Hunter College, CUNY, NY, NY 10021
E-mail address: mbenders@shiva.hunter.cuny.edu

Lehigh University, Bethlehem, PA 18015
E-mail address: dmd1@lehigh.edu

Contemporary Mathematics
Volume **279**, 2001

Homologically exotic free actions on products of S^m

William Browder

ABSTRACT. In this paper we construct homologically exotic free actions of $(\mathbf{Z}/p)^\ell$ on $(S^m)^k$ provided p is a sufficiently large prime number.

INTRODUCTION:

For free actions of a cyclic group $\mathbf{Z}/p$ on a sphere S^m, all the possible homotopy types of such actions are exhausted by the classical Lens spaces, defined by the complex numbers acting by a linear representation. For larger elementary abelian p-groups $G = \overset{l}{\Pi}\, \mathbf{Z}/p$, acting freely on products of spheres $\overset{k}{\Pi}\, S^m$, we will show that there exist actions with a much richer variety of homotopy types, when $l \geq 3$, $k \geq 4$, for "large" primes p.

Our main invariant is the kernel of the cohomology map induced by the classifying map $C : X/G \longrightarrow B_G$ for the principal G-bundle over the quotient space X/G, where G acts freely on X. If $H^i(X) = 0$ for $i < m$, and $H^m(X)$ is $\mathbf{Z}$-free of rank k, then $(\ker C^*)^{m+1}$ is a submodule of $H^{m+1}(B_G)$ with at most k generators, and for $G = \overset{l}{\Pi}\, \mathbf{Z}/p$, $H^*(B_G)$ injects into $H^*(B_G; \mathbf{F}_p) \cong \Lambda(x_1, \ldots, x_\ell) \otimes \mathbf{F}_p[y_1, \ldots, y_\ell]$, $x_i \in H^1$, $y_i \in H^2$.

We set $K = \text{image}\ (\ker C^*)^{m+1} \subset H^{m+1}(B_G; \mathbf{F}_p)$. If the action factors through the action of a connected group, then $K \subset \mathbf{F}_p[y_1, \ldots, y_\ell]$. This is true also for any G action which is the product of actions (not necessarily free) on manifolds homotopy equivalent to S^m. This is shown in §4.

We will construct actions for which $K \not\subset \mathbf{F}_p[y_1, \ldots, y_\ell]$. We call such actions "*homologically exotic.*"

In §3, we construct *smooth* $G = \overset{\ell}{\Pi}\, \mathbf{Z}/p$ actions on $\overset{k}{\Pi}\, S^m$, $\ell \geq 3$, $k \geq 4$, $k \geq \ell$, $p > km/2$, $m \geq 3$ odd, which are homologically exotic. This is done by *perturbing* a non-exotic action in a certain way, and leads to relatively few possibilities for K. We use some constructions of complex n-plane bundles, carried out in §2.

In §5, we show that for any $K_0 \subset H^{m+1}(B_G)$ such that $H^*(B_G; \mathbf{F}_p)/(K)$ is finite dimensional, ($K =$ the mod p reduction of K_0, and (K) denotes the ideal

2000 *Mathematics Subject Classification.* Primary 57P10; 57S17.
Key words and phrases. group action.
Research partially supported by an NSF grant.

generated by K), for $p > n(k-1)/2$, there is a free action of G on a space X, where X/G is dominated by a finite CW complex, and X is homotopy equivalent to $\overset{k}{\Pi}\, S^m$, with K_0 as $\ker C^* \subset H^{m+1}(B_G)$.

§6 is devoted to some differential algebra, used in preceding sections.

We give precise statements of the main results in §1.

The construction of §5 is a geometrical form of a special case of the theorem of [Benson-Carlson]. The obstruction to X/G being of the homotopy type of a finite complex is the class of the Benson-Carlson module in $\tilde{K}_0(\mathbf{Z}G)$, and it is interesting to ask when it is zero.

Likewise, it is interesting to ask whether the Spivak normal fiber space of X/G can be lifted to a topological bundle.

I am indebted to Nick Katz for helpful suggestions.

§1. Statement of Results:

Here $\mathbf{Z}/p$ denotes the integers mod p, $\mathbf{F}_p$ the field with p-elements, (which we distinguish by notation, so that $\mathbf{Z}/p$ is a *transformation group*, and $\mathbf{F}_p$ is the *coefficient* group in cohomology) and $\mathbf{Z}_{(p)}$ denotes the integers localized at p, i.e., rational numbers with denominator prime to p (p a prime).

As in the introduction, let $G = \overset{\ell}{\Pi}\, \mathbf{Z}/p$, and let G act freely on a finite dimensional CW complex X, with $H^i(X;\mathbf{Z}_{(p)}) = 0$ for $i < m$, $C : X/G \longrightarrow B_G$ the classifying map of the principal G-bundle $X \longrightarrow X/G$, B_G the classifying space of G. Let $K_0 = \ker\left(C^* : H^{m+1}(B_G;\mathbf{Z}_{(p)}) \longrightarrow H^{m+1}\left(X/G;\mathbf{Z}_{(p)}\right)\right)$, (called the k-invariant) and $K = im\ K_0 \subset H^{m+1}(B_G;\mathbf{F}_p)$, (mod p k-invariant).

In §3, we will prove:

(3.1) Theorem *Suppose* $H^*(X;\mathbf{Z}_{(p)}) \cong H^*(\overset{k}{\Pi}\, S^m;\mathbf{Z}_{(p)})$ *m odd, and G acts freely on X and trivially on* $H^*(X;\mathbf{Z}_{(p)})$, K_0 *= the k-invariant of this action. Given any* $z \in H^{m+1}(B_G;\mathbf{Z}_{(p)})$, *if* $p > km$, *there exists a free G-action on* $S^m \times X$ *with k-invariant* $= K_0 + (z)$, *which is the restriction to G of a free* $\mathbf{Z}/p \times G$ *action on* $S^m \times X$, *which is smooth if the original G-action on X was smooth.*

The construction in (3.1) comes from a complex n-plane bundle over X/G, ($m = 2n-1$) constructed using the following result from §2:

(2.1) Theorem *For each n, and each $p > n$, and each t such that $n(t-1) < p-1$ there exists a space $B_n(t)$ and a complex n-plane bundle $\xi_n(t)$ over $B_n(t)$ with the properties:*

(i) *$B_n(t)$ has cells only in dimensions*

$$2kn,\ \ 0 \le k < t.$$

(ii) $H^*(B_n(t);\mathbf{Z}_{(p)}) \cong \mathbf{Z}_{(p)}[c_n]/(c_n^t)$ *where* $c_n = c_n(\xi_n(t))$ *is the nth Chern class (localized at p) of* $\xi_n(t) \in H^{2n}(B_n(t);\mathbf{Z}_{(p)})$.

(iii) *Given a CW complex Y and an element of* $y \in H^{2n}(Y;\mathbf{Z}_{(p)})$, *if* $H^i(Y) = 0$ *for* $i \ge 2nt$, *then there exists a map* $\alpha : Y \longrightarrow B_n(t)$ *such that* $\alpha^*(c_n) = qy$ *for some integer q prime to p.*

As a corollary to (3.1) we can construct homologically exotic, smooth actions. For example:

(3.2) Corollary *For $p > 3$, there exists a smooth free action of $G \cong \prod^{\ell} \mathbf{Z}/p$, $\ell =$ 3 or 4 on $(S^3)^4$ with mod p k-invariant not contained in the polynomial part of $H^4(B_G; \mathbf{F}_p)$.*

Calling these actions "homologically exotic" is justified by the following, (proved in §4):

(4.3) Corollary *Let $G \times X_i \longrightarrow X_i$, $i = 1, \dots, q$ be actions of G (not usually free), where X_i are finite dimensional and $H^*(X_i; \mathbf{Z}_{(p)}) \cong H^*(S^n; \mathbf{Z}_{(p)})$. Then the mod p k-invariant of the product action $G \times X \longrightarrow X$, where $X = \prod_1^q X_i$, is contained in the polynomial part of $H^*(B_G; \mathbf{F}_p)$, and is in fact generated by polynomials $P_1, \dots, P_q$, where each P_i is the product of linear forms over $\mathbf{F}_p$.*

In §5 we explore a more general method of constructing actions with a given k-invariant $K \subset H^{n+1}(B_G; \mathbf{Z}_{(p)})$:

(5.1) Theorem *Let $u_1, \dots, u_q \in H^{2n}(B_G; \mathbf{Z}_{(p)})$ such that, if $\bar{u}_i$ = reduction mod p of u_i, then $H^*(B_G; \mathbf{F}_p)/(\bar{u}_1, \dots,, \bar{u}_q)$ is finite dimensional, (where (...) denotes the ideal generated by ...). If $p > \frac{(q-1)(2n-1)}{2} + 1$, then there is a free G action on a finite dimensional CW complex X homotopy equivalent to $\prod^q S^{2n-1}$ with k-invariant generated by $u_1, \dots, u_q$. Further, X/G is a Poincaré space, dominated by a finite CW complex.*

This may be applied to find actions whose mod p k-invariants are in the polynomial part of $H^*(B_G; \mathbf{F}_p)$ but violate the conclusion of (4.3), so are not the product of actions on homology spheres. For example:

Corollary *$\mathbf{Z}/5 \times \mathbf{Z}/5$ can act freely on a finite dimensional CW complex X homotopy equivalent to $S^5 \times S^5$ so that the k-invariant is generated by $u_1 = y_1^3$, $u_2 = y_1^3 + 2y_1^2 y_2 + y_1 y_2^2 + y_2^3$. This is not homotopy equivalent to a product of actions on X_1, X_2 for any finite dimensional X_i homotopy equivalent to S^5, $i = 1, 2$.*

It is not hard to see that $H^*(B_G; \mathbf{F}_p)/(\bar{u}_1, \bar{u}_2)$ is finite dimensional (see §6). One may check directly on the elements of $\mathbf{F}_5$ that <u>no</u> polynomial of the form $\lambda u_1 + u_2$ splits into linear factors over $\mathbf{F}_5$ (set $y_2 = 1$ and verify the resulting polynomial in one variable never splits).

Similarly:

Corollary *There is a smooth free action of $G = (\mathbf{Z}/5)^3$ on $S^5 \times S^5 \times S^5$ which is not homotopy equivalent to products of G actions on homotopy S^5's.*

§2. Constructing Complex Bundles:

In this section we construct a complex bundle such that the base space looks, at the prime p (large), like an Eilenberg-MacLane space $K(\mathbf{Z}, 2n)$ up to a certain dimension, and the sphere bundle looks like a high dimensional sphere (at p). This will be used in the construction of §3.

Specifically we have:

(2.1) Theorem *For each n, each $p > 2n - 1$, and each t such that $2n(t-1) < 2(p-1)$, there exists a space $B_n(t)$ and a complex n-plane bundle $\xi_n(t)$ over $B_n(t)$ with the following properties:*

(i) *$B_n(t)$ has cells only in dimensions $2kn$, $0 \le k < t$.*
(ii) *$H^*(B_n(t); \mathbf{Z}_{(p)}) \cong \mathbf{Z}_{(p)}[c_n]/(c_n^t)$ where $c_n = c_n(\xi_n(t))$.*
(iii) *Given a CW complex X with $H^i(X) = 0$ for $i \ge 2nt$ and $y \in H^{2n}(X; \mathbf{Z}_{(p)})$ there exists a map $\alpha : X \longrightarrow B_n(t)$ such that $\alpha^*(c_n) = my$, for some integer m prime to p.*

Of course for $n = 1$ the infinite dimensional complex projective space gives an example of B_1 for all p with t replaced in (ii) by ∞, and for $n = 2$, we have $B_{SU(2)}$, which is an example of $B_2(t)$ for any $p > 2t$, so we may assume $n > 2$.

Proof of (2.1) According to [Bott], there is a complex n-plane bundle ω over S^{2n} with $c_n(\omega) = (n-1)!\mu$, where μ generates $H^{2n}(S^{2n})$. We will define $B_n(2)$ to be S^{2n} with $\xi_n(2) = \omega$, if $2p > 2n - 1$, and proceed by induction to define $B_n(t)$ for $2 < t$, (see (2.5) below to verify (iii)).

We proceed by a number of lemmas, assuming (2.1) holds for a given t, and defining $B_n(t+1)$, $\xi_n(t+1)$, if $2nt < 2p + 1$.

(2.2) Lemma *In (2.1), (ii) is equivalent to (ii)$'$: If $\pi : E_0(t) \to B_n(t)$ is the (2n-1) sphere bundle over $B_n(t)$ associated to $\xi_n(t)$, then*

$$\bar{H}_i(E_0(t); \mathbf{Z}_{(p)}) = \begin{cases} 0 & \text{for } i \neq 2nt-1 \\ \mathbf{Z}_{(p)} & \text{for } i = 2nt-1. \end{cases}$$

This follows immediately from the Gysin sequence, using the fact that for a $\mathbf{C}^n$ bundle, $c_n = \chi$, the Euler class.

From (2.2) it follows, using the $\mathbf{Z}_{(p)}$-Hurewicz theorem, that

$$\pi_{2nt-1}(E_0(t)) \otimes \mathbf{Z}_{(p)} \longrightarrow H_{2nt-1}(E_0(t); \mathbf{Z}_{(p)})$$

is an isomorphism, so we can find an element $\gamma_t \in \pi_{2nt-1}(E_0(t))$ which generates $H_{2nt-1}(E_0(t); \mathbf{Z}_{(p)})$, $\gamma_t : S^{2nt-1} \longrightarrow E_0(t)$. By (2.3) below,

$$\pi_{2nt-1}(B_{U(n)}) \otimes \mathbf{Z}_{(p)} = 0 \text{ for } 2nt < 2p+1$$

and it follows that for some multiple of γ_t, $m\gamma_t \in \pi_{2nt-1}(E_0(t))$, with $(m,p) = 1$, $\pi \circ (m\gamma_t)^*(\xi_n(t))$ is trivial, so $\pi \circ (m\gamma_t)^*(\xi_n(t))$ extends to an n-plane bundle $\xi_n(t+1)$ over $B_n(t+1) = B_n(t) \underset{m\gamma_t}{\cup} D^{2nt}$.

(2.3) Lemma *$\pi_i(B_{U(n)}) \otimes \mathbf{Z}_{(p)} = 0$ for $2n < i < 2p+1$.*

Proof: We show $\pi_i(U(n)) \otimes \mathbf{Z}_{(p)} = 0$ for $2n \le i < 2p$, (since $\pi_i(B_{U(n)}) \cong \pi_{i-1}(U(n))$ and we proceed by induction on n, using the fibration of $U(n)$ over S^{2n-1} with fibre $U(n-1)$. By induction $\pi_i(U(n-1)) \otimes \mathbf{Z}_{(p)} = 0$, $2n-2 \le i < 2p$ (the case $U(1)$ being trivial) so $\pi_i(U(n)) \otimes \mathbf{Z}_{(p)} \cong \pi_i(S^{2n-1}) \otimes \mathbf{Z}_{(p)}$ in the range $2n \le i < 2p$, so the result follows from the well known result that $\pi_i(S^{2n-1}) \otimes \mathbf{Z}_{(p)} = 0$ for $2n-1 < i < 2n+2p-4$.

(2.4) Lemma *There is a mod p homotopy equivalence $p : B_n(t+1) \longrightarrow T(\xi_n(t))$ where $T(\xi_n(t)) = E(\xi_n(t)) \cup C(E_0(t))$, the Thom complex of the bundle $\xi_n(t)$.*

This is obvious since $S^{2nt-1} \longrightarrow E_0(t)$ is an isomorphism in $H_*(\cdots; \mathbf{Z}_{(p)})$.

Since $H^*(T(\xi_n(t); \mathbf{Z}_{(p)}) \cong \mathbf{Z}_{(p)}[c_n]/(c_n^{t+1})$ from the Thom isomorphism theorem, (2.1) (ii) follows from (2.4). (2.1) (i) is obvious by construction. Finally, we have:

(2.5) Proposition *Suppose B is 1-connected finite dimensional complex with $H^*(B; \mathbf{Z}_{(p)}) \cong \mathbf{Z}_{(p)}[y]$ in dimensions $\leq q < 2n + 2(p-1)$, $\dim y = 2n$. Then there is a map $S : K(\mathbf{Z}, 2n)^q \longrightarrow B$ such that*

$$S_* : H_i(K(\mathbf{Z}, 2n)^q; \mathbf{Z}_{(p)}) \longrightarrow H_i(B; \mathbf{Z}_{(p)})$$

is an isomorphism for $i < q$.

By the calculations of [Cartan], $H^*(K(\mathbf{Z}, 2n); \mathbf{Z}_{(p)}) \cong \mathbf{Z}_{(p)}[\iota]$ in dimensions $< 2n + 2(p-1)$, $\iota \in H^{2n}$, so that if $f : B \longrightarrow K(\mathbf{Z}, 2n)$ is defined by $f^*\iota) = y$, f^* is an isomorphism on $H^i(...; \mathbf{Z}_{(p)})$ for $i < q$.

If F is the fibre of the map f it follows that $\pi_i(F)$ is a finite torsion group of order prime to p for all $i < q$. Therefore, by an easy argument in obstruction theory, for some integer m prime to p, the map mf, defined by $(mf)^*(\iota) = my$ has a section over the q - skeleton $S : (K(\mathbf{Z}, 2n))^{q-1} \longrightarrow B$, (since $mf = f \circ (m\,\iota)$ and the effect of $(m\,i)$ on cohomology is multiplication by powers of m.)

Then (2.1) (iii) follows from (2.1) (ii) and (2.5).

§3. Perturbing Smooth Actions:

(3.1) Theorem *Let $G = \prod^{\ell} \mathbf{Z}/p$ act freely on a finite dimensional CW complex X, where $H^*(X; \mathbf{Z}_{(p)}) \cong H^*(\prod^{k} S^m; \mathbf{Z}_{(p)})$, G acting trivially on $H^*(X; \mathbf{Z}_{(p)})$, and suppose the k-invariant of X/G is $K_0 \subset H^{m+1}(B_G)$, where $m = 2n-1$, and let $z \in H^{m+1}(B_G)$, $z \notin K_0$. If $p > \frac{(k-1)m}{2}$, then there exists a free action of G on $S^m \times X$ with k- invariant $K_0 + (z)$, which is the restriction of a $\mathbf{Z}/p \times G$ action on $S^m \times X$. Further, if the original G action on X was smooth, then the new action of $\mathbf{Z}/p \times G$ on $S^m \times X$ is smooth.*

By (2.1), $X/G \xrightarrow{f} K(\mathbf{Z}, 2n)^q \xrightarrow{s} B_n$ is such that $(sf)^*(c_n) = c^*(rz)$, $(r, p) = 1$, where $c : X/G \longrightarrow B_G$ is the classifying map. It follows that $(sf)^*(c_n) = c^*(z)$ goes to zero in $E_0((sf)^*(\xi_n))$ (since c_n goes to zero in $E_0(\xi_n)$) so that z goes to zero in $E_0((sf)^*(\xi_n))$, and hence the k-invariant of $E_0((sf)^*(\xi_n))$ is $K_0 + (z)$.

If $\pi : X \longrightarrow X/G$ is the covering projection, than $E_0(\pi^*(sf)^*(\xi_n))$ is the covering space of $E_0((sf)^*(\xi_n))$. Now $pz = 0$ in $H^*(B_G)$, and $f^*(\iota) = c^*(z)$, so $pf^*(\iota) = 0$, and we get that $\pi^* f^*(c_n) = 0$ since $H^*(X)$ is torsion free. Hence, the composite map $X \xrightarrow{\pi} X/G \xrightarrow{f} K(\mathbf{Z}, 2n)^q$ is null-homotopic, and hence $sf\pi : X \longrightarrow B_n$ is null-homotopic, and $\pi^*(sf)^*(\xi_n)$ is the trivial bundle and $E_0((sf)^*(\xi_n)) = S^n \times X$.

Now $S^1 \subset \mathbf{C}$ acts freely on the sphere bundle of any complex vector bundle, so this defines a free action of S^1 on $(S^m \times X)/G$. Thus we have fibration (up to homotopy)

$$(S^m \times X)/G \longrightarrow ((S^m \times X)/G)/S^1 \longrightarrow B_{S^1}$$

Since the S^1 preserves fibres we get a diagram

$$\begin{array}{ccc} S^m/S^1 & \longrightarrow & ((S^m \times X)/G)/S^1 \\ \downarrow & & \downarrow \\ B_{S^1} & \xrightarrow{1} & B_{S^1} \end{array}$$

and restricting to $\mathbf{Z}/p \subset S^1$

$$\begin{array}{ccc} S^n/\mathbf{Z}/p & \longrightarrow & (S^n \times X)/G)/\mathbf{Z}_p \\ \downarrow & & \downarrow \\ B_{\mathbf{Z}/p} & \xrightarrow{1} & B_{\mathbf{Z}/p} \end{array}$$

Thus $\mathbf{Z}/p$ splits off from $\pi_1(((S^n \times X)/G)/\mathbf{Z}/p)$ so it follows, since the $\mathbf{Z}/p$ action is a principal $\mathbf{Z}/p$-bundle, that the group is $\mathbf{Z}/p \times G$ which acts on $S^m \times X$.

Since the construction comes from a complex linear n-plane bundle over X/G, it follows that the $\mathbf{Z}/p \times G$ action is smooth on $S^m \times X$ if the original G action was smooth on X.

Now we give the smallest possible example with this approach:

(3.2) Corollary *For $p > 5$, there exists a free smooth action of $G \cong \overset{\ell}{\Pi}\, \mathbf{Z}/p$ on $(S^3)^4$, $l = 3$ or 4 with k- invariant $\not\subseteq$ polynomial part of $H^4(B_G;\, \mathbf{F}_p)$.*

Take a linear action of $G = \overset{3}{\Pi}\, \mathbf{Z}/p$ on $(S^3)^3$, so that the k-invariant $K_0 \subset H^4(B_G)$ injects into the polynomial part of $H^4(B_G;\, \mathbf{F}_p)$, for example, generated by y_1^2, y_2^2 and y_3^2, where $H^*(B_G; \mathbf{F}_p) \cong \Lambda(x_1, x_2, x_3) \otimes \mathbf{F}_p[y_1, y_2, y_3]$. Let $z = \delta(x_1 x_2 x_3) \in H^4(B_G)$, where δ is the Bockstein coboundary associated to the exact coefficient sequence

$$0 \longrightarrow \mathbf{Z} \xrightarrow{\times p} \mathbf{Z} \longrightarrow \mathbf{F}_p \longrightarrow 0.$$

Then the image $z = y_1 x_2 x_3 - x_1 y_2 x_3 + x_1 x_2 y_3 \in H^4(B_G;\, \mathbf{F}_p)$, is not in the polynomial part, and by (3.1) there is a free action of G on $(S^3)^4$ with k-invariant $K_0 + (z)$ which extends to an action of $\mathbf{Z}/p \times G$ on $(S^3)^4$.

Note that for $\ell \leq 2$, $H^{\text{even}}(B_G)$ injects into the polynomial part of $H^*(B_G;\, \mathbf{F}_p)$, so (3.2) is the example of both lowest dimension and lowest rank.

§4 STANDARD k-INVARIANTS

In this paragraph we will show that products of actions of $G \cong \Pi\, \mathbf{Z}/p$ on a single spheres have "standard" k-invariants.

(4.1) Theorem *Let $G \cong \overset{\ell}{\Pi}\, \mathbf{Z}/p$ act on a finite dimensional space X with $H^*(X;\, \mathbf{Z}_{(p)}) \cong H^*(S^m;\, \mathbf{Z}_{(p)})$, preserving orientation, i.e., trivially on $H^*(X;\, \mathbf{Z}_{(p)})$. Then for this action $K \subset \mathbf{Z}_{(p)}[y_1, \ldots, y_\ell] \subset H^*(B_G;\, \mathbf{Z}_{(p)})$, the polynomial part of the group cohomology.*

(4.2) Complement *Under the hypothesis of (4.1) K is generated by a single element which is the product of linear forms in $H^2(B_G;\, \mathbf{Z}_{(p)})$.*

(4.3) Corollary *For G actions on $X_1, X_2; X_q$, X_i as in (4.1), $H^*(X_i;\, \mathbf{Z}_{(p)}) \cong H^*(S^{n_i};\, \mathbf{Z}_{(p)})$, the mod p k- invariant of G acting by the product action of $X_1 \times X_2 \times \ldots \times X_q$ is contained in the polynomial part of $H^*(B_G;\, \mathbf{F}_p)$, and is in fact generated by q elements $P_1; P_q$ where P_i is the product of linear forms over $\mathbf{F}_p$, $P_i \in H^{n_i+1}(B_G;\, \mathbf{Z}_{(p)})$.*

Proof: We simply apply (4.1) and (4.2), noting that the k-invariant of the product action is generated by the k-invariants of the action on each sphere. □

We start to prove (4.1) and (4.2) by proving them for linear actions, i.e., for $G \subset O(n+1)$, or for actions which extend to the torus action.

(4.4) Proposition *Let $T^l = \prod^{\ell} S^1$ act on a finite dimensional space X with $H^*(X;\, \mathbf{Z}_{(p)}) \cong H^*(S^m;\, \mathbf{Z}_{(p)})$. Then for the G action induced by $G \subset T^\ell$, the k-invariant K is contained in the polynomial part of $H^*(B_G;\, \mathbf{Z}_{(p)})$.*

Proof: This follows immediately from the fact that $im\,(H^*(B_{T^\ell};\, \mathbf{Z}_{(p)})) \subset H^*(B_G;\, \mathbf{Z}_{(p)})$ is exactly the polynomial part of $H^*(B_G;\mathbf{Z}_{(p)})$ and the k-invariant of the G action comes from the analogous k-invariant of the T^ℓ action, contained in $H^*(B_{T^\ell};\, \mathbf{Z}_{(p)})$. □

(4.5) Proposition *Let $G \subset SO(m+1)$ acting on S^m. Then the k-invariant of the G action is in the polynomial part of $H^*(B_G;\, \mathbf{Z}_{(p)})$, generated by P = a product of linear forms in $H^2(B_G;\, \mathbf{Z}_{(p)})$.*

Proof: By the usual representation theory of abelian groups, the representation decomposes as the sum of two dimensional representations, and each irreducible representation factors through $\mathbf{Z}/p$, (i.e., the representation is the sum of characters). The k-invariant of each character (as an action on S^1) is a linear form in H^2. The action on S^m is the join of these actions on these S^1's and the k-invariant of the join action is the product of the k-invariants on the S^1's. □

The proof of (4.1) and (4.2) may now be completed by mapping any action as in (4.1) into a linear action by a map of non-zero degree (mod p), which can be done by a result of Dotzel, (to appear). But we may also use an older result of [Dotzel, §3]:

(4.6) Proposition *Let $G \cong \prod^{\ell} \mathbf{Z}/p$ act on a finite dimensional CW complex X with $H^*(X;\, \mathbf{Z}_{(p)}) \cong H^*(S^m;\, \mathbf{Z}_{(p)})$. Then there exists a G-map f of the suspension of X into a linear representation with degree $f \not\equiv 0$ mod p, $f : \Sigma X \longrightarrow S^{m+1}$.*

Now the proof of (4.1) and (4.2) using (4.6) and the following:

(4.7) Proposition *The suspension induces an isomorphism between the spectral sequences of the Borel constructions on (X, X^G) and $(\Sigma X, \Sigma X^G)$. This isomorphism preserves base dimension and shifts fibre dimension by 1.*

Proof: The external cup or cross product induces a G-co-chain map.

$$C^*(X,\, X^G) \longrightarrow C^{*+1}((X,\, X^G) \times ([0,1], \{0,1\})$$

while the second co-chain complex is isomorphic to $C^{*+1}(\Sigma X,\, \Sigma X^G)$. Thus (4.7) follows. □

Now from (4.7) and (4.6) we get that (under the hypothesis of (4.1)) an action on a finite dimensional homology sphere X has the k-invariant of some linear action, which by (4.5) is a product of linear forms in $H^2(B_G\,;\,\mathbf{Z}_{(p)})$. □

§5 General k-invariants for large p.

In this section we show:

(5.1) Theorem *Let $u_1, \dots, u_q \in H^{2n}(B_G\,;\mathbf{Z}_{(p)})$, with the property that $H^*(B_G\,;\,\mathbf{F}_p)/(\bar{u}_1,\,\bar{u}_q)$ is finite dimensional ($\bar{u}_i$ = reduction mod p of u_i). Then there is a free G action on a finite dimensional space X with X homotopy equivalent to $\overset{q}{\Pi}\, S^{2n-1}$, whose k-invariant is generated by $u_1, \cdot, u_q$, provided that*

$$p > \frac{(q-1)\,(2\,n-1)}{2} + 1.$$

Further, X/G is a Poincaré space dominated by a finite complex.

In §6 we will give a few properties of such sets $\{\bar{u}_1\,,\,\dots\,,\bar{u}_q\} \subset H^*(B_G\,;\,\mathbf{F}_p)$.

The idea of the proof of (5.1) is to create the first stage in the Postrikov system of the desired quotient space, and then take a skeleton of it. From our choice of p, if we take the $q(2n-1)+1$ skeleton, this has the right $\mathbf{Z}_{(p)}$-homology to be the quotient, except in dimension $q(2n-1)+1$. We then show that this homology group is spherical (i.e., in the image of the Hurewicz homomorphism) and $\mathbf{Z}_{(p)}\,G$ projective. We may then attach a (possibly infinite) number of free G cells to make the resulting universal covering space mod p homotopy equivalent to a product of $(2n-1)$ spheres. A standard "homotopy mixing" argument (with a product of spheres) completes the construction.

Define a space $\bar{X}_0$ by the pull-back diagram

$$\begin{array}{ccc} \bar{X}_0 & \longrightarrow & \text{(Path Space)} \\ \downarrow & & \downarrow \\ B_G & \underset{\Pi_{u_i}}{\longrightarrow} & \overset{q}{\Pi}\, K(\mathbf{Z},\,2n) \end{array}$$

so that $\bar{X}_0$ has only two non-zero homotopy groups: $\pi_1(\bar{X}_0) = G$ and $\pi_{2n-1}(\bar{X}_0) = (\mathbf{Z})^q$. We let X_0 = the universal covering space of $\bar{X}_0$, so G acts freely on X_0 with $X_0/G = \bar{X}_0$.

Now define a differential graded co-chain algebra A over $\mathbf{F}_p$ by $A = \Lambda(\xi_i, \cdot, \xi_q) \otimes H^*(B_G;\,\mathbf{F}_p)$ with $d(\xi_i) = u_i$, dim $\xi_i = 2n-1$, and $d(H^*(B_G\,;\,\mathbf{F}_p)) = 0$.

(5.2) Lemma *$H^i(A) = 0$ for $i > q\,(2n-1)$.*

We will postpone the proof till §6, (see (6.4)).

According to the calculations of Cartan, $H^i(K(\mathbf{Z};\,2n-1);\,\mathbf{F}_p) = 0$ for $i < 2n-1$ and $2n-1 < i < 2n-1+2(p-1)$, the latter integer being the dimension of the Steenrod operation $\mathcal{P}^1$ applied to the fundamental class in $H^{2n-1}(K(\mathbf{Z},\,2n-1),\,\mathbf{F}_p)$.

It follows that the co-chain algebra A describes the term $E_{2n}^{s,t}$ in the spectral sequence mod p for $\bar{X}_0$ over B_G for $t < 2n-1+2(p-1)$, since the higher dimensional generators of the fibre cohomology $H^*(\Pi\, K(\mathbf{Z},\,2n-1);\,\mathbf{F}_p)$ are all in

higher dimensions and are transgressive, being iterated Steenrod operations on the fundamental class.

It follows that

(5.3) Lemma $E_m^{s,t} = 0$ *for* $m > 2n$, $t + s > q(2n-1)$ *and* $t < 2n - 1 + 2(p-1)$.

For this is the region where $H^i(A) = 0$, and where this homology gives the term E_{2n+1} in the spectral sequence. □

Now let $X_1 =$ the $q(2n-1)+1$ skeleton of X_0. One sees easily that

$$(5.4) \qquad H_i(X_1;\, \mathbf{Z}_{(p)}) \cong \begin{cases} H_i(X_0;\, \mathbf{Z}_{(p)}) & \text{for } i \leq q(2n-1) \\ \text{is } \mathbf{Z}_{(p)} \text{ free} & \text{for } i = q(2n-1)+1 \\ 0 & \text{for } i > q(2n-1)+1 \end{cases}$$

Further we have (if $b = q(2n-1)+1$)

$$(5.5) \qquad \pi_{b+1}(X_0, X_1) \otimes \mathbf{Z}_{(p)} \cong H_{b+1}(X_0, X_1;\, \mathbf{Z}_{(p)}) \cong H_b(X_1; \mathbf{Z}_{(p)})$$

by the relative Hurewicz Theorem, and the fact that $H_{b+1}(X_0; \mathbf{Z}_{(p)}) = H_b(X_0; \mathbf{Z}_{(p)}) = 0$, (using the exact sequence of the pair (X_0, X_1)).

Now we show:

(5.6) Lemma $H^b(X_1;\, \mathbf{Z}_{(p)})$ *is cohomologically trivial as a* $\mathbf{Z}_{(p)}G$ *module.*

From (5.6), it will follow that $H_b(X_1;\, \mathbf{Z}_{(p)})$ is cohomologically trivial as a $\mathbf{Z}_{(p)}G$ module, and since $H_b(X_1;\, \mathbf{Z}_{(p)})$ is $\mathbf{Z}_{(p)}$ free by (5.4), it follows from [Rim] that $H_b(X_1; \mathbf{Z}_{(p)})$ is $\mathbf{Z}_{(p)}G$ projective. Hence, we may add a (possibly infinite) number of free G cells of dimension b and $b+1$ to kill $H_b(X_1; \mathbf{Z}_{(p)})$, with all other homology groups remaining the same, to get a space we call X_2, $X_2 = X_1 \smile \bigcup_{\substack{\alpha \\ g\in G}} g\, D_\alpha^{b+1}$.

It follows that X_2 is the mod p homotopy type of $\Pi\, S^{2n-1}$. We give more detail later.

We deduce (5.6) from (5.7) below. Let $'E_r^{s,t}$ be the spectral sequence for X_1/G over B_G with $\mathbf{Z}_{(p)}$ coefficients, so that $'E_2^{s,t} \cong H^s(B_G;\, H^t(X_1;\, \mathbf{Z}_{(p)}))$.

Since X_1/G is finite dimensional, it follows that $'E_\infty^{s,t} = 0$ for $s+t >$ dimension $X_1/G = b$.

(5.7) Lemma $'E_\infty^{s,b} = {'E_2^{s,b}} \cong H^s(B_G; H^b(X_1;\, \mathbf{Z}_{(p)}))$.

But $'E_\infty^{s,t} = 0$ for $s+t > b$, so that $H^s(B_G; H^b(X_1; \mathbf{Z}_{(p)})) = 0$ for all $s > 0$. The same argument applies to any subgroup $H \subset G$, so that $H^b(X_1; \mathbf{Z}_{(p)})$ is cohomologically trivial.

Proof of (5.7) Since X_1 has dimension $b = q(2n-1)+1$, $H^t(X_1; \mathbf{Z}_{(p)}) = 0$ for $t > b$. Hence

$$d_r : {'E_r^{a,b+r-1}} \longrightarrow {'E_r^{a+r,b}} \text{ is zero,}$$

i.e., $im\, d_r \cap {'E_r^{s,b}} = 0$. It remains to show

$$(5.8) \qquad d_r({'E_r^{s,b}}) = 0, \text{ all } r, \text{ all } s.$$

Assume by induction that:

$$d_t({'E_t^{s,b}}) = 0 \text{ for } s < a, \text{ all } t, \text{ and } d_t({'E_t^{a,b}}) = 0 \text{ for } t < r.$$

We will show that $d_r('E_r^{a,b}) = 0$, and this will conclude the induction step. The inclusion $i : X_1 \subset X_0$ induces a map of the spectral sequences (of the Borel constructions)

$$i_t^* : E_t^{u,v} \longrightarrow {}'E_t^{u,v}$$

and

$$i_2^* : E_2^{u,v} \longrightarrow {}'E_2^{u,v}$$

is an isomorphism for $v < b$, since $H^v(X_0) \longrightarrow H^v(X_1)$ is an isomorphism for $v < b$, $(E_2^{s,t} \cong H^s(B_G;\, H^t(X_0;\, \mathbf{Z}_{(p)}))$.

Let $x \in {}'E_r^{a,b}$. By an argument analogous to (5.3) we have ${}'E_t^{u,v} = 0$ for $u + v > q(2n-1)$, $v < b \leq 2n - 1 + 2(p-1)$ and $t > 2n$. Thus it follows that if $r > 2n$, $d_r(x) = 0 \in {}'E_r^{a+r,b-r+1} = 0$, since $a + b > q(2n-1)$, and $b - r + 1 < b$.

It remains to show that $d_r x = 0$ if $r \leq 2n$, $x \in {}'E_r^{a,b}$, as above.

Since $E_2^{u,v} \longrightarrow {}'E_2^{u,v}$ is onto for $v \neq b$, it follows that it is only $d_{r-1}('E_{r-1}^{u,b})$ which can make ${}'E_r^{u,v}$ different from $E_r^{u,v}$, and since $(d_{r-1}('E_R^{u,b})) \subset {}'E_{r-1}^{u+r,\, b-r+1}$ it follows that

(5.9) $\quad E_r^{u,v}$ is isomorphic to ${}'E_r^{u,v}$ for $v \leq b - r + 1$.

Therefore we get:

(5.10) $\quad d_r x = i^* y,\ y \in E_r^{a+r,\, b-r+1}$

Since

(a) $E_r^{u,v} \cong E_{2n}^{u,v}$,

and

(b) $E_{2n+1}^{u,v} = 0$ for $u + v > q(2n-1)$
$v < b < 2n - 1 + 2(p-1)$ by (5.3),

and it follows $d_{2n}\{y\} = z \neq 0$ if $y \neq 0$, since $E_{2n}^{a+r,\, b-r+1} \cap (im\, d_{2n}) = 0$ (the complementary degree $b - r + 1$ being too large, since $r \leq 2n$).

Now $z \in E_{2n}^{a+r+2n,\, b-r-2n+2}$ and $b - r - 2n + 2 \leq b - 2n$, so $i^* z \neq 0$ in ${}'E_{2n}$ by (5.9), while $i^* y = d_r x$ so $d_{2n} i^* y = 0 = i^*\, d_{2n} y = i^* z$. Hence we have a contradiction unless $d_r x = 0$. So by induction $d_r\, {}'E_r^{a,b} = 0$ all r, all a, so that

$${}'E_2^{a,b} = {}'E_\infty^{a,b} \quad \text{all } a.$$

Since X_1 is finite dimensional, ${}'E_\infty^{a,b} = 0$ for large a, and thus

$$H^a(B_G;\, H^b(X_1)) = {}'E_2^{a,b} = 0$$

for large a, and $H^b(X_1)$ is cohomologically trivial, using the same argument for every subgroup of G.

Since $H^b(X_1)$ is $\mathbf{Z}_{(p)}$ free it follows from [Rim] that $H^b(X_1)$ is $\mathbf{Z}_{(p)}G$ projective, and hence that $H_b(X_1) \cong Hom(H^b(X_1),\, \mathbf{Z}_{(p)})$ is projective.

By the Eilenberg trick, there exists a free module F (possibly infinitely generated) such that $(\pi_{b+1}(X_0, X_1) + F) \otimes \mathbf{Z}_{(p)} \cong H_b(X_1;\, \mathbf{Z}_{(p)}) + F \otimes \mathbf{Z}_{(p)}$ is $\mathbf{Z}_{(p)}G$ free.

Let C_α be a $\mathbf{Z}G$ free basis for F, $\alpha \in A$. Let $\bar{X}_1' = \bar{X}_1 \cup \bigcup_{\alpha\in A} S^b_\alpha$ and $X_1' =$ universal cover of $\bar{X}_1'$. Then define $X_1' \longrightarrow X_1 \longrightarrow X_0$ by simply mapping each S^b_α to the base point (in $\bar{X}_1$). It is easy to see that $\pi_{b+1}(X_0, X_1') \cong \pi_{b+1}(X_0, X_1) + F$, so that $\pi_{b+1}(X_0, X_1') \otimes \mathbf{Z}_{(p)}$ is $\mathbf{Z}_{(p)}G$ free (possibly infinitely generated).

Choose a $\mathbf{Z}_{(p)}G$ basis $\{f_\beta\}$, $\beta \in B$ for $\pi_{b+1}(X_0, X_1')$, and let $\bar{X}_2 = \bar{X}_1 \smile_{f_\beta} \bigcup_{\beta\in B} D^{b+1}_\beta$, and let $X_2 =$ universal covering of $\bar{X}_2$.

Since $\{f_\beta\}$ was a basis for $\pi_{b+1}(X_0, X_1')\otimes\mathbf{Z}_{(p)}$ it follows that $\{\partial f_\beta\}$, $f_\beta : S^b_\beta \longrightarrow \bar{X}_1'$ gives a basis for $H_b(\bar{X}_1'; \mathbf{Z}_{(p)})$. It's easy to see that if $j : X_2 \longrightarrow X_1$ is the natural map, the push out map on each new cell, so that $j_* : H_i(X_2) \longrightarrow H_i(X_1)$ is an isomorphism for $i < b$, and $H_i(X_2) = 0$ for $i \geq b$.

It follows that $H^*(X_2; \mathbf{Z}_{(p)})$ is an exterior algebra, isomorphic to $H^*(X_1; \mathbf{Z}_{(p)})$ in dimensions $< 2n - 1 + 2(p-1)$. Hence, X_2 is the homotopy type at the prime p of a product of $(2n-1)$ spheres, and its G action has the desired k-invariant, generated by $u_1, \dots, u_q$.

To complete the construction, we will simply "mix" the homotopy type of X_2/G at the prime p with the homotopy type of $\Pi S^{2n-1} \times B_G$ at the other primes, "mixing" being the process described by [Sullivan] (see also [Zabrodsky]), but over the free action of G, as follows:

Let $X(\frac{1}{p})$ be the localization of ΠS^{2n-1} away from p, and let $X(\mathbf{Q})$ be the localization of ΠS^{2n-1} at $\mathbf{Q}$, so that $\pi_i(X(\frac{1}{p})) \cong \pi_i(\Pi S^{2n-1}) \otimes \mathbf{Z}[1/p]$ and $\pi_i(X(\mathbf{Q})) \cong \pi_i(\Pi S^{2n-1}) \otimes \mathbf{Q}$, so that ΠS^{2n-1} is homotopy equivalent to the pull back X:

$$\begin{array}{ccc} X & \longrightarrow & X(1/p) \\ \downarrow & & \downarrow \\ X(p) & \longrightarrow & X(\mathbf{Q}) \end{array}$$

($X(p)$ being the localization at p, so that $X(p) \cong X_2$ above).

It follows easily that X_2/G admits a map

$$l_{\mathbf{Q}} : X_2/G \longrightarrow X(\mathbf{Q}) \times B_G$$

so that π_1 is mapped isomorphically and the induced map $X_2 \longrightarrow X(\mathbf{Q})$ is the $\mathbf{Q}$ localization map. Then define X/G as the pull back

$$\begin{array}{ccc} X/G & \longrightarrow & X(1/p) \times B_G \\ \downarrow & & \downarrow \\ X_2/G & \xrightarrow{l_{\mathbf{Q}}} & X(\mathbf{Q}) \times B_G \end{array}$$

The space X/G is a Poincaré duality space, since X/H satisfies Poincaré duality for each prime q, and every subgroup H of G (c.f (6.5)). It follows from [Browder] that X/G is dominated by a finite complex, which completes the proof of (5.1). □

§6 Some differential algebra

In this section we study certain differential algebras of the type which arise in §5. We study the case when the homology is finite dimensional, and prove Poincaré duality in that case. We give the results for fields of odd characteristic and note that, they hold by a similar argument for characteristic 2.

Let $\mathbf{K}$ be a field of characteristic p, and let $A = \Lambda(x_1, \ldots, x_\ell) \otimes \mathbf{K}[y_1, \ldots, y_\ell]$, $\deg x_i = 1$, $\deg y_i = 2$, (e.g. if $\mathbf{K} = \mathbf{F}_p$ p odd, then $A \cong H^*(B_G; \mathbf{F}_p)$) and let

$$B = \Lambda(z_1, \ldots, z_q) \otimes A \text{ be a differential algebra, where}$$

$$d: B \longrightarrow B \text{ is a differential of degree } + 1,$$

$$d(A) = 0 \text{ and } d(z_i) \in A, \text{ all } i.$$

Let $B^t = \Lambda(w_i, \ldots, w_q) \otimes A$ with differential d^t defined by $d^t(A) = 0$ and $d^t(w_i) = (dz_i)^t$ (so that degree $w_i = t$ (degree $z_i + 1) - 1$.

Let $r : A \longrightarrow \mathbf{K}[y_1, \ldots, y_\ell]$ be the natural retraction, $i : \mathbf{K}[y_1, \ldots, y_\ell] \longrightarrow A$ the inclusion, and define

$B_r = \Lambda(z_1, \ldots, z_q) \otimes A$ with differential d_r defined by $d_r(A) = 0$, $d_r(z_i) = r(dz_i) \in A$.

Define

$B_0 = \Lambda(z_1, \ldots, z_q) \otimes \mathbf{K}[y_1, \ldots, y_\ell]$ with $d_0 = d_r$, so that $B_r = B_0 \otimes \Lambda(x_1, \ldots, x_\ell)$ as a differential algebra.

It is clear, using the degree in $\Lambda(z_1, \ldots, z_q)$, that image $(A) \subset H(B, d)$ is $A/(dz_1, \ldots, dz_q)$, where $(u_1, \ldots, u_q)$ denotes the ideal in A generated by $u_1, \ldots, u_q$.

(6.1) Proposition *The following are equivalent*

(a) *$H(B, d)$ finite dimensional*
(b) *$A/(dz_1, \ldots, dz_q)$ finite dimensional*
(c) *There exists an N such that $y_i^N \in (dz_1, \ldots, dz_q)$ for all i.*
(d) *$H(B^t, d^t)$ is finite dimensional for <u>some</u> $t \geq 1$.*
(e) *$H(B^t, d^t)$ is finite dimensional for <u>every</u> $t \geq 1$.*

Further, if char$\mathbf{K} = p \neq 0$ or 2, then (a) - (e) are equivalent to the following:

(f) *$H(B_r, d_r)$ is finite dimensional.*
(g) *$H(B_0, d_0)$ is finite dimensional.*

Proof (a) $\Longrightarrow$ (b) Since $A/(dz_1, \ldots) \subset H(B, d)$.

Now B is an A-module, $d(A) = 0$, so $H(B, d)$ is an A module, and hence a $\mathbf{K}[y_1, \ldots, y_\ell]$ module. Since $\mathbf{K}[y_1, \ldots, y_\ell]$ is Noetherian and B is finitely generated over $\mathbf{K}[y_1, \ldots, y_\ell]$, $H(B, d)$ is also finitely generated over $\mathbf{K}[y_1, \ldots, y_\ell]$, and hence finitely generated over the larger ring A. But the action of A on $H(B, d)$ factors through $A/(dz_1, \ldots, dz_q)$, so $H(B, d)$ is finitely generated over $A/(dz_1, \ldots,)$, so that if the latter is finite dimensional, then so is $H(B, d)$, i.e., (b) $\Longrightarrow$ (a).

If (b), let $M = \max \deg$ in $A/(dz_1, \ldots,)$. Then $2N > M$ implies $y_i^N = 0$ in $A/(dz_1, \ldots)$ so $y_i^N \in (dz_1, \ldots, dz_q)$, i.e., (b) $\Longrightarrow$ (c).

Now we give a lemma which will prove useful several times:

(6.2) Lemma *B as above, let $C = \Lambda(u_1, \dots, u_s) \otimes B$ with differential δ where $\delta | B = d$, and $\delta(u_i) \in (dz_1, \dots, dz_q) \subset A \subset B$. Then $H(B,d)$ is finite dimensional if and only if $H(C,\delta)$ is finite dimensional.*

Proof of (6.2) If we filter C by degree in B, we get a spectral sequence with $E_1 \cong \Lambda(u_1, \dots, u_s) \otimes H(B,d)$ and $d_r \equiv 0$ for $r \geq 1$ since $\delta(u_i) \subset im\, d$, so that $E_1 = E_\infty$. Hence $H(B,d)$ is finite dimensional if and only if E_∞ is finite dimensional. But E_∞ is the associated graded group to a filtration of $H(C,\delta)$, so $H(C,\delta)$ is finite dimensional, if and only if E_∞ is finite dimensional.

This proves (6.2).

Returning to the proof of (6.1), under the hypothesis (c), define $C = \Lambda(u_1, \dots, u_\ell) \otimes B$ with δ defined by $\delta \mid B = d$, and $\delta(u_i) = y_i^N$. By (6.2), $H(C,\delta)$ is finite dimensional if and only if $H(B,d)$ is. But $A/(y_1^N, \dots, y_\ell^N)$ is finite dimensional, so that $A/(\delta u_1, \dots, \delta u_\ell, dz_1, \dots, dz_q)$ is finite dimensional, and $H(C,\delta)$ is finite dimensional by the equivalence of (b) and (a). Hence (c) $\Longleftrightarrow$ (a).

If (d) holds, then

$$A/(d^t w_1, \dots, d^t w_q) = A/((dz_1)^t, \dots, (dz_q)^t)$$

is finite dimensional, and

$$(dz_1, \dots, dz_q) \supset ((dz_1)^t, \dots, (dz_q)^t)$$

so $A/(dz_1, \dots, dz_q)$ is finite dimensional, or (d) $\Longrightarrow$ (b), while (a) $\Longrightarrow$ (d) with $t = 1$, since $B^1 = B$.

If (b) holds, $y_i^N \in (dz_1, \dots, dz_q)$, then $y_i^{Nt} \in ((dz_1)^t, \dots (dz_q)^t)$ so $H(B^t, d^t)$ is finite dimensional, i.e., (b) $\Longrightarrow$ (e). Of course (e) $\Longrightarrow$ (d) $\Longrightarrow$ (b) (as above).

Since $r : A \longrightarrow A$ is an idempotent ($r^2 = r$), it follows that $u = ru + u_0$ where $ru_0 = 0$, any $u \in A$. Also, $\ker r = (x_1, \dots, x_\ell)$.

Note, that if char$\mathbf{K} = p$, and if $\Psi : A \longrightarrow A$ is the Frobenius map $\Psi(u) = u^p$, then Ψ is an algebra homomorphism on A^{even} (the sub-algebra of A of elements of even degree, which is commutative). Further, if p is odd, $\Psi(u) = 0$ if $\deg u$ odd since $u^2 = 0$, so that if $\deg u$ is even, $\Psi(u) = \Psi(ru)$.

On the other hand, we have

(6.3) Lemma *In the differential algebra B, if $p =$ char$\mathbf{K} \neq 2$, and if $\deg z_i$ is even, then $dz_i = 0$, so that we may represent as $B = (\Lambda(z_1, \dots, z_t) \otimes A) \otimes \Lambda(z_{t+1}, \dots, z_q)$ where $\deg z_i$ is odd for $1 \leq i \leq t$, and $d(z_i) = 0$ for $i > t$.*

Proof of (6.3) Since $z_i^2 = 0$ we have $0 = d(z_i^2) = 2z_i\, dz_i$, so that $dz_i = 0$ since multiplication by z_i is an injection on A.

Then $H(B,d)$ is finite dimensional if and only if $H(\Lambda(z_1, \dots, z_t) \otimes A, d)$ is finite dimensional, or in other words, we may assume (if $p \neq 2$) that $\deg z_i$ is odd.

It follows that $B^p \cong (B_r)^p$ as differential algebras, so that (f) is equivalent to (d) with $t = p$. Clearly, $B_r = B_0 \otimes \Lambda(x_1, \dots, x_\ell)$ so that $H(B_r, d_r) = H(B_0, d_0) \otimes \Lambda(x_1, \dots, x_\ell)$ so that (f) $\Longleftrightarrow$ (g). This completes the proof of (6.1).

If char $\mathbf{K} = 0$ we may show that (c) $\Longrightarrow$ (f). We note that the retraction $r : A \longrightarrow \mathbf{K}[y_1, \dots, y_\ell]$ induces a map of graded differential algebras $\hat{r} : B \longrightarrow B_0$. If $y_i^N \in (dz_1, \dots, dz_\ell)$, then $y_i^N \in (rdz_1, \dots, rdz_\ell)$, so that if $H(B,d)$ is finite then $H(B_0, d_0)$ is finite.

Question: If char $\mathbf{K} = 0$ does $H(B_0, d_0)$ finite imply $H(B,d)$ finite.

Definition: A graded algebra A over a field $\mathbf{K}$ is said to *satisfy Poincaré duality* in dimension n if:

1. $A^n \cong \mathbf{K}$ and $A^i = 0$ for $i > n$, and
2. For each $i \leq n$, the pairing given by multiplication $A^i \otimes A^{n-i} \longrightarrow A^n \cong \mathbf{K}$ is non-singular.

This is equivalent to saying that $a \mapsto a\cdot$ is an isomorphism:

$$A^i \longrightarrow (A^{n-i})^* = Hom\,(A^{n-i}, \mathbf{K}).$$

(6.4) Proposition *Let B be as above, and suppose $H(B,d)$ is finite dimensional. Then $H(B,d)$ satisfies Poincaré duality in dimension $m = \sum_{i=1}^{q} \deg z_i$. In particular, $H^s(B,d) = 0$ for $s > m$.*

We need the following for the proof:

(6.5) Proposition

(a) *Suppose U is a differential graded algebra satisfying Poincaré duality in dimension n and suppose $U^n \cong H^n(U)$. Then $H(U)$ also satisfies Poincaré duality in the same dimension n.*

(b) *Suppose U is a filtered algebra such that $F^nU = 0$ for large n, and that the associated graded algebra $\mathcal{G}(U)$ satisfies Poincaré duality in dimension n. Then U satisfies Poincaré duality in dimension n.*

(c) *Let $W = U \otimes V$, tensor product of graded algebras. Then W satisfies Poincaré duality in dimension n if and only if U and V satisfy Poincaré duality in dimensions s and t with $s + t = n$.*

Proof of (6.5) (a) Let μ generate $(U^n)^* \cong H^n(U)^* \cong H_n(U^*) \cong \mathbf{K}$. Then $\mu\cap : U \longrightarrow U^*$ is a chain map $((\mu \cap a)(b) := (a \cdot b)(\mu))$, where $U^* = Hom(U, \mathbf{K})$ is a chain complex, $(\deg d^* = -1)$. Since $\mu\cap$ is an isomorphism $U^i \longrightarrow U_{n-i}$ $(U_{n-i} = Hom(U^{n-i}, \mathbf{K}))$, $(\mu\cap)_* : H^i(U) \longrightarrow H_{n-i}(U^*)$ is also an isomorphism.

To prove (b), we note that the decreasing filtration on the algebra U induces an increasing filtration on the dual U^*, and $\mathcal{G}U \longrightarrow \mathcal{G}U^*$ is the Poincaré duality isomorphism $(\mathcal{G}U^* \cong (\mathcal{G}U)^*)$. The associated graded algebra $\mathcal{G}U$ is a graded algebra for filtration gradation as well as the total degree. It follows that $\mathcal{G}U$ is satisfies Poincaré duality, considered as a graded algebra with either gradation. Suppose $N=$ highest filtration degree of $\mathcal{G}U$, so $\mathcal{G}U^N \cong U^n \cong \mathbf{K}$. Then (b) follows by using the exact sequences:

$$\begin{array}{ccccccccc}
0 \longrightarrow & \dfrac{F^{q+1}U}{F^{q+r}U} & \longrightarrow & \dfrac{F^qU}{F^{q+r}U} & \longrightarrow & \dfrac{F^qU}{F^{q+1}(U)} & \longrightarrow 0 \\
& \downarrow & & \downarrow & & \downarrow \cong & \\
0 \longrightarrow & \dfrac{F_{N-q-1}U^*}{F_{N-q-r}U^*} & \longrightarrow & \dfrac{F_{N-q}U^*}{F_{N-q-r}U^*} & \longrightarrow & \dfrac{F_{N-q}U^*}{F_{N-q-1}U^*} & \longrightarrow 0
\end{array}$$

the Five Lemma and induction on r, $(F^0U = U,\ F^{N+1}U = 0)$

To prove (c) we note first that the tensor product of isomorphisms is an isomorphism, so that if $u\cap : U \longrightarrow U^*$ and $v\cap : V \longrightarrow V^*$ are isomorphisms then $u \otimes v\cap : U \otimes V \longrightarrow U^* \otimes V^* = (U \otimes V)^*$, so that if U and V satisfy Poincaré duality in dimensions s and t (respectively) then $U \otimes V$ satisfies Poincaré duality in dimension $s + t$.

Conversely if $U \otimes V$ satisfies Poincaré duality in dimension m, then for some s, t, $U^s \cong \mathbf{K}$, $U^i = 0$ for $i > s$, and $V^t \cong \mathbf{K}$, $V^j = 0$ for $j > t$, since the highest non-zero dimensional U^i and V^j tensor together to create the highest dimensional $(U \otimes V)^k$. Hence, Poincaré duality in $U \otimes V$ is given by $(u \otimes v)\cap$, where u, v are the generators of U^s, V^t respectively. Since $(u \otimes v)\cap = (u\cap) \otimes (v\cap) : U \otimes V \longrightarrow U^* \otimes V^*$ is an injection, it follows that $u\cap$ and $v\cap$ are also injections. Since $((u \otimes v)\cap)^* = (u \otimes v)\cap$ it follows that they are also surjections, proving (c). This completes the proof of (6.5).

Proof of (6.4) Define a differential algebra $C = \Lambda(\xi_1, \ldots, \xi_\ell) \otimes B$ with differential δ defined by $\delta | B = d$ and $\delta \xi_i = y_i^{N+1}$, where N is chosen so large that $y_i^{N+1} \in d(B)$, which is possible by (6.1).

Then $C = \Lambda(z_1, \ldots, z_q) \otimes B'$ where $B' = \Lambda(\xi_1, \ldots, \xi_\ell) \otimes A$, $d'|B' = \delta|B'$, $\delta(z_i) = d(z_i)$. Filtering C by degree in B' we get a spectral sequence E'_r with $E'_1 = \Lambda(z_1, \ldots, z_q) \otimes H(B')$ and it follows easily from the Kunneth formula that

$$H(B') = \Lambda(x_1, \ldots, x_\ell) \otimes \bigotimes_{j=1}^{\ell} H(\Lambda(\xi_i) \otimes \mathbf{K}[y_i]).$$

Now $H(\Lambda(\xi_i) \otimes \mathbf{K}[y_i]) \cong \mathbf{K}[y_i]/y_i^{N+1}$ which satisfies Poincaré duality in dimension $2N$, so that $H(B')$ satisfies Poincaré duality in dimension $\ell + 2N\ell = \ell(2N+1)$, so that E'_1 satisfies Poincaré duality in dimension $\ell(2N+1) + \sum_{i=1}^{q} \deg z_i$.

By (6.5) (a), $E'_2, \ldots, E'_\infty$ satisfy Poincaré duality in dimension $\ell(2N+1) + \sum \deg z_i$.

It follows from (6.5) (b) that $H(C)$ satisfies Poincaré duality in the same dimension. But we will show

(6.6) $\qquad H(C) \cong \Lambda(\xi_1, \ldots, \xi_\ell) \otimes H(B)$

so that by (6.5) (c) $H(B)$ satisfies Poincaré duality in dimension $\sum \deg z_i$.

Proof of (6.6) Let $w_i \in B$ such that $dw_i = y_i^{N+1}, i = 1, \ldots, \ell$. Define $\xi'_i = \xi_i - w_i$, so that $\delta \xi'_i = 0$ and $(\xi'_i)^2 = 0$, since w_i is odd dimensional. Note, that odd dimensional classes have square zero in C if $p \neq 2$, since $C \cong \Lambda$ (odd dimensional classes) $\otimes$ Polynomials in even dimensions. Then $C \cong \Lambda(\xi'_1, \ldots, \xi'_\ell) \otimes B$ as a B module with $\delta \xi'_i = 0$ so $H(C) = \Lambda(\xi'_1, \ldots, \xi'_\ell) \otimes H(B) \cong \Lambda(\xi_1, \ldots, \xi_\ell) \otimes H(B)$. This concludes the proof of (6.6), and therefore of (6.4).

References

[Be-Ca] Benson, D. & Carlson, J., "Projective resolutions and Poincaré duality complexes," *Trans. AMS*, 342, (1994), 447-488.

[Bo] Bott, R., "The space of loops on a Lie group," *Michigan math. Journal*, 5, (1958), 35-61.

[Br] Browder, W., "Poincaré spaces, their normal fibrations and surgery," *Inv. Math.*, 17, (1972), 191-200.

[Ca] Carlsson, G., "On the rank of abelian groups acting freely on $(S^n)^k$," *Invent. Math.*, 69, (1982), 393-400.

[Ca 2] Carlsson, G., "On the non-existence of free actions of elementary abelian groups on products of spheres," *American Journal of Math.* 102, (1980), 1147-1157.

[Car] Cartan, H., "Seminaire H. Cartan 1954-55," *Algèbres d' Eilenberg-MacLane et homotopice.*

[Do] Dotzel, R., "A converse of the Borel formula," *Trans. AMS*, 250, (1979), 275-287.

[Ri] Rim, D.S., "Modules over finite groups," *Annals of Math.*, 69, (1959), 700-712.

[Su] Sullivan, D., "Genetics of homotopy theory and the Adams conjecture," *Annals of Math.*, 100, (1974), 1-79.

DEPARTMENT OF MATHEMATICS, PRINCETON UNIVERSITY, FINE HALL, WASHINGTON ROAD, PRINCETON, NEW JERSEY

E-mail address: browder@math.princeton.edu

Contemporary Mathematics
Volume **279**, 2001

Surgery formulae for analytical invariants of manifolds

S. E. Cappell, R. Lee, E. Y. Miller

ABSTRACT. This paper discusses some splitting formulae for low eigenmodes and spectral flows of first order, self-adjoint, elliptic operators when the underlying manifolds can be decomposed into two parts along a long cylinder. In studying the analytical invariants of manifolds before and after surgery, these formulae are particularly effective because some of the terms become identical and cancel out each other in a surgery formula. To illustrate this approach, we investigate the Kronheimer-Mrowka Conjecture on the equality of Seiberg-Witten invariants and Casson invariants of three-manifolds.

§1. Introduction.

Recently, there has been a great deal of interest in studying the behavior of analytical invariants of manifolds under the surgery operation. As early as 1990, there was a pioneer effort of Yoshida [**Y**] who presented formulas for the spectral flow of certain operators relevant to Floer homology for 3-manifolds. Nowadays, formulas of this type are known under the general heading of splitting formulas: Roughly, when the underlying manifold can be decomposed into two pieces, such formulas then express the analytical invariant as a sum of the corresponding invariants from the two sides together with an "interaction term" from the cylinder connecting these two pieces. Inspired by Yoshida's effort, our investigations of splitting formulae for spectral flows were in the general setting of a family of first order, self-adjoint, elliptic operators. In the series of three papers [**CLM2**]–[**CLM4**], we obtained several of these splitting formulas, by which spectral flow can be explicitly computed, and then applied them to study the Casson invariant of 3-manifolds and its various generalizations.

As this subject matter of splitting formulae for analytical invariants has become enormous and is still rapidly growing, it would be beyond our scope to present a comprehensive survey of all these different techniques and perspectives. Instead, in the next two sections, we will concentrate on two of our results from [**CLM2**], [**CLM3**] which have had some positive impact on the subsequent development.

2000 *Mathematics Subject Classification.* Primary 55N91,53D12; Secondary 57M27.

Key words and phrases. spectral flow, eta invariant, index theorem, Seiberg-Witten, Casson invariant.

During the research of this paper, all three authors were supported by grants from the National Science Foundation.

The first is the theorem on the exponentially decaying eigenmodes when we stretch the cylinder connecting the two sides of the manifold (see Theorem (2.12)). As it turns out, our gluing techniques for linear, elliptic operators have their counterparts in the study of many moduli spaces, when the underlying manifolds can be decomposed into two pieces. Although the analyses there are nonlinear, the low eigenvalue estimate for the linearized operators is indispensable for the applications of Newton's method (see [**CMW**] [**N3**]).

The second topic is our version of the splitting theorem for spectral flows in [**CLM3**] (see Theorem (3.8)). Despite the fact that it is not as elegant as Nicolaescu's version in [**N1**] [**N2**] and not as complete as Daniel-Kirk in [**DK**], this simple splitting formula has served its purpose as it has found many applications. For example, in [**CLM4**] we applied this splitting formula to study and compare Walker's and other generalizations of Casson's $SU(2)$ -representation theoretic invariant of 3-manifolds, using the η-invariant of certain Dirac operators. In the present paper, we obtain another application by studying a related, though in principle different, invariant of integral homology 3-sphere, known as the Seiberg-Witten Casson invariant (see [**CMW**], [**C**], [**MW**], [**L**]).

Let Y be an oriented, integral homology 3-sphere together with a Riemannian metric g_0. Then, as is well-known, associated to this situation there is a unique spin structure on Y, which in turn gives rise to a bundle S of spinors. Consider the space of all pairs (A, ψ) consisting of an imaginary valued 1-form A on Y and a smooth section ψ of S. Then the 3-dimensional Seiberg-Witten equations are given by:

$$\begin{aligned} D_{g_0}\psi + A\psi &= 0 \\ * \, dA - \tau(\psi, \psi) &= 0 \end{aligned} \tag{1.1}$$

where D_{g_0} is the Dirac operator associated to the spin structure and $\tau(\psi, \psi)$ is a certain bilinear pairing on $\Gamma(S)$.

It will be better to interpret the above equations in terms of the associated *spin*^{c} structure on Y. In this setting, we have in addition to S a $U(1)$-line bundle L whose tensor product $S \otimes L$ with S gives us a $U(2)$-vector bundle. The imaginary valued 1-form A can be viewed a $U(1)$-connection of this line bundle L and so the first equation can be viewed as the twisted Dirac operation ∂_A on the space of sections $\Gamma(S \otimes L)$. As for the second equation, the term $dA = F_A$ is the curvature 2-form associated to the line bundle L and so this equation gives us a restriction on the curvature. From this interpretation, it is not difficult to see that there is a natural gauge group $\mathcal{G} = Map(Y, U(1))$ action on the space of solutions of the above Seiberg-Witten equations. The quotient space of this action is compact and has virtual dimension 0 and is known as the Seiberg-Witten moduli space $\mathcal{M}_{SW}(Y, g, 0)$.

In [**C2**] [**L**] [**MW**], Chen, Lim and Marcolli-Wang developed a systematic method of perturbing the Seibert-Witten equations in (1.1) by varying both the metric g and connection data v:

$$\begin{aligned} \partial_A^g \psi &= 0 \\ * \, F_A + \sigma(\psi, \psi) &= 2i * dv \end{aligned} \tag{1.2}$$

A generic choice of these perturbation data (g, v) is said to be inside a chamber if the moduli space has a unique reducible $(v, 0)$ and the twisted Dirac operator ∂_v^g has no nontrivial harmonic spinors, i.e. $Ker\partial_v^g = 0$. Under such a perturbation, the subspace $\mathcal{M}^*_{SW}(Y,g,v)$ of the moduli space away from the reducible consists of finite number of oriented points and hence we have a well-defined Euler number $\chi(\mathcal{M}^*_{SW}(Y,g,v))$.

Unfortunately, this Euler number by itself is not an invariant of the underlying 3-manifold Y. For, the chambers of admissible perturbations are separated by a codimension-1 subspace $W = \{(g,v)|Ker\partial^g_v \neq 0\}$ called a wall, and when a path of perturbation data (g_t, v_t) connects two points in different chambers, it has to cross the wall at some point $t = t^*$. At such a crossing point some irreducibles in $\mathcal{M}^*_{SW}(Y,g_t,v_t)$ may sink into or emerge from the reducible $(v_{t*},0)$, which in turn leads to a jump in the above Euler characteristic. Indeed, to account for these phenomena there is the following wall crossing formula:

$$\chi(\mathcal{M}^*_{SW}(Y,g_1,v_1)) - \chi(\mathcal{M}^*_{SW}(Y,g_0,v_0)) = SF[\partial^{g_t}_{v_t} : 0 \leq t \leq 1] \tag{1.3}$$

where $SF[\partial^{g_t}_{v_t} : 0 \leq t \leq 1]$ is the spectral flow of a path of Dirac operators connecting $\partial^{g_0}_{v_0}$ to $\partial^{g_1}_{v_1}$. (see [**C2**,page 73], [**MW**, page 53]).

From this wall crossing formula, we will show in Section 4 that the following is a well-defined topological invariant of a 3-manifold:

$$\begin{aligned}\lambda_{SW}(Y) = \chi(\mathcal{M}^*_{SW}(Y,g,v)) + SF[\partial^g_{(1-t)v} : 0 \leq t \leq 1] \\ + 1/8[\eta_{sign}(Y,g) + 4(\eta_{Dirac} + h)(Y,g)]\end{aligned} \tag{1.4}$$

(see Theorem (4.9) and also [**C2**],[**L**],[**MW**]). In fact, a natural conjecture due to Kronheimer and Mrowka is that this invariant coincides with the Casson invariant $\lambda(Y)$. Recently an important step towards a proof of this conjecture has been taken by Carey-Marcolli-Wang, as they established in their paper ([**CMW**,page 61] also [**L**]) the following:

THEOREM 1.5. *Let Y be an integral homology 3-sphere and K be a knot inside. Let Y_1 and Y_0 denote the 3-manifolds obtained from Y by performing $(1,1)$ and $(1,0)$ surgeries on K. Then there exits metric and perturbation data $(g,v),(g_1,v_1)$ on Y, Y_1 and spin^c structure s_k on Y_0 such that*

$$\chi[\mathcal{M}^*_{SW}(Y,g,v)] = \chi[\mathcal{M}^*_{SW}(Y_1,g_1,v_1)] + \sum \chi[\mathcal{M}_{SW}(Y_0,s_k)] \tag{1.6}$$

Note that Y_0 has the same homology as $S^1 \times S^2$ and its *spin^c* structure is completely determined by the first Chern class $c_1(Y_0) = 2k$. According to Theorem 3.2.12 of [**M**], the sum of the Euler characteristic on the right hand side of (1.6) is given by

$$\sum \chi[\mathcal{M}(Y_0,s_k)] = 1/2\Delta''_K(1) \tag{1.7}$$

where $\Delta''_K(1)$ is the second derivative of the Alexander polynomial $\Delta(K)(t)$ of the knot K evaluated at $t = 1$. In short, the above formula (1.6) agrees with the corresponding result of Casson's invariant on performing the same surgery.

At this point, the paper of Carey-Marcolli-Wang is still in a preprint stage. However, granted that Theorem (1.5) is true, the only remaining issue in establishing the conjecture on Seibert-Witten Casson invariant $\lambda(Y) = \lambda_{SW}(Y)$, is the effect of the above surgery on the correction terms in (1.4). This is our goal in Section 5-8, as we will show the following:

THEOREM 1.8. *Let Y be an integral homology 3-sphere and Y_1 be obtained from Y by performing $(1,1)$ surgery along a knot K. Then there exit metric and perturbation data $(Y,g,v),(Y_1,g_1,v_1)$ (see (8.6)) such that the condition of Carey-Marcolli-Wang in (1.5) are satisfied and the sum of the spectral flow and η-invariant correction terms remain the same before and after the surgery, i.e.*

$$\begin{aligned} &SF[\partial^g_{(1-t)v} : 0 \le t \le 1] + 1/8[\eta_{sign}(Y,g) + 4(\eta_{Dirac} + h)(Y,g)] \\ &= SF[\partial^{g_1}_{(1-t)v_1} : 0 \le t] + 1/8[\eta_{sign}(Y_1,g_1) + 4(\eta_{Dirac} + h)(Y_1,g_1)]. \end{aligned} \tag{1.9}$$

Beginning in Section 5, we discuss the Riemannian metric structures of (Y,g) and (Y_1,g_1) in [**CMW**] which requires a positive scalar curvature on a tubular neighborhood $\nu(K)$ of K and is flat near its boundary $\partial\nu(K)$. For this, we follow the method of Kreck and Stolz in [**KS**] to construct a positive scalar metric on the 3-sphere $S^3 = \nu_+(K) \cup \nu_-(K)$ which is flat near an equatorial region (see Proposition 5.10). Then by choosing a fixed metric on the knot complement $Y - \nu(K)$ and identifying $\nu(K)$ with $\nu_+(K)$, we obtain the desired metric g on Y. Since $Y_1 - \nu(K) = Y - \nu(K)$, we can use the same metric on the complement but gluing it this time with the piece $\nu_-(K)$ to get the metric g_1 on Y_1.

In Section 6, we apply a formula of Wojciechowski to study the difference of η-invariants of the Dirac operators before and after surgery. For this we have to refine the choice of the metric and connection data (g,v) on $Y_1 - \nu(K) = Y - \nu(K)$ to make sure that there exit no nontrivial L^2-solutions for the Dirac operators on these manifolds (see (6.5). Then from Wojciechowski's formula it follows that the difference $\eta_{Dirac}(Y,g,v) - \eta_{Dirac}(Y_1,g_1,v_1)$ is the same as the η-invariant $\eta_{Dirac}(S^3)$ of the original 3-sphere with positive scalar curvature (see (6.8)).

In Section 7, we apply a formula of Bunke to study the similar difference in η-invariants but with respect to the signature operator. Combining this with the previous result on Section 6, we prove the combination $[\eta_{sign} + 4\eta_{Dirac}]$ of η-invariants is unchanged by surgery.

Now in addition to the prescribed metrics on $(Y,g),(Y_1,g_1)$, there is an additional requirement in [**CMW**] on the connection v on $\nu(K) \subset Y$, known as the surgery simulated perturbation. The precise description of v is given in Section 8. By applying the Weitzenböck formula, we show that the contribution to the spectral flow $SF[\partial^g_{(1-t)v} : 0 \le t \le 1]$ due to this type of perturbation is zero. Hence, if we take the identical perturbation data v on both $Y_1 - \nu(K)$ and $Y - \nu(K)$, and the surgery simulated perturbation date v on $\nu_+(K) \subset Y$ and the trivial perturbation on $\nu_-(K) \subset Y_1$, then (1.9) follows from our spectral flow decomposition theorem.

During the final stage of preparing this manuscript, a paper of Lim [**L**] on Seiberg-Witten invariants of three manifolds has appeared which contained certain overlaps with our treatment of (1.9). Since Lim's article is a research announcement, not all of the details have been written down. At any rate, on the proof of Kronheimer-Mrowka conjecture, we do not claim to be original on many of the issues. For instance the proof of the surgery formula of (1.5) is entirely the work of Carey-Marcolli-Wang in [**CMW**] and it is also their set up of positive scalar curvature metric and Weitzenböck formula which allows us to calculate the η-invariant correction terms. Our reason for including the material here is that this particular application to Seiberg-Witten invariant is a good illustration of the splitting principle to be reviewed in §1, 2.

Finally, we wish to congratulate Jim Milgram on this celebration of his 60th birthday. For more than 30 years, his work and scientific leadership has been an inspiration to us.

§2. Gluing Construction.

Let M be a closed, oriented, smooth manifold that is decomposed into the union of two submanifolds M_1, M_2 by a codimension -1, oriented submanifold Σ,

$$M = M_1 \cup M_2, \ \Sigma = M_1 \cap M_2 = \partial M_1 = \partial M_2 \tag{2.1}$$

consider a first order, self-adjoint, elliptic differential operators D on the space $\Gamma(E)$ of smooth sections of a real vector $E \to M$,

$$D : \Gamma(E) \to \Gamma(E) \tag{2.2}$$

which have the form

$$D \mid \Sigma \times [-1,1] = \pi^*\sigma \circ (\frac{\partial}{\partial s} + \pi^*\hat{D}) \tag{2.3}$$

near a neighborhood of Σ. More precisely, under an identification $\Sigma \times [-1,1]$ with a collar neighborhood of $\Sigma = \Sigma \times 0$, and an identification of $E \mid \Sigma \times [-1,1]$ with the pullback $\pi^*\hat{E}$ of a vector bundle $\hat{E}$ on Σ via the projection $\pi : \Sigma \times [-1,1]$, the operator $\pi^*\hat{D}$ is the pull back of a self-adjoint, elliptic operator

$$\hat{D} : \Gamma(\hat{E}) \to \Gamma(\hat{E}) \tag{2.4}$$

of $\hat{E}$ and $\frac{\partial}{\partial s}$ is the differentiation with respect to the coordinates of $[-1,1]$ and $\pi^*\sigma$ is the pull back of a bundle isomorphism $\sigma : \hat{E} \to \hat{E}$ of $\hat{E}$.

We note that $\sigma^* = -\sigma$ and $\sigma\hat{D} = -\hat{D}\sigma$. If we denote by $(\ ,\ )$ the inner product on sections of $\hat{E}$, then $\{f,g\} = (f, \sigma g)$ gives rise to a nondegenerate, skew-symmetric pairing. In particular, its restriction to $\mathcal{H} = Ker\hat{D}$ is a finite dimensional, symplectic vector space.

The main point of decomposing the manifold M, the bundle E, the operator D in the above discussion is that this allows us to form a stretched version $M(r)$ of the same manifold

$$M(r) = M_1 \cup \Sigma \times [-r,r] \cup M_2 \tag{2.5}$$

obtained by cutting M open along Σ and then gluing the pieces back to $\Sigma \times [-r,r]$. In a natural manner, the vector bundle E gives rise to vector bundle $E(r)$ over $M(r), E(r) \mid M_j = E \mid M_j$, $E(r) \mid \Sigma \times [-r,r] = \pi^*\hat{E}$, and the operator D to $D(r) : \Gamma(E(r)) \to \Gamma(E(r))$. The key to our approach is that for r sufficiently large there is a relation between the "small" eigenvalues of $D(r)$ and a "gluing" construction of the extended L^2-solutions from the two sides

$$\begin{aligned} M_1(r) &= M_1 \cup \Sigma \times [0,r] \qquad & \partial M_1 &= \Sigma \times \{r\} \\ M_2(r) &= \Sigma \times [-r,0] \cup M_2 \qquad & \partial M_2 &= \Sigma \times \{-r\} \end{aligned} \tag{2.6}$$

of $M(r)$.

Let $L^2(\hat{E})$ and $L^2(E_j(r))$ denote respectively the spaces of L^2-sections of $\hat{E}$ and $E_j(r)$ over $M_j(r)$. Let $\{\phi_k\}$ be a complete orthonormal basis of eigensolutions of $\hat{D}$ with $\hat{D}\phi_k = \mu_k\phi_k$. By the spectral decomposition theorem, there exists a decomposition of $L^2(\hat{E})$ into an orthogonal sum: $L^2(\hat{E}) = \mathcal{H} \oplus P_+ \oplus P_-$, where $P_\pm$

are respectively positive and negative spaces. In addition, by applying separation of variables method to (2.3), a solution ϕ of $D \mid \Sigma \times [-r, r]$ can be written as a sum of three terms

$$\begin{aligned}\phi &= \sum c_k e^{-\mu_k s}\phi_k \\ &= \sum_{\mu_k>0} c_k e^{-\mu_k s}\phi_k + \sum_{\mu_k<0} e^{-\mu_k s}\phi_k \\ &+ \sum_{\mu_k=0} c_k\phi_k.\end{aligned} \tag{2.7}$$

Note that the first term $\sum_{\mu_k>0} c_k e^{-\mu_k s}\phi_k$ converges to an L^2-solution of $D(\infty) \mid M_1(\infty)$ are $r \to \infty$ as the terms $e^{-\mu_k s}\phi_k$, $\mu_k > 0$, decays rapidly on the infinite cylinder $\Sigma \times [0, \infty)$. Similarly the second term $\sum_{\mu_k<0} c_k e^{-\mu_k s}\phi_k$ converges to an L^2-solution of $D(\infty) \mid M_2(\infty)$. As for the third term $\sum_{\mu_k=0} c_k\phi_k$, it is invariant under the time translation along the cylinder, and if we regard $\sum_{\mu_k\le 0} c_k e^{-\mu_k s}\phi_k$, $\sum_{\mu_k\le 0} c_k e^{-\mu_k s}\phi_k$ as extended L^2-solutions from two sides then $\sum_{\mu_k=0} c_k\phi_k$ is the limiting value of these extended L_2-solutions [**APS**].

Given a subspace W in $L^2(\hat{E})$, we denote by $L^2_1(E_j(r); W)$ the Sobolev L^2_1-completion of the smooth sections ϕ such that $\phi \mid \partial M_j(r) = \phi \mid \Sigma$ lies in W. In particular, by letting $W = P_+, P_-, P_+ \oplus \mathcal{H}$, $P_- \oplus \mathcal{H}$, we have four Sobolev spaces $L^2_1(E_1(r), P_+), L^2_1(E_1(r), P_+ \oplus \mathcal{H}), L^2_1(E_2(r), P_-)$ $L^2_1(E_2(r), P_- \oplus \mathcal{H})$. Moreover, as shown in [**APS**], the closures of the operator D induce Freholm operators on these spaces:

$$\begin{aligned}&D_1 : L^2_1(E_1(r), P_+) \to L^2(E_1(r));\\ &\tilde{D}_1 : L^2_1(E_1(r), P_+ \oplus \mathcal{H}) \to L^2(E_1(r)) \text{ over } M_1(r);\\ &D_2 : L^2_1(E_1(r), P_-) \to L^2(E_1(r));\\ &\tilde{D}_2 : L^2_1(E_1(r), P_- \oplus \mathcal{H}) \to L^2(E_1(r)) \text{ over } M_2(r).\end{aligned} \tag{2.8}$$

The kernel of these operators $V_j = Ker\ D_j$, $\tilde{V}_j = Ker \tilde{D}_j$ are finite dimensional and can be identified respectively with the spaces of L^2-solutions and extended L^2-solution.

Note that there are natural homomorphisms

$$\begin{aligned}&P_1 : \tilde{V}_1 \to \mathcal{H} = Ker\hat{D} \qquad (\phi, \hat{\phi}) \to \hat{\phi}\\ &P_2 : \tilde{V}_2 \to \mathcal{H} = Ker\hat{D} \qquad (\psi, \hat{\psi}) \to \hat{\psi}\end{aligned} \tag{2.9}$$

of $\tilde{V}_j$ into $\mathcal{H}$, i.e. sending an extended L^2-solution to its limiting values at $\Sigma \times \infty$. Let L_j be the image of P_j. Then, from the generalized Stokes theorem, it is not difficult to see that L_j is a Lagrangian subspace in the symplectic space $\mathcal{H}$.

We are now in a position to describe the *gluing construction*. Let $\mathcal{W}$ denote the space of pairs of "matching" extended L^2-solution. That is, $\mathcal{W}$ is the Hilbert space of pairs $(\phi, \hat{\phi}) \in \tilde{V}_1$ and $(\psi, \hat{\psi}) \in \tilde{V}_2$ of extended L^2-solutions such that $\hat{\phi} = \hat{\psi}$.

Choose a smooth, nondecreasing function $\rho(t), 0 \leqq t \leqq 1$ such that

$$\rho(t) = \begin{cases} 0 & 0 \leq t \leq \frac{1}{4} \\ 1 & \frac{3}{4} \leq t \leq 1 \end{cases} \tag{2.10}$$

and $|\frac{d\rho}{dt}| \leq 4$. For a matching pair of solutions $((\phi, \hat{\phi}), (\psi, \hat{\psi}))$ in $\mathcal{W}$, we define $h = \Psi_r((\phi, \hat{\phi}), (\psi, \hat{\psi}))$ as a section of $E(r)$ by the formula:

$$\begin{aligned} & h \mid M_1 = \phi \qquad h \mid M_2 = \psi \\ & (h \mid \Sigma \times [-r, -1])(x, s) = \phi(x, s + r) \\ & (h \mid \Sigma \times [-1, 0])(x, s) = [\rho(-s)(\phi - \pi^* \hat{\phi})(x, s + r)] + (\pi^* \hat{\phi})(x, s + r) \\ & (h \mid \Sigma \times [0, 1])(x, s) = [\rho(s) \cdot (\phi - \pi^* \hat{\phi})(x, s - r)] + (\pi^* \hat{\phi})(x, s - r) \\ & (h \mid \Sigma \times [1, r))(x, s) = \phi(x, s - r). \end{aligned} \tag{2.11}$$

The main results of [**CLM1**] is that small eigenvalues (in the range $r^{-(1+\varepsilon)}$) of $D(r)$ are exponentially small ($\sim \exp(-\frac{\delta r}{4})$) and the corresponding eigenmodes can be approximated by the slice construction as in the following [**CLM2**, Theorem A]:

THEOREM (2.12). *Let $N(r, K)$ be the number of linearly independent eigenvectors ψ_r of $D(r)$ with eigenvalues less than K, i.e. $D(r)\psi_i = \lambda_i \psi_i$, $|\lambda_i| < K$. Let $sp(r, K)$ denote the span of these eigenvectors ψ_i. Given $\varepsilon > 0$, there exists an R such that for $r \geqq R$ we have the following:*

$$N(r, r^{-(1+\varepsilon)}) = \dim V_1 + \dim V_2 + \dim(L_1 \cap L_2) \tag{2.13}$$

(2.14) *let $\delta = \min\{|u| : u \neq 0$ is an eigenvalue of $\hat{D}$ over $\Sigma\}$. Then $N(r, \exp(-\frac{\delta}{4} r)) = N(r, r^{-(1+\varepsilon)})$ and $sp(r, \exp(-\frac{\delta r}{4})) = sp(r, r^{-(1+\varepsilon)})$*

(2.15) *Denote the projection of $\Gamma(E(r))$ onto the subspace $sp(r, \exp(-\frac{\delta r}{4}))$ by $\mathbf{P}_r$. Then $\|\Phi_r(x) - \mathbf{P}_r \Phi_r(x)\| < \exp(-\frac{\delta r}{4}) \|\Phi_r(x)\|$ for $x \in \mathcal{W}$.*

By making additional assumption in the above setting, we can provide a sharper version of Theorem (2.12). For example in [**CLM2**,Theorem B], by assuming $\mathcal{H} = Ker\hat{D} = 0$, we show that for some fixed constants $k, \varepsilon, \delta, R$, the eigenvalues of $D(r), r \leqq R$, in the range $[-(k + \varepsilon/2), (k + \varepsilon/2)]$ can be approximated up to exponentially small error

$$|\lambda - \lambda_{jk}| \leqq exp(-\frac{\delta r}{4}), 1 \leq k \leq n(j) \tag{2.16}$$

by eigenvalues $\{\lambda_{j1}, \dots, \lambda_{jn(j)}\}$ coming from $D_j(r)$ on $M_j(\infty)$ and in the range $[-k, k]$. The corresponding eigenvectors of $D(r)$ can also be approximated by elements in the image of a gluing construction

$$\Phi_r : V_1(k) \oplus V_2(k) \to \Gamma(E(r)) \tag{2.17}$$

where $V_j(k)$ are the eigenspaces of $D(j)$ in the range $[-k, k]$. That is,

$$N(r, (k + t/2)) = \dim V_1(k) + \dim V_2(k) \tag{2.18}$$

$$\|\Phi_r(x, y) - \mathbf{P}_r \Phi_r(x, y)\| \leq C[exp(-\frac{\delta}{r}4)] \|(x, y)\| \tag{2.19}$$

for $(x, y) \in V_1(k) \oplus V_2(k)$

$$\mathbf{P}_r \Phi_r \text{ is an isomorphism onto } sp(r, k + t/2) \tag{2.20}$$

In this direction, there is a parallel result by Christian Bär [**B**] who studied the Dirac operator on the connected sum $M_1\#M_2$ of two closed, Riemannian, spin manifolds M_1, M_2 of odd dimension n. Instead of stretching the connecting cylinder $S^{n-1}\times[-1,1]$ of $M_1\#M_2$, Bär constructed a Riemannian metric on $M_1\#M_2$ which is thin and short around $S^{n-1}\times[-1,1]$. Then the spectrum of $M_1\#M_2$ in a small range is very close to that of the disjoint union $M_1\cup M_2$ (see [**B**, Theorem B]).

There also have been a number of other extensions and refinements of the aforementioned results on gluing construction in the literature. For example, Weimin Chen [**C1**] studied an exponentially small, zero-order perturbation of the operator $D(r)$ on the long cylinder. Instead of (2.3) the operator $D(r)\mid\Sigma\times[-r,r]$ is of the form $\pi^*\sigma\circ(\frac{\partial}{\partial s}+\pi^*\hat{D}+P(s))$ where $P(s)$ is of zero order and $\|P(s)\|$ satisfies an exponential decay condition. Together with some additional assumption, he obtained a lower bound of the first eigenvalues and then applied the result to study Seibert-Witten invariants of 3-manifolds [**C2**], [**C3**].

In [**N1**], Liviu Nicolaescu investigated a similar 0-order, exponentially small perturbation of $D(r)\mid\Sigma\times[-r,r]$. Formulated in the elegant language of asymptotically exact sequence, he recovered the aforementioned results in (2.13)-(2.16) as well as those in [**C1**]. In terms of application, this formulation naturally leads to a gluing construction in Seiberg-Witten theory (see Chapter 4 of [**N4**]). Unlike the above discussion, the Seiberg-Witten equations are nonlinear and so the gluing construction there can be regarded as a nonlinear counterpart to the approach of (2.12).

A similar application to Seiberg-Witten theory can also be found in the work of Carey-Marcolli-Wang [**CMW**]. Firstly, the authors used Theorem (2.12) to analyze the small eigensolutions of the linearized Seiberg-Witten equation. Then they applied the Newton's method in combination with Taubes' obstruction theory. An intriguing feature of their work is the analysis on small eigenmodes of $D(r)$ with eigenvalues $\lambda(r)\to 0$ decaying at a rate of $\frac{1}{r}$ (see [**CMW**, Proposition 4.7]) because, as they have shown, these slowly decaying eigenmodes lead to nonvanishing Taubes' obstruction.

Finally, even in the most ideal situation, where $Ker\hat{D}=0$, it is difficult to predict the behavior of the small eigenmodes $\lambda_r\sim\exp(-\delta r/4)$ of $D(r)$ as to whether they are positive, negative, or zero eigenmodes. In particular, a term like $Sign_r(D)=\Sigma\ sign(\lambda_r)$ in Wojciechowski's sum formula (see [**W**, Theorem 0.5], also (6.2)) remains a mystery to us.

§3. Spectral flow.

One of our motivations in the investigation of small eigenmodes via gluing construction is its application to the study of spectral flow $SF\{D(u_:a\le u\le b\}$ of a family $\{D(u): a\le u\le b\}$ of first order, self-adjoint, elliptic operators on M. Roughly, the spectral flow of a family of operators is the count of the number of spectral curves $\{(u,\lambda(u)): D(u)\psi(u)=\lambda(u)\psi(u)$ for some $\psi(u)\}$ crossing a given reference line. A natural choice of this reference line is the zero line $\lambda=0$. However, a technical difficulty arises when there are zero modes for the operator $D(a), D(b)$ at the two ends $u=a,b$. We can bypass this difficulty by perturbing the zero line by an $(\varepsilon_1,\varepsilon_2)$-amount at these ends and this is referred to as $(\varepsilon_1,\varepsilon_2)$-spectral flow.

To give a precise definition of $(\varepsilon_1,\varepsilon_2)$-spectral flows, we need the following:

DEFINITION (3.1). For $u_0\in[a,b]$ a value λ_0 is called an excluded value of $D(u_0)$, if λ_0 is not an eigenvalue of $D(u_0)$. Given two excluded values ε_1 and

ε_2 for $D(a)$ and $D(b)$, a system of excluded values of $D(t)$ consists of a partition $a = u_0 < u_1 < \ldots < u_n = b$ of $[a, b]$ together with values $\varepsilon_1 = \lambda_0, \ldots, \lambda_n = \varepsilon_2$ such that λ_i is an excluded value of $D(u)$ for all u in $[u_{i-1}, u_i], i = 1, \ldots, n$.

By the continuity of eigenvalues of $D(t)$ (see [**BW**,17.1]), the excluded value λ^* of $D(u^*)$ is also an excluded of $D(u)$ for all u in a small neighborhood of u^*. From the compactness of $[a, b]$, it is easy to see that a system of excluded values (u_i, λ_i) always exists connecting two given values (a, ε_1) and (b, ε_2).

DEFINITON (3.2). Given a system of excluded values $(u_i, \lambda_i), i = 1, \ldots n$ of $D(u) : a \leq u \leq b$, from ε_1 to ε_2, we define N_ℓ to be the number (counted with multiplicities) of eigenvalues λ of $D(u)$ between λ_ℓ and $\lambda_{\ell+1}$. Then

$$\begin{aligned}
&(\varepsilon_1, \varepsilon_2) - \text{spectral flow of } \{D(u) : a \leq u \leq b\} \\
&= (\varepsilon_1, \varepsilon_2) - SF\{D(u) : a \leq u \leq b\} \\
&= \sum_{\ell-1}^{n-1} Sign(\lambda_\ell - \lambda_{\ell+1}) N(\ell).
\end{aligned}$$

Conceptually, we can view the above formulas as the intersection number of the spectral curves with the graph of the step functions and vertical lines (with orientation specified by the increasing t-direction) as depicted in the following figure (3.3). From this, it is not difficult to see that $(\varepsilon_1, \varepsilon_2)$-spectral flow $(\varepsilon_1, \varepsilon_2) - SF\{D(u) : a \leq u \leq b\}$ depends only on $\varepsilon_1, \varepsilon_2$ and not on the system (t_i, λ_i) connecting them.

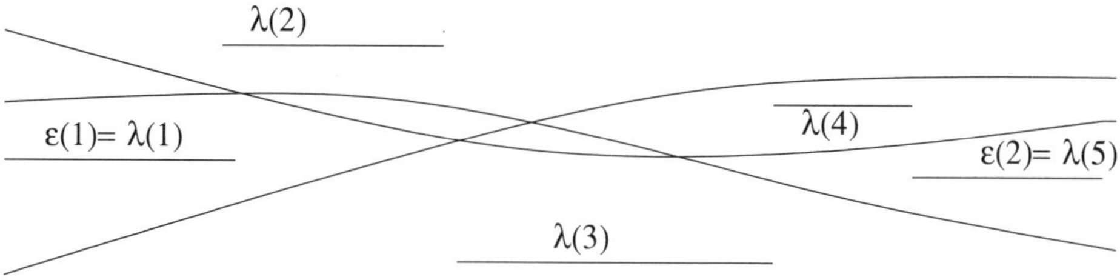

Figure 3.3

In the case $\varepsilon = \varepsilon_1 = \varepsilon_2$, we refer to $(\varepsilon_1, \varepsilon_2)$-spectral flow as the ε-spectral flow. We now consider the family of operator $D(u)$ satisfying the condition

$$D(u) \mid \Sigma \times [-1, 1] = \pi^* \sigma_u \circ (\frac{\partial}{\partial s} + \pi^* \hat{D}(u))$$

over $\Sigma \times [-1, 1]$ where σ_u and $\hat{D}(u)$ are smooth family of bundle automorphisms $\sigma_u : \hat{E} \to \hat{E}$ and self-adjoint, elliptic operators $\hat{D}(u) : \Gamma(\hat{E}) \to \Gamma(\hat{E})$. As explained in the previous section, this condition allows us to form the family of operators

$$D(u)(r) = D(u)(M(r)) : \Gamma(E(r)) \to \Gamma(E(r))$$

over $M(r)$.

As in Theorem (2.12), for r sufficiently large all the eigenvalues λ of $D(u)(M(r))$ in the range $[-\frac{1}{r^2}, \frac{1}{r^2}]$ are exponentially small, $|\lambda| < \exp(-\delta/4r)$. Hence we may fix $R_0 > 0$ such that for all $r > R_0, \pm\frac{1}{r^2}$ is not an eigenvalue of $D(0)(M(r))$ or of $D(1)(M(r))$. As $+\frac{1}{r^2}$ is an excluded value of $D(0)(M(r))$, $D(1)(M(r))$, there is a well-defined $(+\frac{1}{r^2})$-spectral flow of $D(u)(M(r)) : a \leqq u \leqq b$. Our main results in [**CLM3**] give formulas for this $(+\frac{1}{r^2})$-spectral flow, $r \geqq R_0$, in terms of the spectral flows associated with the restriction $D(u) \mid M_1, D(u) \mid M_2$ and a Maslov index term.

To give a precise formulation, we have to overcome a number of technical issues because $Ker\hat{D}(u)$ may have discontinuities and even when $Ker\hat{D}(u)$ varies continuously the Lagrangian subspaces given by the limiting values of extended L^2-solutions may have jumps. To treat this, we partition the parameter space $\{u : a \leqq u \leqq b\}$ into $a = a_0 < a_1 < \ldots < a_n = b$ such that over the subintervals $[a_i, a_{i+1}]$ there are gaps in the spectra of $\hat{D}(u) : a_i \leq u \leq a_{i+1}$. That is, there exists a number $k_i \geq 0$ and $\delta > 0$ such that no eigenvalue λ of $\hat{D}(u)$ for any u with $a_i \leq u \leq a_{i+1}$ lies in the range $(k_i, k_i + \delta)$, $(-k_i - \delta, -k_i)$. Let $\mathcal{H}(u; k_i)$ denote the vector space spanned by the eigensections ϕ_j of $\hat{D}(u)(\phi_j) = \mu_j \phi_j$ with $|u_j| \leq k_j$. By the spectral decomposition theorem, $\mathcal{H}(u, k_i)$ varies smoothly for $a_i \leq u \leq a_{i+1}$.

Let $P_+(u; k_i)$ and $P_-(u; k_i)$ denote respectively the L^2-closures of the spans of eigensections ϕ_j with $\hat{D}(u)\phi_j = \mu_j \phi_j$ where $\mu_j > k_j$ and $\mu_j < k_i$. Then there is an orthogonal sum decomposition:

$$L^2(\hat{E}) = P_-(u; k_i) \oplus \mathcal{H}(u; k_i) \oplus P_+(u; k_i). \tag{3.4}$$

By choosing R_0 large, we may ensure that the operators $D(a_i)(M(r))$, $i = 0, 1, \ldots n$ have no eigenvalues $\pm 1/r^2$ for all $r \geq R_0$. With $r \geq R_0$, the $(+1/r^2)$-spectral flow of $D(u)(M(r)) : a \leq u \leq b$ is then the sum

$$\sum_{i=0}^{n-1} (+1/r^2) - \text{spectral flow of } D(u) : a_i \leq u \leq a_{i+1}.$$

Hence, it suffices to concentrate on a fixed subinterval $a_i \leq u \leq a_{i+1}$ in which the following property holds

$$\begin{cases} \text{For all } a_i \leq u \leq a_{i+1}, \hat{D}(u) \text{ has no eigenvalues} \\ \text{in the range } [k_i, k_i + \delta], [-k_i - \delta, -k_i] \end{cases} \tag{3.5}$$

By replacing $\mathcal{H}(u) = Ker\hat{D}(u)$ by $\mathcal{H}(u, k_i)$, we have a continuous family of symplectic vector space in the interval $a_i \leq u \leq a_{i+1}$. However, there is no natural choice of Lagrangians which are continuous and play the role of $L_1(u), L_2(u)$ in (2.9). Our general spectral flow theorem can be stated in terms of *any choice* of smoothly varying Lagrangian pairs $\mathcal{L}_1(u), \mathcal{L}_2(u), a_i \leq u \leq a_{i+1}$ that satisfy the end point condition:

$$\begin{cases} \mathcal{L}_1(u) = L_1(u) \oplus [P_+(u) \cap \mathcal{H}(u, k_i)] \text{ if } u = a_i, a_{i+1} \\ \mathcal{L}_2(u) = L_2(u) \oplus [P_-(u) \cap \mathcal{H}(u, k_i)] \text{ if } u = a_i, a_{i+1} \end{cases} \tag{3.6}$$

For these choices of Lagrangians, we may introduce the self-adjoint Fredholm operators:

$$\begin{cases} D_1(u, \mathcal{L}_1(u)) : L^2_1(E \mid M_1, \mathcal{L}_(u) \oplus P_+(u, k_i) \to L^2(E \mid M_1) \\ D_2(u, \mathcal{L}_2(u)) : L^2_1(E \mid M_2, \mathcal{L}_2(u) \oplus P_-(u, k_i)) \to L^2(E \mid M_2). \end{cases} \tag{3.7}$$

In [**CLM3**], we proved the following:

THEOREM (3.8). *For the interval $a_i \le u \le a_{i+1}$ with $\hat{D}(u)$ satisfying (3.5), and for k_i and R_0 as above and any choice of smoothly varying Lagrangians $\mathcal{L}_j(u)$ satisfying the endpoint condition (3.6) for all $r \ge R_0$, then the $[(+1/r^2)-$ spectral flow of $D(u)(M(r)) : a_i \le u \le a_{i+1}]$ equals*

$$\sum_{j=1}^{2}[\varepsilon' - \textit{spectral flow of } D_j(u;\mathcal{L}_j(u)) : a_i \le u \le a_{i+1}]$$
$$+ Mas[(\mathcal{L}_1(u), \mathcal{L}_2(u)) : a_i \le u \le a_{i+1}]$$
$$+ \frac{1}{2}[\dim Ker(\hat{D}(a_{i+1})) - \dim Ker(\hat{D}(a_i))]$$

Here ε' is chosen so that the eigenvalues of $D_j(u, \mathcal{L}_j(u))$ at $u = a_i, a_{i+1}$ and in the band $[-\varepsilon', \varepsilon']$ contains at most zero eigenvalues, and $Mas[(\mathcal{L}_1(u), \mathcal{L}_2(u)) : a_i \le u \le a_{i+1}]$ stands for the Maslov index of the path of Lagrangian pairs $(\mathcal{L}_1(u), \mathcal{L}_2(u))$ in $\mathcal{H}(u, k_i)$.

There have been many applications of Theorem (3.8). For example, in [**CLM4**] we used this to study Walker's and other generalizations of Casson $SU(2)$-invariant of 3-manifolds in terms of η-invariant of certain Dirac operators (see §4 for more details). Then in [**CMW**], Carey-Marcolli-Wang used the spectral flow formula to calculate the Floer indices that appeared in the exact triangle of their Seiberg-Witten Floer theory.

There is also the paper of Liviu Nicolaescu [**N1**], in which related, though different, results in the setting of Dirac operators are treated in an elegant manner. Let $\Lambda_1(u) = KerD_1(u)|\Sigma, \Lambda_2(u) = KerD_2(u)|\Sigma$ denote the Cauchy-data spaces, obtained by the restriction to Σ of the space $KerD_j(u)$ of solutions of $D_j(u)\psi = 0$ on M_j. Under the transversality assumption $\Lambda_1(u) \cap \Lambda_2(u) = 0, u = a, b$, the spectral flow $SF\{D(u) : a \leqq u \leqq b\}$ is equal to the Maslov index $Mas\{(\Lambda_1(u), \Lambda_2(u)) : a \leqq u \leqq b\}$. From a technical viewpoint, his methods are quite different from ours because he relies on infinite dimensional Lagrangians whereas we concentrate on the finite dimension ones.

Extending the results of [**N1**], the recent article of M. Daniel and P. Kirk gave a general splitting formula which involved not only the spectral flows on two sides but also twelve more Maslov indices. The advantage of their treatment is that, although they assume a collar neighborhood of the splitting manifold Σ in $M = M_1 \cup M_2$, there is no need to stretch the collar of the splitting submanifold. In addition, there is no transversality condition for the spectral flow of $D(u) : a \le u \le b$ at the two ends $u = a, b$ which, as pointed out before, is much harder to compute than the $1/r^2$-spectral flow.

§4. Seiberg-Witten invariant for 3-manifold.

The sum formula for spectral flow $SF\{D(u) : a \le u \le b\}$ discussed in the previous section has many applications in the study of analytical invariants associated to natural geometrical operators, such as the Dirac and the signature operator. For example, in [**CLM4**], we applied this theory to study the Casson invariant and its various extensions by Walker [**Wa**], Boyer-Nicas [**BN**] and others. From a gauge theory viewpoint (see [**T**]), the Casson invariant $\lambda(Y)$ of an oriented integral homology 3-sphere $Y, H_*(Y, \mathbf{Z}) \cong H_*(S^3, \mathbf{Z})$, can be regarded as the Euler characteristic[1] of the vector field associated to the Chern-Simon functional. More precisely, let $\mathcal{A} = \mathcal{A}(E)$ denote the space of $SU(2)$-connections on a fixed $SU(2)$-bundle $E = Y \times SU(2)$. Then there is an action on $\mathcal{A}(E)$ by the gauge group $\mathcal{G} = Map(Y, SU(2))$ and on the quotient space $\mathcal{A}(E)/\mathcal{G}$ there is a well-defined function $CS : \mathcal{A}/Y \to \mathbf{R}/\mathbf{Z}$ given by the Chern-Simon integral:

$$CS(A) = \frac{1}{2}\int A \wedge F_A \quad \mod \mathbf{Z} \tag{4.1}$$

It turns out that the gradient $\nabla(CS)$ equals F_A and so the critical point of CS coincides with the Donaldson moduli space $\mathcal{M}(Y)$ of flat $SU(2)$-connections $A, F_A = 0$, on Y. In general, this last space has singularities corresponding to points where the connection A becomes reducible. Fortunately, for an integral homology 3-sphere, there exists only one such reducible connection, i.e. the trivial product connection θ. Moreover, this trivial connection is isolated and satisfies the condition $H^1(Y, AdE_\theta) = 0$ throughout the perturbations on the self-dual equations. (These perturbations are necessary in order to ensure a smooth moduli space.) Thus we can ignore the contribution from θ and define:

$$\begin{aligned} \lambda(Y) &= \chi(\mathcal{M}^*(Y)), \text{ where} \\ \mathcal{M}^*(Y) &= \mathcal{M}(Y) - [\theta] \\ &= (A \in \mathcal{A}^*/\mathcal{G} \mid A \text{ is a gauge equivalence class of irreducible} \\ &\qquad SU(2) - \text{connection satisfying } * F_A = 0) \end{aligned} \tag{4.2}$$

In the setting of rational homology 3-spheres Y, $H_1(Y, \mathbf{Q}) = 0$, there are reducible $SU(2)$-connections A other than the trivial one θ where the contribution to the Euler characteristic cannot be ignored. In [**BN**], the method used to overcome this difficulty is to resrict attention to the class of 3-manifolds Y with the property that $H^1(Y, AdE_A) = 0$ at all the reducible, flat, $SU(2)$-connections A. A different approach was provided by Walker [**Wa**] who introduced a correction term $I(A)$ into (3.1) for each of the reducible $SU(2)$-connections A, $A \neq \theta$, i.e.

$$\lambda_{\text{Walker}}(Y) = \chi(\mathcal{M}(Y) - [\theta]) + \sum_{A \neq \theta} I(A). \tag{4.3}$$

In [**CLM4**], we gave an explanation why without the correction terms $I(Y)$ formula (3.2) is not an invariant under perturbations near a singular point. Indeed, the variation in (4.2) due to such a perturbation can be accounted for by a spectral flow from the tangential component of the self-dual operator D_A on $Y \times \mathbf{R}$ to its counterpart after the perturbation. From this, we proved that the Walker correction term $I(A)$ can be identified with an eta invariant η_{D_A} and in the setting of Boyer-Nicas, the sum $\Sigma I(A)$ equals $-\frac{1}{2}$ times the signature defect $Def(Y_{ab} \to Y)$ where $Y_{ab} \to Y$ is the abelian covering associated to the homomorphism $\pi_1(Y) \to H_1(Y; \mathbf{Z}) \to H_1(Y, \mathbf{Z})$/modulo order 2. (For details, see [**CLM4**, Theorem A]).

From a gauge theory viewpoint, the Casson invariant and its companion in Floer homology can be regarded as a descendent of Donaldson's theory - translating the 4-dimension theory to 3-dimension by applying it to the infinite cylinder $Y \times \mathbf{R}$. In the fall of 1994, physicists Seiberg and Witten proposed a new gauge theory

---

[1] In Casson's original treatment, the invariant is actually $\frac{1}{2}$ of the Euler characteristic

which, by now, is known as "the Seiberg-Witten theory". Applied to 4-manifolds, it was used to solve the Thom conjecture, the 10/8-conjecture as well as a number of basic problems on the subject by Kronheimer-Mrowka [**KM**], Morgan-Szabo-Taubes [**MST**] and Furuta [**F**]. In these applications, it is clear that while this theory shared many of the features of the old Donaldson theory, it was "at least a thousand times easier" (Taubes).

Applied to 3-manifolds, the analogy between these two gauge theories has lead to the development of Seiberg-Witten-Floer theory by Carey-Marcolli-Wang, Chen, Lim in a series of preprints [**CMW**], [**MW**], [**C2**], [**L**]. The setting is that of a 3-manifold Y equipped with a Riemannian metric g and a spin-(or spin^c-) structure. Associated to the spin structure, there is the $SU(2)$-vector bundle S of spinors over Y which can be regarded as a Clifford module with the volumn form acting as the identity. Associated to the spin^c-structure, there exists a line bundle L whose first Chern class $c_1(L)$ modulo 2 equals $\omega_2(Y)$.

Consider the tensor product $S \otimes L$ as a $U(2)$-bundle over Y and $\Gamma(S \otimes L)$ as the space of C^∞-sections of this bundle. The Levi Civita connection of the Riemannian metric g lifts to a connection on S. Coupled with an $U(1)$-connection A on L, we have the Dirac operator

$$\begin{aligned} \partial_A &: \Gamma(S \otimes L) \to \Gamma(S \otimes L) \\ \partial_A &= \Sigma e^j \cdot (\nabla_{e_j} + iA_j) \end{aligned} \tag{4.4}$$

given by the composition of the covariant derivative ∇_A and the Clifford multiplication. (In the above formula, $\{e^j\}$ represents a set of local orthonormal frames of $T^*(Y)$).

To put (4.4) in a gauge theory setting, we consider the space $\mathcal{A}_{SW} = \mathcal{A}(L) \times \Gamma(S \otimes L)$ consisting of pairs (A, ψ) where $A \in \mathcal{A}(L)$ is a $U(1)$-connection on L and $\psi \in \Gamma(S \otimes L)$. The gauge group $\mathcal{G}_{SW} = Map(Y, S^1)$ acts on $\mathcal{A}_{SW}$ by $s \cdot (A, \psi) = (A - s^{-1}ds, s\psi)$ and we can form the quotient space $\mathcal{A}_{SW}/\mathcal{G}_{SW}$. As before, this last space has singularities, and to simplify our discussion we assume from now on that Y is an integral homology sphere. Then $\pi_0(\mathcal{G}_{SW}) = H^1(Y, \mathbf{Z}) = 0$ and the singularities of $\mathcal{A}_{SW}/\mathcal{G}_{SW}$ corresponds to the pair (A, ψ) with $\psi \equiv 0$.

Over $\mathcal{A}_{SW}$, there is the Chern-Simons-Dirac functional given by

$$CSD(A, \psi) = -\frac{1}{2} \int_Y A \wedge F_A + \frac{1}{2} \int_Y < \psi, D_A \psi > \tag{4.5}$$

As $\pi_0(\mathcal{G}_{SW}) = 0$, it is easy to see that it descends to a well-defined function on $\mathcal{A}_{SW}/\mathcal{G}_{SW}$. The gradient of CSD at (A, ψ) is given by $(*F_A - \tau(\psi, \psi), D_A \psi)$ with $\tau(\psi, \psi) = \Sigma < e_i \psi, \psi > e_i$. Setting this equal to zero, we obtain the Seiberg-Witten equations:

$$\begin{cases} *F_A = \tau(\psi, \psi) \\ \partial_A \psi = 0 \end{cases}$$

Thus the counterpart of $\mathcal{M}(Y)$ in the Seiberg-Witten theory is the moduli space $\mathcal{M}_{SW}(Y)$ of equivalence classes of solutions (A, ψ) of the above equation.

Properties of this moduli space $\mathcal{M}_{SW}(Y)$ have been investigated in the aforementioned articles [**CMW**], [**MW**], [**C2**], [**L**]. In particular, to ensure that the moduli space $\mathcal{M}_{SW}(Y)$ is smooth, they provided a perturbation term in the above

Seiberg-Witten equation:

$$\begin{cases} *F_A = \tau(\psi, \psi) + 2i\rho \\ \partial_A \psi = 0. \end{cases} \tag{4.6}$$

When Y is an integral homology sphere, the perturbation ρ can be written as $\rho = *dv$ and v can be regarded as a $U(1)$-connection in $\Omega^1(Y, i\mathbf{R})$. Under a generic choice of metric g and connection v, there exists a unique isolated reducible solution $\theta = [v, 0]$ of (4.6) in $\mathcal{M}_{SW}(Y, g, v)$. Furthermore, away from this reducible point, the space

$$\mathcal{M}^*_{SW}(Y, g, v) = \mathcal{M}_{SW}(Y, g, v) - [\theta]$$

is smooth (i.e. cutting out transversally by the equation) and as it is compact and zero-dimensional $\mathcal{M}^*_{SW}(Y, g, v)$ consists of a finite number of points (see Proposition 2.2, 2.3 of [**MW**]).

In view of the definition of the Casson invariant in (4.2), it would be natural to consider the Euler characteristic $\chi(\mathcal{M}^*_{SW}(Y, g, v))$ as the counter part of the Casson invariant in Seiberg-Witten theory. Unfortunately, this scheme does not work because $\chi(\mathcal{M}^*_{SW}(Y, g, v))$ depends on both the metric and perturbation data (g, v) and thus by itself is not a well-defined topological invariant. This can be explained by comparing the difference between the Zariski tangent space $T\mathcal{M}(Y)_\theta$ of the Donaldson Moduli space $\mathcal{M}(Y)$ at the reducible connection $[\theta]$ and its counterpart $T\mathcal{M}_{SW}(Y, g, v) \mid_{[\theta]}$ in the Seiberg-Witten theory. In the Donaldson case, this tangent space $\mathcal{M}(Y)_\theta$ is given by the kernel of the linearized operator $\delta_A \oplus d_A$

$$\Omega^1(Y, AdE_A) \to \Omega^0(Y, AdE_A) \oplus \Omega^2(Y, AdE_A)$$

and in particular at the trivial connection $A = \theta$, this kernel can be identified with the ordinary cohomology $H^1(AdE_0) \cong H^1(Y) \otimes AdE_0 \cong 0$, which is independent of the underlying metric. Thus, given two moduli spaces $\mathcal{M}(g_0), \mathcal{M}(g_1)$, we can connect up g_0, g_1 by a path $\{g_t\}$ of metrics, which in turn gives rise to a smooth cobordism (after some perturbation) connecting $\mathcal{M}(g_0), \mathcal{M}(g_1)$. In this cobordism, the reducibles $\{\theta_t\}$ form a path and are isolated from the set of irreducible points, as $H^1(Y, AdE_{\theta_t}) = 0$. Hence $\chi(\mathcal{M}^*_{g_0}) = \chi(\mathcal{M}^*_{g_1})$ and the invariant is well-defined.

Similar analysis for the Seiberg-Witten moduli space $\mathcal{M}_{SW}(Y, g, v)$ was carried out in [**MW**], [**C2**], [**L**]. This time, at a point (A, ψ) of $\mathcal{M}_{SW}(Y, g, v)$, the Zariski tangent space is given by the following deformation complex:

$$L : \Omega^0(Y) \oplus \Omega^1(Y) \oplus \Gamma(Y, S \otimes L) \to \Omega^0(Y) \oplus \Omega^1(Y) \oplus \Gamma(Y, S \otimes L)$$

$$L(A, \psi)(f, \alpha, \phi) = \begin{cases} T \mid_{(A,\psi)} (\alpha, \phi) + G \mid_{(A,\psi)} (f) \\ G^* \mid_{(A,\psi)} (\alpha, \phi) \end{cases}$$

Here T is the linearization of the Seiberg-Witten equation

$$T \mid_{(A,\phi)} (\alpha, \phi) = \begin{pmatrix} - * d\alpha + 2\mathrm{IM}(< e_i\psi, \phi >)e^i \\ -\partial_A \phi - i\alpha \cdot \psi \end{pmatrix},$$

and $G \mid_{(A,\psi)} (f) = (-df, f\phi)$ is induced by the infinitesimal action of the gauge group and G^* is its adjoint.

It is not hard to see that the above operator L has index zero and so the virtual dimension of the moduli space is zero. Following the usual perturbation argument as in the Donaldson case, there exists a generic choice of perturbation data such that the moduli space $\mathcal{M}^*(Y, g, v)$ is a smooth manifold, i.e. cut out transversely by the perturbed equation. As $\mathcal{M}^*(Y,g,v)$ is compact and zero-dimensional, it consists of a finite number of points. Hence we have a well-defined number $\chi[\mathcal{M}^*(Y,g,v)]$ by counting with sign these oriented points. The sign (or orientation) is given by the mod-2 spectral flow of a path of the linearized operators connecting the irreducible point in question to the reducible point, as explained in [**C2**].

In fact, for a path (A_t,ψ_t) of irreducible points, the same perturbation method will give us a cobordism connecting up the two endpoints. The difficulty comes in when a path of irreducibes encounters some reducible point. At such a reducible point $[\theta]=[v,0]$, the Zariski tangent space can be identified with the kernel $Ker\partial_v^g$ of the twisted Dirac operator $\partial_v^g : S\otimes L \to S\otimes L$ with the Levi-Civita connection on S associated to g and the $U(1)$-connection v on L. This last space is sensitive to both the metric and perturbation date (g,v). For a generic metric and perturbation (g,v) in $Met\times\Omega^1(Y,i\mathbf{R})$, we have $Ker\partial_v^g=0$ but there are codimension one stratified subspaces for which $Ker\partial_v^g\neq 0$. In the literature, the latter

$$W=\{(g,v)\in Met\times\Omega^1(Y,i\mathbf{R}) \mid Ker\partial_v^g\neq 0\}$$

is referred to as a *wall* and a connected component in the complement as a *chamber*.

If we stay inside a chamber of metric and perturbation data, the previous argument for Donaldson moduli space still works and so $\chi(\mathcal{M}^*_{SW}(g,v))$ is well-defined inside such a chamber. For, given two such moduli spaces $\mathcal{M}_{SW}(g_0,v_0),\mathcal{M}_{SW}(g_1,v_1)$, we can connect up $(g_0,v_0),(g_1,v_1)$ by a path (g_t,v_t) without crossing the wall. Hence, we have a family of moduli spaces $\cup M_{SW}(g_t,v_t)$ with the trivial connections θ_t forming an isolated path inside this family.

On the other hand, suppose (g_0,v_0) and (g_1,v_1) lie in two different chambers. Then they can be connected up by a path (g_t,v_t) which crosses the codimension-1 wall $\{(g,v) \mid Ker\partial_\nu^g=\mathbf{C}\}$ transversely at a finite number of points. At each of these crossing points $t=t^*$, there is a jump in dimension of the Zariski tangent space at θ_t. Analogous to the situation of a reducible nontrivial connection in the Donaldson moduli space $\mathcal{M}(Y)$, such a jump is caused by an irreducible point sinking into or emitting from $[\theta]$. As in the treatment of Casson invariant of rational homology sphere (see [**CLM4**]), this change in the number of irreducible point can be accounted for by a spectral flow $SF\{\partial_{g_t}^{v_t} : 0\le t\le 1\}$. Indeed there is the following *wall crossing formula* (see [**C**, page 73], [**MW**, page 53]):

$$\begin{aligned}&\chi[\mathcal{M}^*_{SW}(g_1,v_1)]-\chi[\mathcal{M}^*_{SW}(g_0,v_0)]\\&=SF\{\partial_{v_t}^{g_t} : 0\le t\le 1\}.\end{aligned}\tag{4.7}$$

In view of (4.7), we define the Seiberg-Witten Casson invariant $\lambda_{SW}(Y)$ by adding correction terms to $\chi[\mathcal{M}^*_{SW}(Y,g,v)]$ which compensate for the possible discrepancy:

$$\begin{aligned}\lambda_{SW}(Y)=&\chi[\mathcal{M}^*_{SW}(Y,g,v)]\\&+SF[\partial^g_{(1-t)v} : 0\le t\le 1]\\&+\frac{1}{8}[\eta_{sign}(Y,g)+4(\eta_{Dirac}+h)(Y,g)]\end{aligned}\tag{4.8}$$

Here $SF[\partial^g_{(1-t)v}]$ is the $-\epsilon$ spectral flow of the Dirac operator $\partial^g_{(1-t)v}$ associated to the fixed Riemannian metic g and coupled with the family of connections $\{(1-t)v : 0\le t\le 1\}$. This last family starts from v at $t=0$ and ends with the trivial connection at $t=1$. The terms $\eta_{sign}(Y,g)$ and $\eta_{Dirac}(Y,g)$ are respectively the

η-invariants of the signature and Dirac operator of Y, using the Riemannian metric g and the trivial connection.

In (4.11), we will give another formulation of the above correction terms, replacing the spectral flow by a Chern-Simons integral.

THEOREM (4.9). *Let Y be an integral oriented homology 3-sphere. Then the above formula (4.8) gives a well-defined topological invariant $\lambda_{SW}(Y)$ of Y.*

PROOF. In view of the wall-crossing formulas, it suffices to show that the difference of the correction terms at $(g_0, v_0), (g_1, v_1)$ is the same as $-SF\{\partial^{g_t}_{v_t} : 0 \leq t \leq 1\}$:

$$\begin{aligned} &- SF\{\partial^{g_0}_{(1-t)v_0} : 0 \leq t \leq 1\} + SF\{\partial^{g_1}_{(1-t)v_1} : 0 \leq t \leq 1\} \\ &- \frac{1}{8}[\eta_{sign} + 4(\eta_{Dirac} + h)](Y, g_0) + \frac{1}{8}[\eta_{sign} + 4(\eta_{Dirac} + h)](Y, g_1) \\ &= -SF\{\partial^{g_t}_{v_t}\} \end{aligned} \tag{4.10}$$

For this we choose a path of metrics $\{h_s : 0 \leq s \leq 1\}$ connecting $h_0 = g_0$ to $h_1 = g_1$. Then associated to $\{h_s\}$, there is the following path $\{\partial^{g_t}_{v_t} : 0 \leq t \leq 3\}$ of metric and perturbation data (g_t, v_t) connecting (g_0, v_0) to (g_1, v_1):

(i) $(g_t, v_t) = (g_0, (1-t)v_0)$ on the interval $[0, 1]$ i.e. keeping the metric fixed at g_0 and shrinking v_0 to 0

(ii) $(g_t, v_t) = (h_{t-1}, 0)$ on the interval $[1, 2]$, i.e. keeping the connection fixed at 0 and changing the metric from g_0 to g_1.

(iii) $(g_t, v_t) = (g_1, (t-2)v_1)$ on the integral $[2, 3]$ i.e., keeping the metric fixed at g_1 and changing the connection from 0 to v_1.

Note, by the additivity of spectral flow, we have

$$\begin{aligned} SF\{\partial^{g_t}_{v_t}\} &= SF\{\partial^{g_0}_{(1-t)v_0} : 0 \leq t \leq 1\} \\ &+ SF\{\partial^{h_s}_0 : 0 \leq s \leq 1\} - SF\{\partial^{g_1}_{(1-t)v_1} : 0 \leq t \leq 1\} \end{aligned}$$

Comparing this with (4.10), we see that the proof of (4.10) is reduced to the following:

$$\begin{aligned} &- \frac{1}{8}[\eta_{sign} + 4(\eta_{Dirac} + h)](Y, g_0) + \frac{1}{8}[\eta_{sign} + 4(\eta_{Dirac} + h)](Y, g_1) \\ &= -SF\{\partial^{h_s}_0 : 0 \leq s \leq 1\}. \end{aligned}$$

Now consider the spin 4-manifold $W = Y \times I$ with the product metric structure $(ds)^2 + g_s^2$. Then the above $-\epsilon$ spectral flow $-SF\{\partial^{g_t}_0\}$ can be expressed as the sum of the index of the Dirac operator over W and the dimension of the zero mode $h(Y, g_0)$ at $t = 0$. On the other hand, we can also consider the signature operator on W whose index is zero, as it can be identified with the signature of W. Thus we have

$$SF\{\partial^{h_s}_0 : 0 \leq s \leq 1\} = \text{index } W + \frac{1}{8}\text{Sign}(W) + h(Y, g_0).$$

The above combination of index W and $\frac{1}{8}SignW$ is chosen because when we apply the Atiyah-Patodi-Singer index theorem

$$\begin{aligned} \text{Sign}(W) &= \int \frac{1}{3}P_1(W) - \eta_{sign}(\partial W) \\ \text{Index}(W) &= \int \frac{-1}{24}P_1(W) - \frac{1}{2}[h + \eta_{Dirac}(\partial W)] \end{aligned}$$

the integral terms cancel out each other in the sum. Hence we have

$$\begin{aligned} SF\{\partial_0^{h_s} : 0 \le s \le 1\} &= -\frac{1}{8}[\eta_{sign}(\partial W) + 4(h + \eta_{Dirac}(\partial W))] + h(Y, g_0) \\ &= -\frac{1}{8}[\eta_{sign} + 4(h + \eta_{Dirac})](Y, g_1) \\ &+ \frac{1}{8}[\eta_{sign} + 4(h + \eta_{Dirac})](Y, g_0) \end{aligned}$$

This proves (4.9).

By applying the Atiyah-Patodi-Singer index theorem as in the above, we can express the spectral flow $SF[\partial^g_{(1-t)v} : 0 \le t \le 1]$ as a sum of η-invariants and Chern-Simons integral:

$$\begin{aligned} &- SF[\partial^g_{(1-t)v} : 0 \le t \le 1] = Index(Y \times I) + h(Y, g_0) \\ &= -\int \frac{1}{24} P_1(Y \times I) - \frac{1}{2}[(\eta_{Dirac} + h)(\partial(Y \times I)] + h(Y, g_0) \\ &= -\int \frac{1}{24} P_1(Y \times I) - \frac{1}{2}[\eta_{Dirac}(Y, g, v)] + \frac{1}{2}[\eta_{Dirac}(Y, g, 0) + h] \end{aligned}$$

Substituting this into (4.8), we obtain the following alternative definition of λ_{SW}:

$$\begin{aligned} \lambda_{SW}(Y) &= \chi[\mathcal{M}^*_{SW}(Y, g, v)] \\ &+ \frac{1}{24}\int P_1(Y \times I) \\ &+ \frac{1}{8}[\eta_{sign}(Y, g) + 4\eta_{Dirac}(Y, g, v)] \end{aligned} \tag{4.11}$$

As it turns out in §6, this alternative will be more convenient to use.

§5. Kronheimer-Mrowka Conjecture.

With $\lambda_{SW}(Y)$ defined as before, a natural conjecture due to Kronheimer and Mrowka is that $\lambda_{SW}(Y)$ is the same as the Casson invariant $\lambda(Y)$. For some Brieskorn homology 3-spheres, this was verified by Kronheimer-Mrowka [**KM2**] and a formulation for rational homology spheres as well as verification of this conjecture in some other interesting cases has been treated by Nicolaescu [**N2**], [**N3**].

An important step towards the study of this conjecture and also the Seiberg-Witten Floer homology is provided by the recent work of Carey-Marcolli-Wang in which they established the following *surgery formula*:

THEOREM 5.1. *Let Y be an integral homology 3-sphere and K be a knot in Y. Let Y_1 and Y_0 denote the 3-manifolds obtained from Y by performing $(1,1)$ and $(1,0)$ surgeries on K. Then there exits metric and perturbation data $(g,v),(g_1,v_1)$ on Y,Y_1 and spin^c structure s_k on Y_0 such that*

$$\chi[\mathcal{M}^*_{SW}(Y, g, v)] = \chi[\mathcal{M}^*_{SW}(Y_1, g_1, v_1)] + \sum \chi[\mathcal{M}_{SW}(Y_0, s_k)] \tag{5.2}$$

Note that Y_0 has the homology of $S^1 \times S^2$ and so $H_2(Y_0) \cong H_2(S^1 \times S^2) \cong \mathbf{Z}[S^2]$. In particular its spin^c-structure (Y_0, s_k) is defined by specifying its first Chern class $c_1(Y_0) = 2k$. In the case of the trivial spin^c-structure $s_0, c_1(Y_0) = 0$, the moduli space $\mathcal{M}_{SW}(Y_0, s_0)$ is the space of gauge equivalent classes of solutions of a perturbed equation $F_A = *\sigma(\psi, \psi) + f'(T_A)\mu$, $\partial_A \psi = 0$. (See [**CMW**, page 61]).

According to [**M**, Theorem 3.2.12], the contribution of $\mathcal{M}_{SW}(Y_0, s_k)$ to the Euler characteristic can be computed in terms of the coefficients a_j of the Alexander polynomial $\Delta_K(t) = a_0 + a_1(t^{-1} + t) + \ldots + a_r(t^{-r} + t^r)$:

$$\chi(Y_0, s_k) = \sum_{k>0} j a_{j+|k|}. \tag{5.3}$$

Putting these terms together into (5.2), we obtain

$$\begin{aligned}
\sum_{k \in \mathbf{Z}} \chi(Y_0, s_k) &= \sum_{k} \sum_{j>0} j a_{j+|k|} \\
&= \sum_{j>0} \frac{j(j-1)}{2} a_j \\
&= \frac{1}{2} \Delta''_K(1) \\
\chi(\mathcal{M}_{SW}(Y)) &= \chi(\mathcal{M}_{SW}(Y_1)) + \frac{1}{2}\Delta''_K(1).
\end{aligned} \tag{5.4}$$

Comparing with the surgery formula $\lambda(Y) - \lambda(Y_1) = \frac{1}{2}\Delta''_K(1)$ for Casson's invariant [**AM**], we see the obvious similarity. Thus, to establish the conjecture of Kronheimer-Mrowka, it remains to study the changes in the spectral flow and η-invariant terms in (4.8) before and after surgery:

$$\begin{aligned}
&SF[\partial^g_{(1-t)v} : 0 \le t \le 1] + 1/8[\eta_{sign}(Y, g) + 4(\eta_{Dirac} + h)(Y, g)] \\
&= SF[\partial^{g_1}_{(1-t)v_1} : 0 \le t \le 1] + 1/8[\eta_{sign}(Y_1, g_1) + 4(\eta_{Dirac} + h)(Y_1, g_1)].
\end{aligned} \tag{5.5}$$

We begin the proof by explaining our choice of the Riemannian matrices (Y, g), (Y_1, g_1) on Y, Y_1. Let $\nu(K) \cong D^2 \times K \cong D^2 \times S^1$ denote the tubular neighborhood of the knot K where the surgery is to take place. As in [**CMW**], the basic requirement for the metric (Y, g) on Y is that its restriction to $\nu(K)$ has positive curvature and is flat on a cylindrical neighborhood $[-r, r] \times T^2$ of the boundary $T^2 = \partial\nu(K)$. Of course, we can choose such a metric on $\nu(K)$ and then extend to the complement $Y - \nu(K)$.

Now the manifold Y_1 after $(1,1)$-surgery can be obtained by gluing $Y - \nu(K)$ and $\nu(K)$ via a diffeomorphism $h_1 : \partial\nu(K) \to \partial\nu(K)$,

$$Y_1 = (Y - \nu(K)) \bigcup_{\substack{\nwarrow_{h_1} \quad \nearrow_{id} \\ T}} \nu_-(K) \tag{5.6}$$

With respect to the natural basis of $H_1(T) = \mathbf{Z}[m] \oplus \mathbf{Z}[\ell]$ given by the longitude $[\ell]$ and meridian $[m]$, the induced homomorphism $h_1 : H_1(T) \to H_1(T)$ has the following matrix presentation:

$$[h_1] = \begin{bmatrix} 1 & 1 \\ 0 & 1 \end{bmatrix} \tag{5.7}$$

In order to calculate the combination of η-invariants, it is necessary for us to choose a metric (Y_1, g_1) so that it is the same metric on the complement $(Y_1 - \nu(K))$, i.e. $g_1 \mid Y_1 - \nu(K) = g \mid Y - \nu(K)$, and with positive scalar curvature on $\nu(K)$. Note that because h_1 is of infinite order it can not be an isometry of the flat torus T. Hence there is a problem of extending the metric $h_1^*(g \mid T)$ to $\nu_-(K)$ directly.

To construct these metrics, it will be convenient to keep in mind the following "Y"-shaped configuration in [**CLM 1**]:

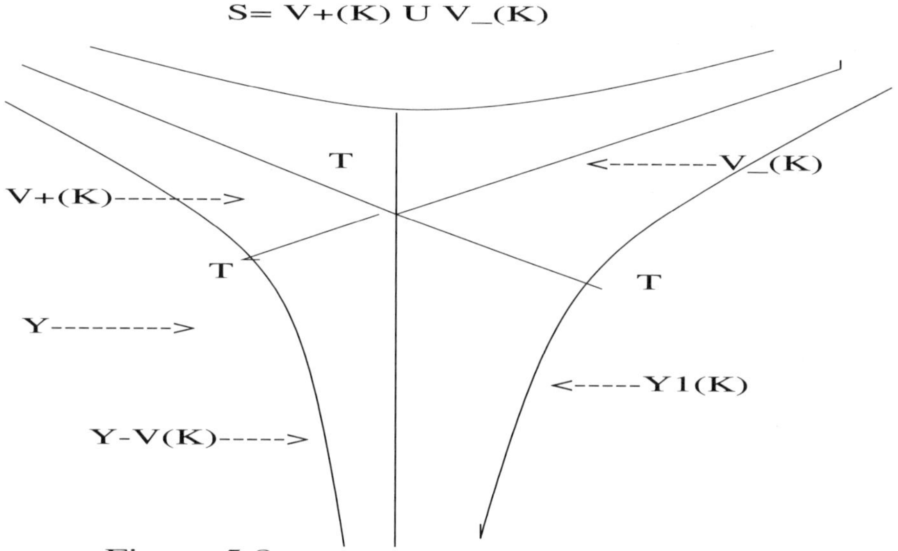

Figure 5.8

Note that the manifolds Y and Y_1 have the following decomposition:

$$\begin{aligned} Y &= (Y - \nu(K)) \bigcup \nu_+(K) \\ Y_1 &= (Y - \nu(K)) \underset{\substack{\nwarrow \quad \nearrow \\ h_1 \quad id \\ T}}{\bigcup} \nu_-(K) \end{aligned} \tag{5.9}$$

both of which have the submanifold $Y - \nu(K)$ in common. By putting these two identical pieces together, we obtain the lower leg of the"Y"-shaped figure and then attach two copies of $\nu(K)$ by id and h_1 as the upper arm. After thickening these pieces by their product with an interval, we obtain a piecewise linear 4-manifold. Furthermore, we can smooth out the corners in a canonical fashion so that the result is a smooth, compact 4-manifold W with three boundary components $\partial W = Y \cup Y_1 \cup S$. More detailed description of W as well as its relation to the Maslov triple index can be found in [**CLM 1**].

Note that two of these boundary components of W are our manifolds Y and Y_1. As for the third component S, it is not difficult to see that S is obtained from gluing two copies of $\nu_+(K) \cong \nu_-(K)$ together,

$$S = \nu_-(K) \underset{\substack{\nwarrow \quad \nearrow \\ h_1 \quad id \\ T}}{\bigcup} \nu_+(K).$$

In fact, if we write T as the product $T = S^1 \times S^1 = \{(z_1, z_2) \mid |z_1| = |z_2| = 1\}$, then the map h_1 can be realized by the diffeomorphism $(z_1, z_2) \mapsto (z_1, z_1 z_2)$. Since the last diffeomorphism is also the clutching function of the well-known Hopf fibration, it follows that S is diffeomorphic to the 3-sphere.

In view of (5.8), the problem of extending the pull back metric $h_1^*(g \mid T)$ to $\nu_-(K)$, can be reformulated as finding a Riemannian metric on S with the following properties:

PROPOSITION 5.10. *There exists a positive scalar curvature metric on the sphere* $S = \nu_+(K) \bigcup_T \nu_-(K)$ *which coincides with the flat metric* $dx^2 + dy^2 + const.dr^2$ *on* $T \times [-r, r]$.

This is because, once this metric is constructed on S, we can identify $\nu_+(K)$ with $\nu(K)$ and extend the flat metric near $\partial\nu(K)$ to a metric $Y - \nu(K)$. Using this as the metric on $Y_1 - \nu(K)$, we can glue it to the piece $\nu_-(K)$ to get the metric g_1 on Y_1.

PROOF. To prove (5.10), we follow a method of Kreck-Stolz in [**KS**]. Consider the setting of a principal G-bundle $\pi : P \to B$ with a metric g_B on the base B, a G-invariant metric g_F on the fiber, and a principal connection θ on P. In this situation, there exists one and only one metric g_P on P such that the projection π is a Riemannian submersion from (P, g_P) to (B, g_B) with totally geodesic fiber (F, g_F) and horizontal distribution induced by θ. ts scalar curvature can be computed by a formula of Besse [**Be**]. In fact by rescaling g_P in the fiber direction, i.e. replacing g_F by tg_F for some $t > 0$, we obtain a family of metrics g_t on P whose scalar curvature s_t is given by the formula:

$$s_t = \frac{1}{t} s_F + s_B \circ \pi - t|A|^2. \tag{5.11}$$

Here s_F and s_B are respectively the scalar curvature and $|A|^2$ is an invariant defined by O'Neill. (Essentially A is the curvature of θ and in particular $|A| = 0$ over the region where the connection θ is flat.)

Now, we apply the above to the Hopf fibration with the principal bundle $P = SU(2) = S^3$ and with the base space $B = SU(2)/U(1) = S^2$. Consider on S^3 the $SU(2)$-invariant metric with constant curvature 1. From this, we have an induced metric (the Killing metric) on S^2 with constant curvature 4. Then the Hopf fibration is a Riemannian submersion with totally geodesic fiber.

We can also view the above 2-sphere as obtained from revolving a circular arc of radius $\frac{1}{2}$ (as $K = \frac{1}{R_0^2} = 4$). Recall that for a surface of revolution $r = f(z)$, the Gaussian curvature is of the form $f''/f(1 + f'^2)^{3/2}$ and the scalar curvature equals to 2 times the Gaussian curvature K. Hence for our radius $\frac{1}{2}$ sphere, we take $f(z) = \sqrt{(\frac{1}{2})^2 - z^2}$ and its scalar curvature R_{1212} equals to 8.

Deform the above 2-sphere by replacing $\sqrt{(\frac{1}{2})^2 - z^2}$ by a function $f(z)$ such that $f''(z) \geq 0$ and agrees with $\sqrt{(\frac{1}{2})^2 - z^2}$ near the pole $z = \pm\frac{1}{2}$ and equals to $\frac{1}{2} - \epsilon$ for z in an annular region containing the equation $z = 0$. At the same time, we can choose the connection on the S^1-bundle so that it is the trivial product connection on a slightly bigger annular region.

Using the above deformed metric on S^2 and connection data A, we have the scalar curvature s_t of the 3-sphere given by $s_{S^2} \circ \pi - t|A|^2$ as the curvature of the fiber circle s_F is zero in (5.11). By making suitable choice of the scaling factor t, we can be sure that the term $s_B \circ \pi - t|A|^2 \geq 0$ and is equal to zero over the region near the equtor. Thus we have a Riemannian metric on S^3 which has positive scalar curvature and is the product flat metric near the equator. This proves (5.10).

§6. Formula of Wojciechowski.

With the Riemannian metric over the tubular neighborhood $\nu(K)$ chosen as in the last section, it remains to choose a generic metric on the complement $Y - \nu_+(K) = Y_1 - \nu_-(K)$ so that we can combine them to form metrics on Y and Y_1 satisfying the condition of [**CMW**]. Furthermore, we would like to calculate the η-invariants by a formula of Wojciechowski [**W**] and so the metrics have to satisfy the needed condition there. As we will see in (6.6), all these can be achieved by choosing a generic metric with a certain vanishing condition on its L_2-harmonic spinors.

Let us return to the settting of our previous discussion of spectral flow $D(u): \Gamma(E) \to \Gamma(E)$, with the underlying manifold M decomposed into two pieces $M = M_1 \cup M_2$, $\Sigma = M_1 \cap M_2 = \partial M_1 = \partial M_2$. Over a collar neighborhood $\Sigma \times [-1,1]$, the first order elliptic operator D of M can be written in the form $(\pi^*\sigma)\circ(\frac{\partial}{\partial s} + \pi^*\hat{D})$. In this situation, we can stretch the manifold M to $M(r) = M_1 \cup \Sigma \times [-1,1] \cup M_2$, the bundles E, F to $E(r)$ and $F(r)$, and the operator D to $D(r): \Gamma(E(r)) \to \Gamma(F(r))$. As we have seen in our previous discussion, there are certain addition formulae for spectral flow of $M(r)$ after making r sufficiently large. Similar results for the eta invariants $\eta(D)$ modulo integers also hold,

$$\begin{aligned} \eta(D) = \eta(D_1; L_1) + \eta(D_2; L_2) \\ + \eta(L_1, L_2) \mod \mathbf{Z} \end{aligned} \tag{6.1}$$

where L_1, L_2 are Lagrangian boundary conditons, $\eta(D_1, L_1), \eta(D_2, L_2)$ are operators $D_1 = D \mid M_1, D_2 = D \mid M_2$ with these boundary conditions. As for the term $\eta(L_1, L_2)$, this is the the counterpart to the Maslov index and its definiiton can be found in [**Bu**], [**LW**]. The proofs of (6.1) have been achieved by quite a few authors: Bunke [**Bu**], Wojciechowski [**W**], Muller [**M**], Brüning-Lesch [**BL**], Mazzero-Melrose [**MM**] just to name a few.

However, for our applications, it is necessary to consider not just the η-invariant modulo $\mathbf{Z}$, but the actual invariant itself. Fortunately, there is the following theorem of Wojciechowski [**W**, Theorem 0.5].

THEOREM (6.2). *Let $M(r)$ and $D(r)$ be given as above. Assume that $Ker\hat{D} = 0$. Then for r sufficiently large*

$$\eta(D(r)) = \eta(D_1(r)) + \eta(D_2(r)) + Sign_r(D)$$

where $Sign_r(D) = \sum_{\lambda_r \neq 0} sign(\lambda_r)$ is the sum of the signs of all exponentially small eigenvalues λ_r.

The precise definition of exponentially small eigenvalues $|\lambda_r| < a_2 r^{-a_3 r}$ is the same as in §2. As reviewed there, the term $\sum_{\lambda_r \neq 0} \text{Sign}\lambda_r$ is difficult to compute, and to overcome this difficulty we have to restrict to the case when the L^2-kernels from both sides are zero. In this situation there exists no exponentially small eigenvalues and in particular $\text{Sign}_r(D) = 0$.

COROLLARY (6.3). *If* $ker_{L^2}D_1(\infty) = ker_{L^2}D_2(\infty) = 0$, *then* $\eta_{D(r)} = \eta_{D_1(r)} + \eta_{D_2(r)}$.

Our object is to choose convenient metrics on both $Y = (Y - \nu(K)) \cup \nu_+(K)$ and $Y_1 = (Y_1 - \nu(K)) \cup \nu_-(K)$ so that we can apply (6.3) to calculate the η-invariants of their Dirac operators. Note that the splitting surface $T = \partial\nu(K) = \partial(Y - \nu(K))$ is a spin torus $(T, w \mid T)$ whose Kervaire-Arf invariant is zero because it bounds, i.e. $[T] = 0$ in Ω_2^{spin}. According to a result of Hitchin [**H**], we have the following:

PROPOSITION (6.4). *For a Riemann surface of genus* < 3, *the dimension of the space of harmonic spinors* $Ker\partial_\theta^g$ *is independent of the metric* g. *In particular, for a bounding spin torus, this dimension is equal to zero.*

On the other hand, as in (5.10), we have chosen on the tubular neighborhood $\nu_+(K)$, a positive scalar metric. For a closed spin manifold with a positive scalar curvature metric, it is a well-known result of Lichnerowicz that there exit no harmonic spinors on such manifolds (see [**H**]). The argument, based on the Weitzenböck formula, can be extended to the solid torus case (see §8) to deduce the vanishing of the L^2-kernel of ∂_θ^g on $\nu(K)$, $Ker_{L^2}(\partial_\theta^g \mid \nu(K)) = 0$. To apply (6.3) it will require a similar vanishing condition for the complement, $Y - \nu_+(K)$.

ASSUMPTION (6.5). *We choose the Riemannian metric* g *and perturbation data* v *on* $Y - \nu_+(K)$ *so that the metric* g *over a collar neighborhood of* $\partial\nu_+(K)$ *coincides with the flat metric in (5.10) and* $v = 0$ *over this neighborhood. Furthermore the* L^2*-kernel,* $Ker_{L^2}[\partial_v^g \mid (Y - \nu_+(K)]$, *of the Dirac operator is zero.*

From now on we choose the metric and perturbation data (g, v) on $Y - \nu(K)$ so that (6.5) is satisfied. Then, because of the condition on the collar neighborhood near the boundary, they can be glued to $\nu_+(K), \nu_-(K)$ to form the metric and perturbation data $(Y, g, v), (Y_1, g_1, v_1)$ on Y, Y_1.

COROLLARY (6.6). *Let* $(Y, g, v), (Y_1, g_1, v_1)$ *be the metric and perturbation data obtained by gluing those on* $Y - \nu(K) = Y_1 - \nu(K)$ *satisfying (6.5) with a positive scalar curvature metric on* $\nu_+(K), \nu_-(K)$ *and trivial connection* $v = 0$. *Then, after stretching the cylinder* $\partial\nu(K) \times [-r, r]$ *sufficiently long, the Dirac operators satisfy the vanishing condition* $Ker\partial_v^g = Ker\partial_{v_1}^{g_1} = 0$.

The proof of (6.6) follows immediately from Theorem (2.12).

With the stretched metrics given as above, we can now apply (6.3) to calculate the η-invariant of the Dirac operators on $(Y(r), g(r), v(r))$ and $(Y_1(r), g_1(r), v_1(r))$:

$$\begin{aligned} \eta_{Dirac}(Y(r), g, v) &= \eta_{Dirac}(Y(r) - \nu_+(K)) + \eta_{Dirac}(\nu_+(K)) \\ \eta_{Dirac}(Y_1(r), g_1, v_1) &= \eta_{Dirac}(Y(r) - \nu_-(K)) + \eta_{Dirac}(\nu_-(K)) \\ \eta_{Dirac}(S(r)) &= \eta_{Dirac}(\nu_+(K)) - \eta_{Dirac}(\nu_-(K)) \end{aligned} \tag{6.7}$$

By adding up the above three equations, we obtain

$$\eta_{Dirac}(Y(r), g, v) = \eta_{Dirac}(Y_1(r), g, r) + \eta_{Dirac}(S(r))$$

or in other words, before and after surgery the difference of η-invariant terms in (4.11) is the same as the corresponding η-invariant of the sphere:

$$\eta_{Dirac}(Y(r), g, v) - \eta_{Dirac}(Y_1(r), g_1, v_1) = \eta_{Dirac}(S(r)) \tag{6.8}$$

As for the Chern-Simons integral in (4.11), the difference is zero when we take the connection v to be trivial over $\nu_+(K), \nu_-(K)$. This is because as $Y = (Y - \nu(K)) \cup \nu_+(K)$ and $Y_1 = (Y_1 - \nu(K)) \cup \nu_-(K)$ the corresponding integrals over Y, Y_1 are the sum

$$\int P_1(Y \times I) = \int P_1((Y - \nu(K)) \times I) + \int P_1(\nu_+(K) \times I)$$

$$\int P_1(Y_1 \times I) = \int P_1((Y_1 - \nu(K)) \times I) + \int P_1(\nu_-(K) \times I)$$

Since over $\nu_+(K), \nu_-(K)$ the connections $v = v_1 = 0$, the corresponding Chern-Simons integrals $\int P_1(\nu_+(K) \times I) = \int P_1((\nu_-(K)) \times I) = 0$. Likewise, over $Y - \nu(K) = Y_1 - nu(K)$, they have identical metric and perturbation data and so $\int P_1((Y - \nu_+(K)) \times I) = \int P_1((Y_1 - \nu_-(K)) \times I)$. From this it follows that

$$\frac{1}{24} \int P_1(Y \times I) = \frac{1}{24} \int P_1(Y_1 \times I). \tag{6.9}$$

To conclude this section, we show that the metric and perturbation data satisfying (6.5) are generic.

PROPOSITION (6.10). *Let Met be the space of Riemannian metrics g on $Y - \nu(K)$ which coincides with a fixed flat metric in the neighborhood on a neighborhood of $\nu(K)$. Let $\Omega^1_{comp}(Y - \nu_+(K), i\mathbf{R})$ denote the space of $U(1)$-connection v which are the trivial connection $v = 0$ near the boundary $\partial \nu_+(K)$. Then there exits a Baire set of metric g and connection data v in $Met \times \Omega^1_{comp}(Y - \nu_+(K), i\mathbf{R})$ such that $Ker_{L^2} \partial^g_v = 0$*

PROOF. In the case of a closed manifold, the proof can be found in [**MW**]. Hence we will only give an outline of some of the basic ideas and indicate the necessary modification for our noncompact situation. Let (g_0, v_0) be a pair of metric and perturbation data on Y such that $Ker_{L^2} \partial^{g_0}_{v_0} \neq 0$. We consider the problem of perturbing the data near (g_0, v_0) to make this kernel zero. Denote by Met^0 the subspace of metrics in Met which have the same volume element as the metric g_0. Note any metric in Met is conformally equivalent to one in Met^0. Conformally equivalent metrics have Dirac operators which are conjugate to each other and in particular have isomorphic kernel spaces. Thus we can replace Met by Met^0 and show that for generic metrics in Met^0 the space of L^2-harmonic spinors is zero.

For metric and perturbation data (g, v) sufficiently close to (g_0, v_0), there is a natural isometry h from the spaces of spinors associated to g to those associated to g_0. Under this identification, the Dirac operator for (g, v) can be conjugated to an operator $h^{-1} \partial^g_v h$ on the spinor space of g_0. For lack of good notation, we will denote this by ∂^g_v. The advantage of choosing g in Met^0 is that this last operator is self-adjoint.

Consider the real Hilbert bundle E over the Banach manifold

$$B = Met^0 \times \Omega^1_{comp}(Y - \nu_+(K), i\mathbf{R}) \times (L^2_{APS}(S))$$

whose fiber at (g, v, ψ) is given by

$$E_{(g,v,\psi)} = \{\phi \in (L^2_{APS}(S) | Re < \phi, i\psi >= 0\}$$

Here $L^2_{APS}(S)$ denotes the space of spinors which satisfy the Atiyah-Patodi-Singer boundary condition.

Because of the self-adjointness property, we can define a section s of E by assigning to (g, v, ψ) the element $\partial_v^g \psi$. Our object is to prove that this section is transverse to the zero section of E. Suppose $(g, v, \psi) \in s^{-1}(0)$; then the differential of s at this point (g, v, ψ) in the direction (V, Ψ) tangent to $\Omega^1_{comp}(Y - \nu(K), i\mathbf{R}) \times L^2_{APS}(S)$ is given by

$$\delta s_{(}g, v, \psi)(V, \Psi) = \partial_{v_0}^{g_0}(\Psi) + \frac{1}{2} V \cdot \psi.$$

It follows that if $\phi \in E_{(g_0, v_0, \psi_0)}$ is orthogonal to the image of δs, then ϕ satisfies the conditions:

$$Re < \phi, i\psi >_{g_0} = 0$$
$$Re < \phi, V \cdot \psi_0 > = 0 \text{ for all } V \in \Omega^1_{comp}(Y - \nu(K), i\mathbf{R})$$
$$\partial_{v_0}^{g_0} \phi = 0$$

From the second equation we see that there exits a function $f : Y \to \mathbf{R}$ such that $\phi = i f \psi_0$. Substituting into the third equation, using $\partial_{v_0}^{g_0} \psi_0 = 0$ and the unique continuation property of twisted Dirac operators [Bo], we obtain $df = 0$, which implies $f = C$ is a constant function. Then the first equation $Re < iC\psi_0, \psi_0 > = C|\psi_0|^2$ implies $C = 0$. This proves that $\phi = 0$ and δs is surjective. In other words, s is transverse to the zero section and $s^{-1}\{0\}$ is a Banach manifold.

It also follows that the projection $pr : s^{-1}(0) \to Met^0 \times \Omega^1_{comp}(Y - \nu(K), i\mathbf{R})$ is a Fredholm map of index 1. This last assertion is because the Dirac operator with the Atiyah-Patodi-Singer boundary condition has index zero and as we apply this to $s : B \to E$ this index is cut down by 1, due to the orthogonal condition in the definition of E. On the other hand, ∂_v^g is complex linear and so by Sard-Smale, for generic elements (g, v) in $Met^0 \times \Omega^1_{comp}(Y - \nu(K), i\mathbf{R})$, its preimage $pr^{-1}(g, v)$ is empty. Thus for a Baire set of metric and perturbation data (g, v) the L^2-kernel $Ker_{L^2} \partial_v^g = 0$. This proves (6.10).

§7. Formula of Bunke.

The sum formula in (6.2) and (6.3) are very effective tools in studying the η-invariants of Dirac operators. However, we cannot apply them to study the signature operator because over the splitting submanifold $T = \partial\nu(K)$ the signature operator has nontrivial kernel. The remedy for this situation is the following result of U. Bunke [**Bu**, page 423].

PROPOSITION (7.1). *Let $(M_i, \Sigma), i = 1, 2, 3$ be three compact oriented Riemannian manifolds of dimension $4k - 1$ with each boundary isometric to Σ and a product metric near the boundary. Let $V = H_{2k-1}(\Sigma, \mathbf{R})$ be the symplectic vector space given by the intersection pairing and let $L_i = Ker(V \to H_{2k-1}(M_i))$ denote the Lagrangian subspace in V given by the natural induced homomorphism $H_{2k-1}(\Sigma) \to H_{2k-1}(M_i)$. Then*

$$\begin{aligned} &\eta_{sign}(M_1 \cup -M_2) + \eta_{sign}(M_2 \cup -M_3) + \eta_{sign}(M_3 \cup -M_1) \\ &= \tau_v(L_1, L_2, L_3) \end{aligned} \tag{7.2}$$

*where $\tau_V(L_1, L_2, L_3)$ is the triple Maslov index associated to the three Lagrangians L_1, L_2, L_3 in V (see [**CLM1**]).*

By applying (7.2) to the manifolds $M_1 = Y - \nu_+(K), M_2 = \nu_+(K), M_3 = \nu_-(K)$ with $T = \partial M_1 = \partial M_2 = \partial M_3$, we have

$$\eta_{sign}(Y) + \eta_{sign}(-Y_1) + \eta_{sign}(-S) = \tau_V(L_1, L_2, L_3). \tag{7.3}$$

Here the triple Maslov index can be computed in the following manner. First of all, the symplectic vector space $V = H_1(T)$ is isomorphic to the direct sum $\mathbf{C} = \mathbf{R} \oplus \mathbf{R}i \cong \mathbf{R}\ell \oplus \mathbf{R}m$ where the first component is generated by the longitude ℓ and the second by the meridian m with $< \ell, m >= 1$. It is not difficult to see that the three Lagrangians are:

$$\begin{aligned} L_1 &= \mathrm{Ker}(V \to H_1(Y - \partial\nu_+(K)) = \mathbf{R}\ell = \mathbf{R} \\ L_2 &= \mathrm{Ker}(V \to H_1(\nu_+(K)) = \mathbf{R}m = \mathbf{R}i \\ L_3 &= \mathrm{Ker}(V \to H_1(v_-(K)) = \mathbf{R}(\ell + m) = \mathbf{R}(i+1) \end{aligned}$$

Using the normalization condition of the triple Maslov index (page 163 [**CLM1**]), we have

$$\tau_V(L_1, L_2, L_3) = -\tau_V(\mathbf{R}, \mathbf{R}(1+i), \mathbf{R}i) = -1.$$

Thus, we have

$$\eta_{sign}(Y) - \eta_{sign}(Y_1) = \eta_{sign}(S) - 1. \tag{7.4}$$

Combining (6.8) and (7.4) together, it is clear that the difference of the correction term $[\eta_{sign} + 4\eta_{Dirac}]$ for the Seiberg-Witten invariant before and after surgery is given by:

$$\begin{aligned} &[\eta_{sign}(Y) + 4\eta_{Dirac}(Y)] \\ &= [\eta_{sign}(Y_1) + 4\eta_{Dirac}(Y_1)] + [\eta_{sign}(S) + 4\eta_{Dirac}(S)] - 1. \end{aligned} \tag{7.5}$$

To establish the η-invariant part of (5.5), it remains to show that the above sum is zero and this is a consequence of the following:

PROPOSITION (7.6). *Let S be the 3-sphere with the positive scalar curvature metric g defined as before and with $v = 0$. Then the correction term for the Seiberg-Witten invariant $[\eta_{sign} + 4\eta_{Dirac}](S, g, \nu) = 1$.*

PROOF. For the proof, we use the fact that S is the boundary of the disk bundle $N = D(E)$ over S^2 with the positive scalar curvature metric as constructed in §5. Note that this positive scalar curvature metric extends to the disk bundle N as $S_F \geq 0$ and the decrease in radius t can only make the scalar curvature more positive than (5.11). In addition, we add a collar with product metric to the boundary so that we can apply the Atiyah-Patodi-Singer index theorem (see [**N3**]).

Note that the orientation on S is opposite to $\partial N, S = -\partial N$. It follows that

$$\begin{aligned} &[\eta_{sign}(S) + 4\eta_{Dirac}(S)] = -[\eta_{sign}(\partial N) + 4\eta_{Dirac}(\partial N)] \\ &= 8 \text{ Index } N + \mathrm{Sign}(N). \end{aligned}$$

Here Index N and Sign N are computed in terms of the Dirac and signature operators coupled with the trivial flat connection on the line bundle. In this case, Sign N is computed by the signature of the intersection pairing $H_2(N, \partial N) \times H_2(N) \to \mathbf{Z}$ and is equal to 1.

To prove (7.6), it remains to show that Index $N = 0$. For this we consider the Weitzenböck formula as in [**Ok-T**]:

$$\partial_A^2 \phi = \nabla_A^* \nabla_A \phi + \Gamma(F_A)\phi + (s(N)/4) \cdot \phi. \tag{7.7}$$

Here we consider $\Gamma(F_A)$ as an element in $\Omega^0(\mathrm{End}(S^+ \oplus S^-) \otimes L)$ induced by the curvature F_A and $s(N)$ is the scalar curvature of N. Suppose ϕ is an L^2-solution of $\partial_A = 0$. Rewriting (7.7) in terms of positive and negative spinors ϕ_+, ϕ_- of ϕ, we have

$$0 = \int (\partial_A^2 \phi, \phi) = \int (\nabla_A^* \nabla_A \phi, \phi) + \int (\Gamma(F_A^+)\phi_+, \phi_+) \qquad (7.8)$$
$$+ \int (\Gamma(F_A^-)\phi_-, \phi_-) + \int (s(N)/4) \cdot |\phi|^2.$$

As $\Gamma(F_A) = \Gamma(F_A^+) = \Gamma(F_A^-) = 0$, we can omit the middle two terms. By Stokes' theorem and the exponential decay condition of ϕ along the boundary, we can write $\int(\nabla_A^* \nabla_A \phi, \phi)$ as $\int(\nabla_A \phi, \nabla_A \phi)$, which is positive. It follows that every term on the left is positive and so this is impossible unless all the terms are zero. In particular, $\int(s(N)/4) \cdot |\phi|^2 = 0$ and so $\phi = 0$ over an open region where $s(N) > 0$. By the unique continuity property of solutions of Dirac operators, $\phi \equiv 0$. This proves $\mathrm{Ker}\partial_A|S^+ = \mathrm{Ker}\partial_A|S^- = 0$. As explained in [**APS**], the index of the Dirac operator Index N is the same as $\dim\mathrm{Ker}(\partial_A|S^+) - \dim\mathrm{Ker}(\partial_A|S^-) - h(\partial N)$. Here $h(\partial N)$ is the dimension of the space of harmonic spinors along ∂N, and so by Lichnerowicz $h(\partial N) = 0$. It follows that Index $N = 0$ and completes the proof of (7.6).

§8. Surgery simulated perturbation.

In the previous section, we have established that there exists chambers of metric and perturbation data (g, v) on $Y - \nu_+(K) = Y_1 - \nu_-(K)$ such that, with respect to the positive scalar curvature metric on $\nu_+(K), \nu_-(K)$ and $v \mid \nu_+(K) = v \mid \nu_-(K) = 0$, the correction terms in (4.11)

$$\frac{1}{24}\int P_1(Y \times I) + \frac{1}{8}[\eta_{sign}(Y, g) + 4\eta_{Dirac}(Y, g, v)]$$
$$= \frac{1}{24}\int P_1(Y_1 \times I) + \frac{1}{8}[\eta_{sign}(Y_1, g_1) + 4\eta_{Dirc}(Y_1, g_1, v_1)]$$

coincide. However in [**CMW**], Carey-Marcolli-Wang relied on a surgery simulated perturbation v which is not trivial over $\nu_+(K)$. Thus, to complete the proof of (5.5), it remains to discuss the effect of this type of perturbation on the correction terms.

Let us explain the surgery simulated perturbation in [**CMW**]. First of all this perturbation v^* is concentrated on $\nu_+(K)$ and not at all on $\nu_-(K)$. As for the complement $Y - \nu_+(K) = Y_1 - \nu_-(K)$, they are the same as given in (6.7). The effect of this perturbation produces a new reducible solution $(v^*, 0)$ with curvature $F_A, A = v \mid \nu_+(K)$ over $\nu_+(K)$ satisfying the formula $F_A = f'(u)\mu$. Here μ is a compactly supported 2-form representing the generator of $H^2_{cpt.}(\nu_+(K)) = H^2_{cpt.}(D^2 \times S^1) = \mathbf{Z}$ such that

$$\int_{D^2 \times \{u\}} \mu = 1 \text{ for any } u \in S^1, \qquad (8.1)$$

$4\sqrt{2}|\mu| - s < 0$ and s is the scalar curvature associated to the Riemannian metric in (5.10) (8.2) (see Lemma 3.18, and 6.1 of [**CMW**]). As for the function $f(u)$, it is the derivative of a continuously differentiable, periodic function $f : \mathbf{R} \to \mathbf{R}$, $f(u+1) = f(u)$ such that the range of $f'(u)$ is $[0,1]$, $f'(0) = 0$, and $\sup_{u\in[0,1]} |f'(u) - u| < \varepsilon$ for a fixed $\varepsilon > 0$.

Combining the above with (g, v) over $Y - \nu_+(K)$, we obtain a new perturbation data (g, v^*) on $Y = (Y - \nu_+(K)) \cup \nu_+(K)$. On the other hand, on $Y_1 = (Y_1 - \nu_-(K)) \cup \nu_-(K))$ we have the data (g_1, v_1) which is the same as (g, v) on $Y_1 - \nu_-(K) = (Y - \nu_+(K))$ but with the connection trivial over $\nu_-(K)$.

Using the definition (4.8) and the fact that the metric data g is unchanged, it is easy to see that the difference in the correction terms between (Y, g, v^*) and (Y, g, v) is in the difference of the two spectral flows, $SF\{\partial^g_{(1-t)v}\} - SF\{\partial^g_{(1-t)v^*}\}$. Hence to complete the proof of (5.5) it suffices to show that these two spectral flows are the same.

Recall that by the result of Hitchin [**H**], there exit no zero modes $\mathcal{H} = \mathrm{Ker}[\partial^g_0 \mid \partial\nu_+(K)] = 0$ along the splitting torus. Hence in calculating the above spectral flows, there is no contribution from the Maslov index and the result is a sum formula of the L^2-spectral flows from two sides:

$$SF\{\partial^g_{(1-t)v}\} = SF\{\partial^g_{(1-t)v} \mid Y - \nu(K)\} + SF\{\partial^g_{(1-t)v} \mid \nu_+(K)\} \tag{8.3}$$

$$SF\{\partial^g_{(1-t)v^*}\} = SF\{\partial^g_{(1-t)v^*} \mid Y - \nu(K)\} + SF\{\partial^g_{(1-t)v^*} \mid \nu_+(K)\} \tag{8.4}$$

Since the perturbation data $(g, v), (g, v*)$ are identical over $Y - \nu(K)$, the first two terms in (8.3) and (8.4) are the same. On the other hand, over $\nu_+(K)$ the connection $v \mid \nu_+(K)$ is the trivial connection and so is the interpolation $(1-t)v \mid \nu_+(K) = 0$. This implies that the operator $\partial^g_{(1-t)v} \mid \nu_+(K)$ remains constant and so $SF\{\partial^g_{(1-t)v} \mid \nu_+(K)\} = 0$.

It remains to show that the L^2-spectral flow $SF\{\partial^g_{(1-t)v^*} \mid \nu_+(K)\}$ is zero. At $t = 1$, the connection is flat and as we have seen there exists no L_2-solution by the Weitzenböck formula because of the scalar curvature $s \mid \nu_+(K) \geq 0$ condition. At $t = 0$, the connection of the line bundle L is no longer flat but Carey-Marcolli-Wang applied a similar argument to deduce the vanishing of the L^2-harmonic spinors (see p. 50 of [**CMW**]). The key observation is that by applying the Weitzenböck formula (7.8) we obtain in this case for L^2-solutions ϕ:

$$\begin{aligned} -\int (\nabla_A\phi, \nabla_A\phi) &= \int (\Gamma(F_A)\phi, \phi) + \int s(\nu_+(K))/4|\phi|^2 \\ &\geq \int -\sqrt{2}|F_A|\, |\phi|^2 + s(\nu_+(K))/4|\phi|^2. \end{aligned} \tag{8.5}$$

Condition (8.2) implies that the right-hand side is positive and so on an open region $\phi \equiv 0$. Unique continuation implies that ϕ is zero over the entire $\nu_+(K)$.

Now to calculate the spectral flow $SF\{\partial^g_{(1-t)v^*} \mid \nu_+(K)\}$ we observe that the $U(1)$-connection $|F_{(1-t)A}| = |(1-t)F_A|$ is getting smaller and so the same inequality

$$-\int \sqrt{2}|F_{(1-t)A}| + s(\nu_+(K))/4 \geq 0$$

holds for all $t \in [0,1]$. Thus, by the Weitzenböck formula, we conclude that there exist no nontrivial L^2-zero modes for the operator $\partial^g_{(1-t)v^*}$ and so the spectral flow $SF\{\partial^g_{(1-t)v^*} \mid \nu_+(K)\}$ is zero. This completes the proof of (5.5).

REMARK 8.6. To conclude this paper, we list the conditions on the metric and perturbation data $(Y, g, v), (Y_1, g_1, v_1)$ under which Theorem (1.7) holds:

(1) On the two tubular neighborhoods of the knot, the metrics $(\nu_+(K), g)$ $(\nu_-(K), g)$ have positive scalar curvature and are flat near their boundaries so that they can be glued together to form the metric on the 3-sphere (S, g) in Proposition (5.10).

(2) On the complement $Y - \nu(K) = Y_1 - \nu_-(K)$, the metric and perturbation data are the same. They are chosen so that the space $Ker_{L^2}(\partial^g_v|(Y - \nu(K))) = 0$, as in (6.5).

(3) Both the metric and connection data on $(Y, g, v), (Y_1, g_1, v_1)$ have to be stretched sufficiently long so that Wojciechowski's sum formula in (6.2) can be applied.

(4) Over $\nu_+(K)$, the perturbation $v = v^*$ satisfy conditions (8.1) and (8.2), i.e. the same as the surgery simulated perturbation of [**CMW**] and over $\nu_-(K), v = 0$.

References

[AM] Akbulut, S. and McCarthy, J., *Casson's Invariant for Oriented Homology* 3*-spheres - an exposiiton*, Mathematical Notes 36, Princeton University Press, Princeton, New Jersey, USA, 1990.

[APS] Atiyah, M.F., Patodi, V.K. and Singer, I.M., *Spectral symmetry and Riemannian geomtry I*, Math. Proc. Camb. Phil. Soc. **77** (1975), 43-69.

[B] Bär, C., *Metrics with harmonic spinors*, Geometry And Functional Analysis **6** (1996), 899-942.

[Be] Besse, A. L., *Einstein Manifolds*, Springer-Verlag, Berlin and New York, 1986.

[BN] Boyer, S. and Nicas, A., *Variety of group representations and Casson invariant for rational homology* 3*-spheres*, Trans. Amer. Math. Soc. **322** (1990), 507-572.

[BF] Booss, B. and Furutani, K., *The Maslov index - a functional analytical definition and the spectral flow formula*, Tokyo J. Math. **21** (1998), 1-34.

[BL] Brüning, J. and Lesch, M., *On the eta-invariant of certain non-local boundary value problem*, Preprint dg-ga/96090001.

[BN] Boyer, S. and Nicas, A., *Varieties of group representations and Casson's invariant for rational homology 3-spheres*, Trans. Amer. Math. Soc. **322** (1990), 507-522.

[Bo] Booss, B., *Boundary reduction of spectral invariants and unique continuation property*, Preprint Roskilde Universitetscenter **367** (1999).

[Bu] Bunke, U., *On the gluing problem for the η-invariant*, J. Diff. Geom. **41** (1995), 397-448.

[BW] Booss, B. and Wojciechowski, K.P., *Elliptic Boundary Problems for Dirac Operators*, Mathematics: Theory and Applications, Birkhauser, 1993.

[C1] Chen, W., *A lower bound of the first eigenvalue of certain self-adjoint operators on manifolds containing long necks*, Turkish J. Math. **21** (1997), 93-98.

[C2] Chen, W., *Casson's invariant and Seiberg-Witten gauge theory*, Turkish J. Math. **21** (1997), 61-81.

[C3] Chen, W., *The Seiberg-Witten theory of homology 3-spheres*, Preprint, dg-ga/9703009.

[CMW] Carey, A.L., Marcolli, M. and Wang, B.L, *Exact triangles in Seiberg-Witten Floer theory Part I: the geometric triangle*, Preprint, math.DG/9907065.

[CLM1] Cappell, S., Lee, R., and Miller, E., *On the Maslov index*, Comm. Pure and Appl. Math. **47** (1994), 121-186.

[CLM2] Cappell, S., Lee, R., and Miller, E., *Self-adjoint operators and manifold decomposition, Part* I*: Low eigenmodes and stretching*, Commm. Pure and Appl. Math. **49** (1996), 825-866.

[CLM3] Cappell, S., Lee, R., and Miller, E., *Self-adjoint operators and manifold decomposition, Part II: Spectral flow and Maslov index*, Commm. Pure and Appl. Math. **48** (1996), 869-909.
[CLM4] Cappell, S., Lee, R. and Miller, E., *Self-adjoint operators and manifold decomposition, Part III: Determinant line bundles and Lagrangian intersection*, Commm. Pure and Appl. Math. **52** (1999), 543-611.
[DK] Daniel, M. and Kirk, P., *A general splitting theorem for spectral flow, with an appendix by K.P. Wojciechowski*, Michigan Math. Journal **46** (1999), 589-617.
[F] Furuta, M., *Monopole equations and 11/8 conjecture*, Preprint.
[H] Hitchin, N., *Harmonic spinors*, Advances in Math. **14** (1974), 1-55.
[KK] Kirk, P. and Klassen, E., *Computing spectral flow via cup products*, J. Diff. Geom. **40** (1994), 505-562.
[L] Lim, Y., *The equivalence of Seiberg-Witten and Casson invariants for homology 3-spheres*, Math. Research Letters **6** (1999), 631-643.
[L1] Lim, Y., *Seiberg-Witten invariants for 3-manifolds in the case* $b_1 = 1$, To appear in Pacific J. Math..
[LW] Lesch, M. and Wojciechowski, K.P., *On the* η*-invariant of generalized Atiyah-Patodi-Singer problems.*, Illinois J. Math. **40** (1996), 30-46.
[KM] Kronheimer, P. and Mrowka, T., *The genus of embedded surfaces in the projective plane*, Math. Research Letters **1** (1994), 797-808.
[KM] Kronheimer, P. and Mrowka, T., *The 1st International Press Lectures, UC Irvine,* (1996).
[KS] Kreck, M. and Stolz, S., *Nonconnected Moduli spaces of positive curvature metrics*, J. Amer. Math. Soc. **6** (1993), 825-850.
[M] Marcolli, M., *Seiberg-Witten Gauge Theory*, Texts and Readings in Mathematics, Hindustan Book Agency, New Delhi, India, 1999.
[Mu] Müller, W., *Eta invariants and manifolds with boundary*, J. Diff. Geom. **40** (1994), 311-377.
[MM] Mazzeo, R.R. and Melrose, R.B., *Analytic surgery and the eta invariant*, GAFA **5** (1995), 14-75.
[MOY] Mrowka, T., Ozsvath, P. and Yu, B., *Seiberg-Witten Monopoles on Seifert fibered spaces*, Communications in Analysis and Geometry **5** (1997), 685-793.
[MST] Morgan, J., Szabo, Z., and Taubes, C.H., *A product formula for the Seiberg-Witten invariants and the generalized Thom conjecture*, J. Diff. Geom. **44** (1996), 706-788.
[MW] Marcolli, M. and Wang B.L., *Equivariant Seiberg-Witten Floer Homology*, Preprint arXiv.dg-ga/9606003.
[N1] Nicolaescu, L., *The Maslov index, the spectral flow, and decomposition of manifolds*, Duke Math. J. **80** (1995), 485-533.
[N2] Nicolaescu, L., *On the Cappell-Lee-Miller glueing theorem*, Preprint, McMaster University.
[N3] Nicolaescu, L., *Lattice points, Dedekind-Rademacher sums and a conjecture of Kronheimer and Mrowka*, Preprint, math.DG/9801030.
[N4] Nicolaescu, L., *Notes on Seiberg-Witten Theory*, Preprint, University of Notre Dame.
[Ok-T] Okonek, C. and Teleman, A., *The coupled Seiberg-Witten equations, vortices, and moduli spaces of stable pairs*, Internat. J. Math. **6**, 893-910.
[W] Wojciechowski, K. P., *The additivity of the* η*-invariant: The case of an invertible tangential operator.*, Houston J. Math. **20** (1994), 603-621.
[W1] Wojciechowski, K. P., *The additivity of the* η*-invariant: The case of a singular tangential operator*, Commun. Math. Phys. **169** (1995), 315-327.
[W2] Wojciechowski, K. P., *The* ζ*-determinant and the additivity of the* η*- invariant on the smooth, self-adjoint Grassmannian*, Commun. Math. Phys. **201** (1999), 432-444.
[Wa] Walker, K., *An Extension of Casson's Invariant*, Annals of Mathematics Studies, 126, Princeton University Press, Princeton, New Jersey, USA, 1992.
[T] Taubes, C. H., *Casson's invariant and gauge theory*, J. Diff. Geom. **31** (1990), 547-599.
[Y] Yoshida, T., *Floer homology and splittings of manifolds*, Ann. Math. **134** (1992), 277-324.

Courant Institute, 251 Mercer Street, New York, NY 10012
E-mail address: cappell@cims.nyu.edu

Yale University, Department of Mathematics, New Haven, CT 06520

E-mail address: rlee@math.yale.edu

POLYTECHNIC UNIVERSITY OF NEW YORK, 33 JAY STREET, BROOKLYN, NY 11215
E-mail address: emiller@math.poly.edu

Contemporary Mathematics
Volume **279**, 2001

On genus one mapping class groups, function spaces, and modular forms

Frederick R. Cohen*

ABSTRACT. The purpose of this article is to record the homology of the space of n unordered distinct particles on the surface of a punctured torus modulo the natural $SL(2,\mathbb{Z})$ action on the punctured torus. The associated Borel construction for this space admits a description as a $K(\pi,1)$ where π is given in terms of mapping class groups, as well as an analogue of Artin's braid group.

One feature here is that summing the answers over all punctures gives a single easily described graded object which itself is given in terms of modular cusp forms based on the the standard $SL(2,\mathbb{Z})$-action on the upper half-plane by fractional linear transformations. The results obtained by assigning a sign as particles are interchanged also follow.

The view here is that the homology of these mapping class groups are naturally assembled as those of a certain choice of function space on the torus. The specific computations give (1) the rational cohomology of the "pointed mapping class group" with punctures for genus one surfaces with either trivial coefficients or coefficients in the sign representation, or (2) the rational cohomology for the mapping class group of genus one surfaces with punctures, and with coefficients in the sign representation.

1. Introduction

The purpose of this article is to give a calculation of the rational cohomology for certain choices of mapping class groups obtained from genus one surfaces by either (1) restricting to diffeomorphisms which leave a fixed set of k points invariant as well as fixing a single point, or (2) using coefficients in the sign representation and restricting to diffeomorphisms which leave a fixed set of k points invariant. The answers in these cases assemble into a single easily described bigraded group which is given in terms of modular cusp forms based on the standard $SL(2,\mathbb{Z})$-action on the upper half plane by fractional linear transformations. The results where one assigns a sign as "particles are interchanged" also follows, is simpler, and more direct.

2000 *Mathematics Subject Classification.* Primary: 20J99, 20F36, 55N25, Secondary: 11F99, 32N99.

Key words and phrases. mapping class groups, function spaces, modular forms.

*Partially supported by the National Science Foundation.

Exact sequences of $SL(2,\mathbb{Z})$-modules arise naturally from the classical Koszul complex, and consequently there is a long exact sequence relating the cohomology of the mapping class group with punctures to the full diffeomorphism group of the torus. After addressing this point, the cohomology calculations for the "pointed mapping class groups" follow directly from (1) techniques of [**5**], as well as (2) work of Furusawa, Tezuka, and Yagita [**9**] which in turn is based on classical work of Eichler [**7**], and Shimura [**14**]. The modifications of [**9**] given here are an analysis of the cohomology of the $SL(2,\mathbb{Z})$-Borel construction for the space of continuous functions on an elliptic curve with target given by a sphere.

The results here are related to certain earlier calculations due to E. Getzler [**10**] which address the cohomology of a related moduli space. However, the work here addresses a somewhat different problem concerning invariants for the "pointed mapping class group" that is described below.

In the cases considered here, the answers can be assembled into a single bigraded group which admits a simple global "picture" where the answers are in terms the tensor product with a polynimial ring which is tensored with 6 copies of the ring of modular forms together with an additional contributions due to a single cusp. The case of cohomology with coefficients in the sign representation is cleaner, and is given in terms of one copy of the ring of modular forms together with some additional contributions due to a single cusp tensored with an exterior algebra on a single generator.

The methods here also apply to the torsion in the cohomology of these discrete groups. These types of calculations will not be addressed here.

More precisely, let T denote a fixed complex curve which is topologically isomorphic to the torus $S^1 \times S^1$. Consider the group of orientation preserving homeomorphisms which leave a set Q_k of k distinct points invariant $Top^+(T; Q_k))$. Define Γ_1^k to be the the group of path-components $\pi_0(Top^+(T; Q_k))$.

Similarly, let $Top^+(T; Q_k, *)$ denote the group of pointed, orientation preserving homeomorphisms which

1. leaves a chosen point $*$, say the identity, fixed, and
2. leaves a set Q_k of k other distinct points invariant.

Define $\Gamma_1^{k,*}$ as the group of path-components $\pi_0(Top^+(T; k, *))$.

These mapping class groups themselves are a kind of half-way house between mapping class groups which leave a set of k+1 points invariant, or k+1 points fixed. The particular choice here of the "pointed mapping class group" is useful in the sense that the answers for the resulting homology groups assemble themselves cleanly, and systematically. In addition, analogous results for the mapping class group Γ_1^k with coefficients in the sign representation are also direct and are listed below. This is one of the motivations for this article.

In addition, the cohomology groups given below are listed in terms of vector spaces of modular forms. The dimensions of these spaces are well-known. The Euler-Poincaré series for these cohomology groups is also listed. However, it is more natural to give the answers in terms of modular forms as these admit further topological interpretations.

There are similar features satisfied for surfaces of higher genus where the resulting homology groups share some features with cyclic homology [**3**]. The results here are applications of those techniques. Bundle theoretic interpretations related to elliptic curves, and which provide some explanation for these calculations might be given elsewhere.

The mathematics here represents part of a larger project on course notes which are in preparation with Jonathan Pakianathan [**5**].

We would like to congratulate Jim on this happy occasion of his birthday. It has been a real pleasure both to know him, and to learn some of the beautiful mathematics which he has done. He has been an inspiration.

We would also like to thank the organizers for the opportunity to participate in a very enjoyable, and stimulating conference.

CONTENTS

2. Statement of results

Recall that $SL(2,\mathbb{Z})$ acts on the upper 1/2-plane $\mathbb{H}$ via fractional linear transformations with $M(z) = (az+b)/(cz+d)$ where z is a complex number with strictly positive pure imaginary part, and where M is the element in $SL(2,\mathbb{Z})$ given by

$$M = \begin{pmatrix} a & b \\ c & d \end{pmatrix}.$$

In addition, $SL(2,\mathbb{Z})$ acts on the plane $\mathbb{R}^2$ by the formula $M\binom{x}{y} = \binom{ax+by}{cx+dy}$. Both of these actions are used below.

It is well-known [**11, 13**] that $SL(2,\mathbb{Z})$ is isomorphic to the amalgamated free product

$$\mathbb{Z}/6\mathbb{Z} *_{\mathbb{Z}/2\mathbb{Z}} \mathbb{Z}/4\mathbb{Z}$$

with generators given by the elements

$$\sigma = \begin{pmatrix} 1 & 1 \\ -1 & 0 \end{pmatrix},$$

and

$$\tau = \begin{pmatrix} 0 & 1 \\ -1 & 0 \end{pmatrix}.$$

The element σ generates a cyclic group of order 6 while the element τ generates a cyclic group of order 4. Furthermore, $\sigma^3 = \tau^2 = \lambda = -Identity$ is the generator of $\mathbb{Z}/2\mathbb{Z}$ the center of $SL(2,\mathbb{Z})$ which is given by the matrix

$$\begin{pmatrix} -1 & 0 \\ 0 & -1 \end{pmatrix}.$$

The group $SL(2,\mathbb{Z})$ also acts on the standard integral lattice $\mathbb{Z}\oplus i\mathbb{Z}$ in $\mathbb{C}$. Write the elements in the lattice as $\binom{x}{y}$ in $\mathbb{C}$ regarded as a topological space homeomorphic to $\mathbb{R}^2$. Let u $= \binom{1}{0}$, and v $= \binom{0}{1}$. Then

1. $\sigma(u) = u - v$,
2. $\sigma(v) = u$,
3. $\tau(u) = -v$, and
4. $\tau(v) = u$.

There is an exact sequence of groups $1 \to S^1 \tilde{\times} S^1 \to Diff^+(S^1 \times S^1) \to SL(2,\mathbb{Z}) \to 1$ where $S^1 \tilde{\times} S^1$ is homotopy equivalent to $S^1 \times S^1 = T$ [**6**]. Furthermore, this sequence is split via an action $SL(2,\mathbb{Z}) \times T \to T$ which sends a matrix M to the function which itself sends (u,v) to $(u^a v^b, u^c v^d)$. Thus $SL(2,\mathbb{Z})$ fixes the point $(1,1)$, and hence acts on the complement of the point $(1,1)$ denoted T'. In addition, this action induces the standard action of $SL(2,\mathbb{Z})$ on $\mathbb{Z}\oplus\mathbb{Z}$ after passage to the first homology group of T.

The $SL(2,\mathbb{Z})$ action on T descends to an action on the configuration spaces

$$F(T,k) = \{(x_1,...,x_k)\epsilon T^k | x_i \neq x_j, i \neq j\},$$

and

$$F(T',k) = \{(x_1,...,x_k)\epsilon (T')^k | x_i \neq x_j, i \neq j\}.$$

This action on T also descends to quotients of the configuration space modulo the natural action of the symmetric group $F(N,k)/\Sigma_k = B(N,k)$ for a manifold N, and where Σ_k denotes the symmetric group on k letters.

Let $Br_k(S)$ denote the k-stranded braid group for the surface S given by the fundamental group $\pi_1 F(S,k)/\Sigma_k$, and let $PBr_k(S)$ denote the pure k-stranded braid group for the surface S given by $\pi_1 F(S,k)$ [**8**]. If $k \geq 1$, then spaces $B(T,k)$, and $B(T',k)$ are $K(\Pi,1)'s$ where Π is the k stranded braid for the surfaces T, and T' respectively.

The Borel construction

$$ESL(2,\mathbb{Z}) \times_{SL(2,\mathbb{Z})} B(T',k)$$

gives a $K(\Gamma,1)$ for the group Γ in a group extension

$$1 \to \; Br_k(T') \to \; \Gamma \to \; SL(2,\mathbb{Z}) \to \; 1.$$

The group Γ is isomorphic to $\Gamma_1^{k,*}$ if $k \geq 1$. The analogue of the pure braid group $P\Gamma_1^{k,*}$ is the fundamental group of

$$ESL(2,\mathbb{Z}) \times_{SL(2,\mathbb{Z})} F(T',k).$$

In addition, the action of $SL(2,\mathbb{Z})$ on the upper upper 1/2-plane $\mathbb{H}$ has finite isotropy groups. Thus the natural map

$$ESL(2,\mathbb{Z}) \times_{SL(2,\mathbb{Z})} F(T',k) \to \; \mathbb{H} \times_{SL(2,\mathbb{Z})} F(T',k)$$

is singular fibration which induces an isomorphism in rational cohomology. Similar statements apply in case T' is replaced by T or spaces are replaced by their natural quotients via the natural action of the symmetric group.

In addition, consider $Top^+(S_g;k,*)$ the group of orientation preserving homeomorphisms of a closed Riemann surface S_g of genus g which leave a set of k distinct points invariant, and which fix a $(k+1)-st$ point. Define $\Gamma_g^{k,*} = \pi_0 Top^+(S_g;Q_k,*)$. If $g \geq 1$, then $B(S_g,k)$ is a $K(Br_k(S_g),1)$ where $Br_k(S_g)$ is the k-stranded braid group for the surface S_g [**8**]. If g = 0, then a $K(Br_k(S_g),1)$ is gotten by quotienting by an action of $SO(3)$ on the product of the configuration space, and $ESO(3)$ [**4**]. The next fact is certainly a "folk-theorem" and a proof is given in [**5**].

PROPOSITION 2.1. *Let M be a 2 dimensional manifold.*

1. *If M is a surface of genus g without boundary (possibly with punctures), and $g \geq 2$, the spaces*

   $$ETop(M)^+ \times_{Top(M)^+} F(M,q),$$

   and

   $$ETop(M)^+ \times_{Top(M)^+} F(M,q)/\Sigma_q$$

   are respectively $K(P\Gamma_g^q,1)$, and $K(\Gamma_g^q,1)$.
2. *Let M be a closed orientable surface of genus g with p a point of M, with N given by $M-\{p\}$. The spaces*

   $$ETop(N)^+ \times_{Top(N)^+} F(N,q),$$

   and

   $$ETop(N)^+ \times_{Top(N)^+} F(N,q)/\Sigma_q$$

   are respectively $K(P\Gamma_g^{q,*},1)$, and $K(\Gamma_g^{q,*},1)$.
3. *If $q \geq 1$, then $ESL(2,\mathbb{Z}) \times_{SL(2,\mathbb{Z})} B(T',q) = K(\Gamma_1^{q,*},1)$.*
4. *If $q \geq 3$, then $ESO(3) \times_{SO(3)} F(S^2,q)/\Sigma_q$ is a $K(\Gamma_0^q,1)$.*

The next theorem describes how the homology of certain choices of function spaces encodes the homology of the mapping class groups above into a single bi-graded object. The point of this theorem is that it gives a way to assemble all answers "at once" into a single object that is computable in a few instances for g = 0,1 or 2, and where computations for individual groups appear to be complicated.

Let $\mathbb{F}$ denote a field which can be considered as a $\Gamma_1^{k,*}$-module in two natural ways:

1. The trivial $\Gamma_1^{k,*}$-module denoted $\mathbb{F}$ where each element of $\Gamma_1^{k,*}$ fixes any element in the field.
2. The group $\Gamma_1^{k,*}$ maps surjects to the symmetric group Σ_k, and $\mathbb{F}$ inherits the structure of a $\Gamma_1^{k,*}$-module via the sign of a permutation. This $\Gamma_1^{k,*}$-module is denoted $\mathbb{F}(\pm 1\ )$.

The following theorem is given in [**5**]. Here, $map(M,T)$ denotes the space of continuous functions from M to T with the compact-open topology for a locally compact Hausdorff space M while $map_*(M,T)$ denotes the subspace of pointed functions.

THEOREM 2.2. *Assume that homology is taken with coefficients in a field $\mathbb{F}$.*

1. *Let M be a closed orientable surface of genus g with p a point of M, and with N given by $M - \{p\}$. Let G denote $Top(N)^+$. Then for all $n > 0$,*
$$H_*(EG \times_G map_*(M, S^{2n+2}); \mathbb{F})$$
is a bigraded group $H_{s,k}$ which is isomorphic to
$$H_{s-2nk}(\Gamma_g^{k,*}; \mathbb{F}).$$
For fixed s, $\oplus_{k\geq 0} H_{s,k}$ is isomorphic to
$$H_s(EG \times_G map_*(M, S^{2n+2}); \mathbb{F}).$$
2. *Let M be a closed orientable surface of genus g with p a point of M, and with N given by $M - \{p\}$. Let G denote $Top(N)^+$. Then for all $n > 0$,*
$$H_*(EG \times_G map_*(M, S^{2n+1}); \mathbb{F})$$
is a bigraded group $H_{s,k}$ which is isomorphic to
$$H_{s-(2n-1)k}(\Gamma_g^{k,*}; \mathbb{F}(\pm 1\ )).$$
For fixed s, $\oplus_{k\geq 0} H_{s,k}$ is isomorphic to
$$H_s(EG \times_G map_*(M, S^{2n+1}); \mathbb{F}).$$

Specialize to the case of genus 1 with $G = SL(2,\mathbb{Z})$.

1. *For all $n > 0$, $H_*(EG \times_G map_*(T, S^{2n+2}); \mathbb{F})$ is a bigraded group $H_{s,k}$ which is isomorphic to $H_{s-2nk}(\Gamma_1^{k,*}; \mathbb{F})$. There is an isomorphism for fixed s given by $\oplus_{k\geq 0} H_{s,k} \to\ H_s(EG \times_G map_*(T, S^{2n+2}); \mathbb{F})$.*
2. *For all $n > 0$, $H_*(EG \times_G map_*(T, S^{2n+1}); \mathbb{F})$ is a bigraded group $H_{s,k}$ which is isomorphic to $H_{s-tk}(\Gamma_1^{k,*}; \mathbb{F}(\pm 1\ ))$. There is an isomorphism for fixed s given by $\oplus_{k\geq 0} H_{s,k} \to\ H_s(EG \times_G map_*(T, S^{2n+1}); \mathbb{F})$ where t = 2n-1.*

There is an analogue of the above theorem which applies to the mapping class groups Γ_g^k in which the space of "free maps" $map(S,T)$ is used instead of the pointed function space $map_*(S,T)$ [**3**].

Furthermore in the case of genus 1, the bigrading above corresponds to weights of modular forms, as listed in this section. As a corollary, the calculation of the rational homology for the "pointed mapping class group" with coefficients in the sign representation is given additively next while the proof occupies section 5. The vector space M^0_{2n} denotes the vector space of modular cusp forms of weight $2n$ based on the standard $SL(2,\mathbb{Z})$ action on the upper 1/2-plane $\mathbb{H}$ which will be recalled in section 3 [**14**]. These modular forms are used as language to give the cohomology of the mapping class groups $\Gamma_1^{k,*}$ addressed here with coefficients in either the sign representation $\mathbb{Q}(\pm 1\ )$, or the trivial representation $\mathbb{Q}$. The "weight convention" used here is that of [**14**] rather than that of [**12**].

THEOREM 2.3. 1. *There are isomorphisms for $s > 0$ given by*

$$H_s(ESL(2,\mathbb{Z}) \times_{SL(2,\mathbb{Z})} map_*(T, S^3); \mathbb{Q}) \cong \Sigma_{k>0} H_{s-k}(\Gamma_1^{k,*}; \mathbb{Q}(\pm 1\ )).$$

2. *Furthermore, there are isomorphisms of real vector spaces for $k \geq 1$ given as follows.*

$$H^i(\Gamma_1^{k,*}; \mathbb{R}(\pm 1\ )) = \begin{cases} M^0_{2n+2} \oplus \mathbb{R} & \textit{if } k = 2n, \textit{ and } i = 2n+1, \\ M^0_{2n+2} \oplus \mathbb{R} & \textit{if } k = 2n+1, \textit{ and } i = 2n+1, \\ 0 & \textit{otherwise} \end{cases}$$

One could ask whether $B\Gamma_1^{k,*}$ is a stable summand of $B\Gamma_1^{k+s,*}$ for all $s \geq 0$ as in the case of ordinary configuration spaces. The above theorem shows that such stable splittings do not exist as the classes given in this theorem are not "stable".

Recall $map(T, X)$, the space of "free maps" from T to X, and that $SL(2,\mathbb{Z})$ acts naturally on the space of maps $map(T, X)$.

THEOREM 2.4. 1. *There is an isomorphism*

$$H^*(ESL(2,\mathbb{Z}) \times_{SL(2,\mathbb{Z})} map(T, S^3); \mathbb{Q}) \cong H^*(BDiff^+(T); \mathbb{Q}) \otimes \mathbb{E}[c] \otimes \mathbb{E}[g]$$

where c is of degree 1, and g is of degree 3.

2. *Furthermore, there are isomorphisms of real vector spaces given as follows.*

$$H^i(\Gamma_1^k; \mathbb{R}(\pm 1\ )) = \begin{cases} (M^0_{2n} \oplus \mathbb{R}) \oplus (M^0_{2n+2} \oplus \mathbb{R}) & \textit{if } k = 2n, \textit{ and } i = 2n+1, \\ M^0_{2n+2} \oplus \mathbb{R} & \textit{if } k = 2n+1, \textit{ and } i = 2n+1, \\ M^0_{2n+2} \oplus \mathbb{R} & \textit{if } k = 2n+1, \textit{ and } i = 2n+3, \\ 0 & \textit{otherwise} \end{cases}$$

The real cohomology of $\Gamma_1^{k,*}$ with trivial coefficients is given in the next theorem.

THEOREM 2.5. *There are isomorphisms of real vector spaces given as follows.*

1. *There are isomorphisms for $s > 0$ given by*

$$H_s(ESL(2,\mathbb{Z}) \times_{SL(2,\mathbb{Z})} map_*(T, S^4); \mathbb{Q}) \cong \Sigma_{k>0} H_{s-2k}(\Gamma_1^{k,*}; \mathbb{Q}).$$

2. $H^0(\Gamma_1^{k,*}; \mathbb{R})$:
   (a) *This group is isomorphic to $\mathbb{R}$.*
3. $H^{4j}(\Gamma_1^{k,*}; \mathbb{R})$ *for $j > 0$:*
   (a) *If $k = 1,2$, then $H^4(\Gamma_1^{k,*}; \mathbb{R})$ is isomorphic to $\{0\}$.*
   (b) *If $k = 3$, then $H^4(\Gamma_1^{k,*}; \mathbb{R})$ is isomorphic to $(M^0_4 \oplus \mathbb{R})$.*

(c) *If $k = 4$, then $H^4(\Gamma_1^{k,*};\mathbb{R})$ is isomorphic to $(M_4^0 \oplus \mathbb{R}) \oplus \mathbb{R}$.*
(d) *If $k \geq 5$, then $H^4(\Gamma_1^{k,*};\mathbb{R})$ is isomorphic to $(M_4^0 \oplus \mathbb{R}) \oplus_2 \mathbb{R}$.*
(e) *If $k < 4j-1$, with $j > 0$, then $H^{4j}(\Gamma_1^{k,*};\mathbb{R})$ is isomorphic to $\{0\}$.*
(f) *If $k = 4j-1$, with $j > 1$, then $H^{4j}(\Gamma_1^{4j-1,*};\mathbb{R})$ is isomorphic to*

$$(M_{2j+2}^0 \oplus \mathbb{R}) \oplus (M_{2j}^0 \oplus \mathbb{R}).$$

(g) *If $k \geq 4j$, with $j > 1$, then $H^{4j}(\Gamma_1^{k,*};\mathbb{R})$ is isomorphic to*

$$H^{4j}(\Gamma_1^{4j-1,*};\mathbb{R}) \oplus (M_{2j}^0 \oplus \mathbb{R})$$

which is isomorphic to

$$(M_{2j+2}^0 \oplus \mathbb{R} \oplus M_{2j}^0 \oplus \mathbb{R}) \oplus (M_{2j}^0 \oplus \mathbb{R}).$$

4. $H^{4j+1}(\Gamma_1^{k,*};\mathbb{R})$*:*
(a) *If $k \geq 5$, then $H^5(\Gamma_1^{k,*};\mathbb{R})$ is isomorphic to*

$$M_4^0 \oplus \mathbb{R}.$$

(b) *If $k < 4j-1$ and $j \geq 1$, then $H^{4j+1}(\Gamma_1^{4j,*};\mathbb{R})$ is isomorphic to $\{0\}$.*
(c) *If $4j \leq k \leq 8j+4$, and $j \geq 1$, then $H^{4j+1}(\Gamma_1^{k,*};\mathbb{R})$ is isomorphic to*

$$M_{2j+2}^0 \oplus \mathbb{R}.$$

(d) *If $k \geq 8j+5$, and $j \geq 1$, then $H^{4j+1}(\Gamma_1^{k,*};\mathbb{R})$ is isomorphic to*

$$(M_{2j+2}^0 \oplus \mathbb{R}) \oplus (M_{4j+4}^0 \oplus \mathbb{R} \oplus M_{4j+2}^0 \oplus \mathbb{R}).$$

5. $H^{4j+2}(\Gamma_1^{k,*};\mathbb{R})$*:*
(a) *If $j \geq 0$, this group is isomorphic to $\{0\}$.*
6. $H^{4j+3}(\Gamma_1^{k,*};\mathbb{R})$*:*
(a) *If $k = 3$, $H^3(\Gamma_1^{k,*};\mathbb{R})$ is isomorphic to $\mathbb{R}$.*
(b) *If $k \geq 4$, $H^3(\Gamma_1^{k,*};\mathbb{R})$ is isomorphic to $\oplus_2\mathbb{R}$.*
(c) *If $k < 8j+9$, and $j > 0$, then $H^{4j+3}(\Gamma_1^{k,*};\mathbb{R})$ is isomorphic to $\{0\}$.*
(d) *If $k \geq 8j+9$, and $j > 0$, then $H^{4j+3}(\Gamma_1^{k,*};\mathbb{R})$ is isomorphic to*

$$(M_{4j+6}^0 \oplus \mathbb{R}) \oplus (M_{4j+4}^0 \oplus \mathbb{R}).$$

7. *All other groups $H^i(\Gamma_1^{j,*};\mathbb{R})$ vanish.*

REMARK 2.6. A special case of Theorem 2.5 is listed next. There are the isomorphisms of real vector spaces given as follows.

$$H^i(\Gamma_1^{4,*};\mathbb{R}) = \begin{cases} \mathbb{R} & \text{if i = 0,} \\ \{0\} & \text{if i = 1,2} \\ \oplus_2\mathbb{R} & \text{if i = 3} \\ M_4^0 \oplus \mathbb{R} & \text{if i = 4,5} \\ \{0\} & \text{if } i \geq 6 \end{cases}$$

Theorem 2.5 corresponds to a bigrading of the modules

$$[\{\oplus_2 H^*(SL(2,\mathbb{Z});P)\} \oplus \{\oplus_2 H^*(SL(2,\mathbb{Z});V \otimes P)\}] \otimes \mathbb{R}[a]$$

where

1. the fundamental 2 dimensional representation of $SL(2,\mathbb{Z})$ is denoted V,
2. P denotes the polynomial ring in 2 indeterminates generated by the fundamental representation V of $SL(2,\mathbb{Z})$ so that $P = \mathbb{R}[V]$, and
3. $\mathbb{R}[a]$ denotes the polynomial ring on one indeterminate.

These cohomology groups are given in terms of the direct sum of 6 copies of the graded vector space of modular cusp forms tensored with $\mathbb{R}[a]$, a statement which is made precise in section 6.

The rational cohomology for the space of free maps $map(T, S^{2n+2})$ is given in the thesis of K.K.Tse [**15**]. These calculations are complicated as well as delicate. Furthermore, these calculations give one indication of why the "pointed mapping class group" is more convenient to handle rather than the "free" version. Namely, Tse's calculations are considerably more delicate than those given here. It appears that the rational cohomology of the Borel construction $ESL(2,\mathbb{Z}) \times_{SL(2,\mathbb{Z})} map(T, S^{2n+2})$ ought to follow from Tse's calculations while these last computations give the rational cohomology of the groups Γ_1^k. It also appears that a computation of the invariants of cohomology classes given in [**10**] should give the results in Theorem 2.5 directly.

3. Recollections for $SL(2,\mathbb{Z})$ and modular forms

The purpose of this section is to recall some well-known information concerning modular forms based on the $SL(2,\mathbb{Z})$ action on the upper 1/2-plane $\mathbb{H}$ by fractional linear transformations. This information is then used as language to describe the rational cohomology of certain braid groups associated to a punctured torus, as well as the mapping class group for a punctured torus.

The main input required here consists of results due to Eichler [**7**], and Shimura [**14**] together with additional refinements to group cohomology given by Furusawa, Tezuka, and Yagita [**9**]. Some definitions are recalled from [**14, 12**] first for coherence, and secondly to give the author a fighting chance to understand a little bit.

Recall that $SL(2,\mathbb{R})$ acts on the upper-1/2 plane $\mathbb{H}$ by fractional linear transformations. It is well-known that $PSL(2,\mathbb{R})$ is the group of all analytic isomorphisms of $\mathbb{H}$. In addition, the action of $SL(2,\mathbb{Z})$ extends to one on $\mathbb{C} \cup \infty$.

There is a second action of $SL(2,\mathbb{Z})$ on $\mathbb{R}^2$ obtained by extending the natural action through linear transformations on $\mathbb{Z} \oplus \mathbb{Z}$. Both of these actions are used below.

For each positive integer k, a complex holomorphic self-map of $\mathbb{H}$ is called an (integral) modular form of weight k with respect to $SL(2,\mathbb{Z})$ provided

1. $f(\{az+b\}/\{cz+d\}) = (cz+d)^k f(z)$ for every matrix A in $SL(2,\mathbb{Z})$, with A given by
$$\begin{pmatrix} a & b \\ c & d \end{pmatrix}.$$

2. The function f is holomorphic everywhere including ∞.
3. Furthermore, f is called a cusp form if $f(\infty) = 0$.

Write M_k for the complex vector space of all modular forms of weight k, and M_k^0 for the subspace of all cusp forms. These vectors spaces admit an algebra structure for M where

$$M = \Sigma_{0 \leq k} M_k$$

denotes the direct sum of the modules M_k. Let G_k denote the Eisenstein series given by a modular form of weight 2k with

$$G_k(z) = \Sigma_{(m,n) \neq (0,0)} 1/(mz+n)^{2k}$$

where (m, n) run over the non-zero points in the lattice $\mathbb{Z} + i\mathbb{Z}$. The functions G_k are given in terms of the Laurent series expansion for the Weierstrass $\wp$-function. Some standard facts concerning modular forms are as follows:

1. The vector space M_k is of dimension 0 for all odd integers k.
2. The vector space M_k^0 is the kernel of the homomorphism $M_k \to \mathbb{C}$ given by sending f to $f(\infty)$.
3. If $k = 0, 2, 4$, or 6, the complex dimension of M_k is one, and a basis is given by the Eisenstein series G_k .
4. If $k \geq 6$, there is an isomorphism of vector spaces $M_{k-6} \to M_k^0$ given by multiplication by $\Delta = g_2^3 - 27g_3^2$ with $g_2 = 60G_2$, and $g_3 = 140G_3$.
5. By [**14**] Proposition 2.27, or [**12**], pages 83-89, M is a polynomial ring over $\mathbb{C}$ with generators g_2, and g_3
6. The real dimension of M_k [**14**] is given as follows where $k \geq 0$ and [x] denotes the Gauss function which gives the integer part of x with [x]= b if b is an integer such that $b \leq x < b+1$:

$$dim_{\mathbb{R}} M_k = \begin{cases} 2[k/12] & \text{if k is } 1(\text{mod } 12) \\ 2[k/12]+1 & \text{if k not equal to } 1(\text{mod } 12) \end{cases}$$

Before proceeding further, it should be pointed out that analogous results are satisfied for subgroups Γ of $PSL(2, \mathbb{Z})$ such that $PSL(2, \mathbb{R})/\Gamma$ have finite volume [**14**], and where automorphic forms with respect to the action of Γ are used. Reference to such groups will be interspersed in this article, but the focus will specifically be for $SL(2, \mathbb{Z})$. The work in [**9**] applies to all such groups Γ.

Note: Some authors omit the case for which k is odd as these modules are trivial. In addition, the modular forms $G_k(z) = \Sigma_{(m,n) \neq (0,0)} 1/(mz+n)^{2k}$ are given weight k with this convention. This last convention concerning weights is not the one used in this article. The convention used here is that of Shimura [**14**] for which $G_k(z)$ has weight 2k.

Additional standard properties are listed next where M is regarded as a graded algebra, with Euler-Poincaré series for M given by $\Sigma_{k \geq 0}(dimension_{\mathbb{R}} M_k)t^k$, and where this series keeps track of the weights of modular forms.

1. The Euler-Poincaré series for M regarded as a real vector space is given by

$$\Sigma_{k \geq 0} dimension_{\mathbb{R}}(M_k) t^k = 2/(1-t^4)(1-t^6).$$

2. $M_k = M_k^0 \oplus \mathbb{C} \cdot G_k$ for $k \geq 2$, and k even.
3. A non-zero modular form weight 0 is given by $j = 1728g_2^3/\Delta$.
4. Define $M^0 = \Sigma_{0\leq k} M_k^0$ with Euler-Poincaré series $\Sigma_{i\geq 0} dim_{\mathbb{R}}(M_k^0)t^k$. Notice that this series is equal to $2/(1-t^4)(1-t^6) - 2 - [2t^4/(1-t^2)]$.
5. In what follows below, it is convenient to double the degrees here. The associated Euler-Poincaré obtained from this doubling process is $2/(1-t^8)(1-t^{12}) - 2 - [2t^8/(1-t^4)]$

A recapitulation of some work in [**9**] that is useful for the computations below is given next: Consider the polynomial ring on two indeterminates $\mathbb{Z}[x,y]$ where x and y have degree 2 with the natural $GL(2,\mathbb{Z})$-action. This algebra is the cohomology of $\mathbb{CP}^\infty \times \mathbb{CP}^\infty$ with the $GL(2,\mathbb{Z})$-action on the cohomology algebra extending the natural action on $\mathbb{Z} \oplus \mathbb{Z}$.

The ring $\mathbb{Z}[x,y]$ is naturally graded by the elements of degree 2k, denoted V^{2k} which are given by the linear span of the elements $\{x^k, x^{k-1}y, ..., x^{k-i}y^i, ...\}$. That is the linear span of monomials having classical degree k, or the homogeneous polynomials in x and y of homological degree 2k. Notice that V^{2k} inherits the structure of $GL(2,\mathbb{Z})$-module.

Facts concerning the cohomology of $SL(2,\mathbb{Z})$ with coefficients in the modules V^{2k} are as follows where the goal is to determine the cohomology of $SL(2,\mathbb{Z})$ with coefficients in the cohomology of $\mathbb{CP}^\infty \times \mathbb{CP}^\infty$. The main point for recapitulating these computations is that the method will apply to other function spaces below replacing $\mathbb{CP}^\infty \times \mathbb{CP}^\infty$. These replacements will be a subspace of the space of continuous functions on an elliptic curve, and will inform on the cohomology of the mapping class group.

1. Consider the group extension
$$1 \to \mathbb{Z}/2\mathbb{Z} \to SL(2,\mathbb{Z}) \to PSL(2,\mathbb{Z}) \to 1,$$
together with the Lyndon-Hochschild-Serre spectral sequence of this extension which abuts to $H^*(SL(2,\mathbb{Z}); H^*(\mathbb{CP}^\infty \times \mathbb{CP}^\infty))$. The E_2-term is given by
$$H^*(PSL(2,\mathbb{Z}); H^*(\mathbb{Z}/2\mathbb{Z}; \mathbb{Z}[x,y])).$$
In case the integers are replaced by any ring R in which 6 is a unit, the E_2-term is equal to the E_∞-term which is given by
$$H^*(PSL(2,\mathbb{Z}); \Sigma_{k\geq 0} V^{2k} \otimes_{\mathbb{Z}} R).$$
Work of Eichler, and Shimura is used next to identify these cohomology groups in terms of modular forms.
2. The Eichler-Shimura isomorphism [**7, 14**] is an $\mathbb{R}$-linear isomorphism
$$H^1_{Eichler}(PSL(2,\mathbb{Z}); V^{2k} \otimes \mathbb{R}) \to M^0_{k+2}.$$
Thus $H^1_{Eichler}(PSL(2,\mathbb{Z}); V^{2k} \otimes \mathbb{R})$ as a real vector space is of dimension twice the complex dimension of M^0_{k+2}. The complex dimension of M^0_{k+2} is stated above, and is taken from [**14**].

3. By [**9**], Lemma 2.1, there is a split short exact sequence of real vector spaces if $k \geq 1$ where k is the classical degree of the polynomial, 2k is the homological degree, and where the right-hand copy of $\mathbb{R}$ arises from a single cusp at ∞:
   $0 \to H^1_{Eichler}(PSL(2,\mathbb{Z}); V^{2k} \otimes \mathbb{R}) \to H^1(PSL(2,\mathbb{Z}); V^{2k} \otimes \mathbb{R}) \to \mathbb{R} \to 0$
4. Thus there are isomorphisms of real vector spaces given as follows.
   (a) $H^1(PSL(2,\mathbb{Z}); V^{4k} \otimes \mathbb{R}) \to M^0_{2k+2} \oplus \mathbb{R}$ for $k \geq 1$,
   (b) $H^1(PSL(2,\mathbb{Z}); H^{4k}(\mathbb{CP}^\infty \times \mathbb{CP}^\infty; \mathbb{R})) \to M^0_{2k+2} \oplus \mathbb{R}$ for $k \geq 1$,
   (c) $H^i(PSL(2,\mathbb{Z}); H^{4k}(\mathbb{CP}^\infty \times \mathbb{CP}^\infty; \mathbb{R})) = 0$ for $i \neq 1, k \geq 1$,
   (d) $H^i(PSL(2,\mathbb{Z}); H^{4k+2}(\mathbb{CP}^\infty \times \mathbb{CP}^\infty; \mathbb{R})) = 0$,
   (e) $H^1(SL(2,\mathbb{Z}); V^{4k} \otimes \mathbb{R}) \to M^0_{2k+2} \oplus \mathbb{R}$ for $k \geq 1$,
   (f) $H^i(SL(2,\mathbb{Z}); V^{4k} \otimes \mathbb{R}) = 0$ for $i \neq 1$, for $k \geq 1$, and
   (g) $H^i(SL(2,\mathbb{Z}); V^{4k+2} \otimes \mathbb{R}) = 0$ for $i \geq 0$, and

In addition,

1. these results in [**14, 12, 9**] also apply to subgroups Γ such that $PSL(2,\mathbb{R})/\Gamma$ has finite volume. The cohomological results reduce to a description of the automorphic forms with respect to these subgroups Γ together with the number of inequivalent cusps.
2. The authors give two proofs of this lemma 2.1, one of which is "bare-hands", while the second proof follows from Shapiro's lemma together with Borel-Serre duality.

These remarks are collected below where results of [**9**] are stated.

THEOREM 3.1. 1. *There are isomorphisms of real vector spaces*
   (a) $H^1(SL(2,\mathbb{Z}); V^{4k} \otimes \mathbb{R}) \to M^0_{2k+2} \oplus \mathbb{R}$ *for* $k \geq 1$,
   (b) $H^i(SL(2,\mathbb{Z}); V^{4k} \otimes \mathbb{R}) = 0$ *for* $i \neq 1$, *and* $k \geq 1$.
   (c) $H^i(SL(2,\mathbb{Z}); \mathbb{R}) = 0$ *for* $i \geq 1$, *and is* $\mathbb{R}$ *for* $i = 0$.
   (d) $H^i(SL(2,\mathbb{Z}); V^{4k+2} \otimes \mathbb{R}) = 0$ *for* $i \geq 0$.
2. *There is an isomorphism* $H^*(BDiff^+(S^1 \times S^1); \mathbb{Q}) \cong H^*(SL(2,\mathbb{Z}); H^*(\mathbb{CP}^\infty \times \mathbb{CP}^\infty; \mathbb{Q}))$ *of graded abelian groups.*
3. *The Euler-Poincaré series for* $H^*(SL(2,\mathbb{Z}); H^*(\mathbb{CP}^\infty \times \mathbb{CP}^\infty; \mathbb{Q}))$ *is given by* $1+2t^2(t^6+t^4-1)/(1-t^4)(1-t^6)$ $1+\{[1/(1-t^8)(1-t^{12})]-1\}t^{-3}+(1/(1-t^4))t^5$ *which is equal to* $(1+2t^4+2t^9-t^{21})/(1-t^8)(1-t^{12})$.

4. On the Koszul complex and $SL(2,\mathbb{Z})$-modules

One problem that arises naturally in the computation above is to obtain information concerning the cohomology of $SL(2,\mathbb{Z})$ with coefficients in the tensor product of the polynomial ring on 2 indeterminates with the module of exterior 1-forms in the exterior algebra on 2 indeterminates. The case of coefficients in the polynomial ring for the case $n = 2$ was given in [**7**], [**14**], [**9**], as collected in section 3 above.

The information below is a reinterpretation of that work to the questions in group cohomology considered here. The main point of this section is that the classical Koszul complex provides a clean way to give the computation with the input of [**14, 12, 9**].

Fix a strictly positive integer q, and consider the graded free abelian groups V_{2q-1} and W_{2q} which are given by $\mathbb{Z}\oplus\mathbb{Z}$ in degrees 2q-1, and 2q respectively, and are zero otherwise. Let $\{v_1, v_2\}$ be a basis for V_{2q-1} in degree 2q-1, and $\{w_1, w_2\}$ be a basis for W_{2q} in degree 2q. Thus V_{2q-1} and W_{2q} are copies of the fundamental representation of $SL(2,\mathbb{Z})$ with odd and even degrees attached.

The Koszul complex $\Lambda[V_{2q-1}]\otimes Z[W_{2q}]$ here is a cochain complex that is an acyclic differential graded algebra with $dv_i = x_i$ where $\Lambda[V_{2q-1}]$ denotes the exterior algebra generated by V_{2q-1}, and $Z[W_{2q}]$ denotes the polynomial algebra generated by W_{2q}. Furthermore, the differential is $SL(2,\mathbb{Z})$-linear (as checked below).

Notice that both σ, and τ fix the product $v_1\cdot v_2$, and $d(v_1\cdot v_2) = v_2\cdot x_1 - v_1\cdot x_2$ since

1. $\sigma(v_1\cdot v_2) = (v_1 - v_2)\cdot(v_1) = -(v_2\cdot v_1) = v_1\cdot v_2$,
2. $\tau(v_1\cdot v_2) = -(v_2\cdot v_1) = v_1\cdot v_2$, and
3. $\sigma(v_2\cdot x_1 - v_1\cdot x_2) = \tau(v_2\cdot x_1 - v_1\cdot x_2) = v_2\cdot x_1 - v_1\cdot x_2 = d(v_1\cdot v_2)$.

Let $\Lambda_j(q)$ denote the $SL(2,\mathbb{Z})$-modules of $\Lambda[V_{2q-1}]$ given by the exterior j-forms. Let $\bar{\mathbb{Z}}[W_{2q}]$ denote the augmentation ideal, the kernel of the augmentation $\epsilon : \mathbb{Z}[W_{2q}] \to \mathbb{Z}$. The classical Koszul complex is of course a resolution, and is given by the chain complex of $SL(2,\mathbb{Z})$-modules

$$0\to \Lambda_2(q)\otimes\mathbb{Z}[W_{2q}] \to \Lambda_1(q)\otimes\mathbb{Z}[W_{2q}] \to \Lambda_0(q)\otimes\mathbb{Z}[W_{2q}] \to \mathbb{Z}\to 0.$$

This sequence is given more finely by a direct sum of complexes which are indexed by polynomial degree as follows. Let S_m denote the submodule of $\mathbb{Z}[W_{2q}]$ given by the linear span of polynomials of classical degree m (and thus homological degree 2mq). Regard the Koszul complex as a direct sum of short exact sequences where the natural grading below is deliberately ignored. Then for $m\geq 0$, there are short exact sequences of $SL(2,\mathbb{Z})$-modules

$$0\to \Lambda_2(q)\otimes S_m \to \Lambda_1(q)\otimes S_{m+1} \to S_{m+2}\to 0.$$

The application here is to give the cohomology of $SL(2,\mathbb{Z})$ with coefficients in $\Lambda_j(q)\otimes\mathbb{Z}[W_{2q}]$, a calculation required for the cohomology of the mapping class group for elliptic curves with punctures. The resulting answers are given in terms of the cohomology of $SL(2,\mathbb{Z})$ with coefficients in $\mathbb{Q}[W_{2q}]$ which was described in the previous section in terms of modular forms.

Let χ denote the Euler-Poincaré series for $H^*(SL(2,\mathbb{Z});\mathbb{Q}[W_{2q}])$, and let $\bar{\chi}$ denote the Euler-Poincaré series for $H^*(SL(2,\mathbb{Z});\bar{\mathbb{Q}}[W_{2q}])$.

PROPOSITION 4.1. *For $q > 0$, and after reduction to the rational numbers $\mathbb{Q}$, there are isomorphisms of vector spaces given as follows:*

1. $H^{k+2q-2}(SL(2,\mathbb{Z});\Lambda_2(q)\otimes M)\cong H^k(SL(2,\mathbb{Z});M)$ *for any $SL(2,\mathbb{Z})$-module M.*
2. $H^1(SL(2,\mathbb{Z});\Lambda_1(q)\otimes S_{m+1})\cong H^1(SL(2,\mathbb{Z});\Lambda_2(q)\otimes S_m)\oplus H^1(SL(2,\mathbb{Z}); S_{m+2})$ *for all $m\geq 0$.*

3. *Regarding S_m as a graded $SL(2,\mathbb{Z})$-module concentrated in dimension 2qm, there are isomorphisms*

$$H^i(SL(2,\mathbb{Z});S_m) = \begin{cases} 0 & \textit{if } m \textit{ is odd,} \\ 0 & \textit{if } m \textit{ is even, and } i \textit{ not equal to } 1+2qm, \\ M^0_{2s+2}\oplus\mathbb{R} & \textit{if } m=2s, \textit{ and } i=1+2qm=1+4qs. \end{cases}$$

4. *There are induced isomorphisms where $G = SL(2,\mathbb{Z})$:*
   (a) $H^{k+2q-2}(G;\Lambda_2(q)\otimes\mathbb{Q}[W_{2q}]) \cong H^k(G;\mathbb{Q}[W_{2q}])$.
   (b) $H^k(G;\Lambda_1(q)\otimes S_{m+1}) \cong H^k(G;\Lambda_2(q)\otimes S_m)\oplus H^k(G;S_{m+2})$ *for all* $m\geq 0$.
   (c) $H^k(G;\Lambda_1(q)\otimes\mathbb{Q}[W_{2q}]) \cong H^k(G;\Lambda_2(q)\otimes\mathbb{Q}[W_{2q}])\oplus H^k(G;\bar{\mathbb{Q}}[W_{2q}])$.
   (d) $H^k(G;\Lambda_1(q)\otimes\mathbb{Q}[W_{2q}]) \cong H^{k-2q+2}(G;\mathbb{Q}[W_{2q}])\oplus H^k(G;\bar{\mathbb{Q}}[W_{2q}])$.
   (e) $H^k(G;W_{2q}\otimes\mathbb{Q}[W_{2q}]) \cong H^{k-1}(G;\Lambda_1(q)\otimes\mathbb{Q}[W_{2q}])$
   (f) $H^{k-1}(G;\Lambda_1(q)\otimes\mathbb{Q}[W_{2q}]) \cong H^{k-2q+1}(G;\mathbb{Q}[W_{2q}])\oplus H^{k-1}(G;\bar{\mathbb{Q}}[W_{2q}])$.

Furthermore, $\chi = 1+\bar{\chi}$.

PROOF. By the above remarks, $\Lambda_2(q)$ is a trivial one dimensional $SL(2,\mathbb{Z})$-module concentrated in degree 2q-2. Thus part 1 is clear.

Define $SL(2,\mathbb{Z})$-modules
1. $A(m,q) = \Lambda_2(q)\otimes S_m$,
2. $B(m,q) = \Lambda_1(q)\otimes S_{m+1}$, and
3. $C(m,q) = \Lambda_0(q)\otimes S_{m+2}$

To check the second part, recall that there is an exact sequence of $SL(2,\mathbb{Z})$-modules

$$0\to A(m,q)\to B(m,q)\to C(m,q)\to 0$$

for all $m\geq 0$ together with an induced long exact sequence in cohomology as follows: $\cdots\to H^k(SL(2,\mathbb{Z});A(m,q))\to H^k(SL(2,\mathbb{Z});B(m,q))\to H^k(SL(2,\mathbb{Z});C(m,q))\to H^{k+1}(SL(2,\mathbb{Z});A(m,q))\to\cdots$. Furthermore, there is a second exact sequence obtained by tensoring with $\mathbb{Q}$, and where $\bar{Q}[W_{2q}]$ is the kernel of the augmentation $\epsilon:\mathbb{Q}[W_{2q}]\to\mathbb{Q}$:

$$0\to\Lambda_2(q)\otimes\mathbb{Q}[W_{2q}]\to\Lambda_1(q)\otimes\mathbb{Q}[W_{2q}]\to\bar{\mathbb{Q}}[W_{2q}]\to 0.$$

This short exact sequence gives a long exact sequence in cohomology

$$\cdots\to H^k(G;A)\to H^k(G;B)\to H^k(G;C)\to H^{k+1-t}(G;A)\to\cdots$$

for which the following notation is used:
1. $G = SL(2,\mathbb{Z})$,
2. $A = \Lambda_2(q)\otimes\mathbb{Q}[W_{2q}]$,
3. $B = \Lambda_1(q)\otimes\mathbb{Q}[W_{2q}]$,
4. $C = \bar{\mathbb{Q}}[W_{2q}]$, and
5. t = 2q-2.

By the remarks in Theorem 3.1, the groups $H^k(SL(2,\mathbb{Z});\bar{\mathbb{Q}}[W_{2q}])$ are non-zero in degrees where $k = 4nq+1$ with $n>0$, and are zero otherwise. Thus the boundary maps in this long exact sequence are all trivial, and part 2 follows while part 3 is a restatement of parts 1, and 2. Part 4 follows from the definition.

Part 5 follows as W_{2q} is isomorphic to $\Lambda_1(q)$ as an $SL(2,\mathbb{Z})$-module with a degree shift by one. □

5. On $H^*(SL(2,\mathbb{Z}); H^*map_*(T, S^{2n+1}))$

The group $SL(2,\mathbb{Z})$ acts on T as above fixing a point, and thus on the space of continuous functions $map(T,X)$ as well as on the space of pointed continuous functions $map_*(T,X)$ for any pointed space X. Thus the cohomology of $map_*(T,X)$ is naturally an $SL(2,\mathbb{Z})$, and one could ask about the cohomology groups

$$H^*(SL(2,\mathbb{Z}); H^*(map_*(T,X);\mathbb{Q})).$$

Notice that if $G = SL(2,\mathbb{Z})$, and coefficients are taken in the a field of characteristic zero, then the Serre spectral sequence for $EG \times_G map_*(T,X) \to BG$ collapses at E_2. The purpose of the next two sections is to work out $H^*(G; H^*(map_*(T,S^n);\mathbb{Q}))$ as the resulting answer gives the cohomology of the "pointed mapping class group" with punctures by Theorem 2.2 in this article. The first step is to work out the cohomology ring of $map_*(T,S^n)$ as an algebra over the group ring of $SL(2,\mathbb{Z})$ with this section restricted to the case $map_*(T, S^{2q+1})$. Throughout this section, assume that n is odd. The case for which n is even is addressed in section 6.

The cohomology of the space of free maps $map(T,S^n)$ as well as the cohomology groups $H^*(SL(2,\mathbb{Z}); H^*(map(T,S^n);\mathbb{Q}))$ are worked out in this section. This last computation gives the cohomology of the mapping class group for genus 1 with punctures, and with coefficients in the sign (Fermionic) representation.

Consider the fibration $map_*(T,X) \to (\Omega X)^2 \to \Omega X$ where

1. the map $(\Omega X)^2 \to \Omega X$ is induced by the commutator map by regarding ΩX as a group, and
2. the function $map_*(T,X) \to (\Omega X)^2$ is induced by inclusion of the axes S^1 in T.

If X is an odd dimensional sphere of dimension 2n + 1 for $n \geq 0$, which is localized away from 2, $S^{2n+1}_{(1/2)}$, then this localized sphere is an H-space, and the commutator map $(\Omega X)^2 \to \Omega X$ is null. Thus the space $map_*(T, S^{2n+1}_{(1/2)})$ is homotopy equivalent to a product $\Omega S^{2n+1}_{(1/2)})^2 \times \Omega^2 S^{2n+1}_{(1/2)}$, and the cohomology of $map_*(T, S^{2n+1}_{(1/2)})$ over a ring R where 2 is a unit is isomorphic to $H^*(\Omega S^{2n+1}_{(1/2)})^2 \times \Omega^2 S^{2n+1}_{(1/2)}; R)$.

All coefficients will be taken in either $\mathbb{Q}$ or $\mathbb{R}$ for the rest of this section. The cohomology algebra of $(\Omega S^{2n+1}_{(1/2)})^2 \times \Omega^2 S^{2n+1}_{(1/2)}$ is given by $\mathbb{Q}[x_1,x_2] \otimes E[c]$ where the x_i, are of degree 2n, and c is of degree 2n-1. Furthermore, the action of $SL(2,\mathbb{Z})$ is specified by setting $x_1 = \binom{1}{0}$, and $x_2 = \binom{0}{1}$, and the action on c is trivial. These observations are recorded next.

THEOREM 5.1. *If $n \geq 1$, then*

1. *The rational cohomology ring of $map_*(T, S^{2n+1})$ is isomorphic to $\mathbb{Q}[x_1, x_2] \otimes E[c]$ where x_i, are of degree 2n, and c is of degree 2n-1.*
2. *Furthermore, this ring is an algebra over the integral group ring of $SL(2,\mathbb{Z})$ where*
   (a) *the action of $SL(2,\mathbb{Z})$ is trivial on c, and*
   (b) *is specified by setting $x_1 = \binom{1}{0}$, and $x_2 = \binom{0}{1}$.*
3. *Let $G = SL(2,\mathbb{Z})$. There are isomorphisms*

$$H^k(EG \times_G map_*(T, S^{2n+1}); \mathbb{Q}) \cong \Sigma_{s+t=k} H^s(SL(2,\mathbb{Z}); H^t(map_*(T, S^{2n+1}); \mathbb{Q})),$$

and

$$H^*(SL(2,\mathbb{Z}); H^* map_*(T, S^{2n+1})) \cong H^*(SL(2,\mathbb{Z}); \mathbb{Q}[x_1, x_2]) \otimes E[c].$$

4. *If n = 1, then $H^*(EG \times_G map_*(T, S^{2n+1}); \mathbb{Q})$ is isomorphic to*

$$H^*(BDiff^+(T); \mathbb{Q}) \otimes E[c]$$

where $G = SL(2,\mathbb{Z})$, and there are isomorphisms

$$H_s(EG \times_G map_*(T, S^3); \mathbb{Q}) \cong \Sigma_{k>0} H_{s-k}(\Gamma_1^{k,*}; \mathbb{Q}(\pm 1\ )).$$

5. *Furthermore, there are isomorphisms of real vector spaces for $k \geq 1$ given by*

$$H^i(\Gamma_1^{k,*}; \mathbb{R}(\pm 1\ )) = \begin{cases} M^0_{2n+2} \oplus \mathbb{R} & \text{if } k = 2n, \text{ and } i = 2n+1, \\ M^0_{2n+2} \oplus \mathbb{R} & \text{if } k = 2n+1, \text{ and } i = 2n+1, \text{ and} \\ 0 & \text{otherwise.} \end{cases}$$

Before proving this theorem, a digression concerning the methods is given. Let S_g be a surface of genus g. If $n \geq 3$, and coefficients are taken in a field $\mathbb{F}$, the homology of the function spaces $map_*(S_g, S^n)$, and $map(S_g, S^n)$ as well as $EG \times_G map_*(S_g, S^n)$, and $EG \times_G map(S_g, S^n)$ for $G = Top^+(S_g)$ are filtered by "configuration space degree" which is that of "weights" as in the weight filtration of configuration spaces [**2, 5**].

For $n >> j$, with α any element in $H^j(map_*(S_g, S^n); \mathbb{F})$, the "configuration space degree" of α is k where $k(n-2) \leq j < (k+1)(n-2)$. The "configuration space degree" is well-defined for sufficiently large n, and will be used in the remainder of this section as well as section 6. The integer k corresponds roughly to the "degree" filtration of the homology of the function space divided by $(n-2)$, at least in low degrees with respect to n, and is obtained by filtering via numbers of points in a configuration. Since the function spaces in question stably split, the filtration satisfies the additional property that it is a gradation. Furthermore, elements α of "configuration space degree" k give elements in the homology of the configuration space as follows by [**2, 5**]. Let ${S_g}^*$ denote the punctured surface $S_g - Q_1$ where Q_1 is a subset of S_g consisting of a single. The following example is listed for use below where the coefficients $\mathbb{F}$ will be assumed to be the rational numbers $\mathbb{Q}$ for the remainder of this article:

1. In case $n = 2t+1$ with $n >> j$, and $k > 0$, there are isomorphisms:
   (a) $H_j(map_*(S_g, S^n); \mathbb{Q}) \to H_{j-(n-2)k}(F({S_g}^*, k)/\Sigma_k; \mathbb{Q}(\pm 1\ ))$.
   (b) $H_j(map(S_g, S^n); \mathbb{Q}) \to H_{j-(n-2)k}(F(S_g, k)/\Sigma_k; \mathbb{Q}(\pm 1\ ))$.

(c) $H_j(EG \times_G map_*(S_g, S^n); \mathbb{Q}) \rightarrow H_{j-(n-2)k}(EG \times_G F({S_g}^*, k)/\Sigma_k; \mathbb{Q}(\pm 1))$.

(d) $H_j(EG \times_G map(S_g, S^n); \mathbb{Q}) \rightarrow H_{j-(n-2)k}(EG \times_G F(S_g, k)/\Sigma_k; \mathbb{Q}(\pm 1))$.

2. In case $n = 2t$ with $n >> j$, there are isomorphisms:

(a) $H_j(map_*(S_g, S^n); \mathbb{Q}) \rightarrow H_{j-(n-2)k}(F({S_g}^*, k)/\Sigma_k; \mathbb{Q})$.

(b) $H_j(map(S_g, S^n); \mathbb{Q}) \rightarrow H_{j-(n-2)k}(F(S_g, k)/\Sigma_k; \mathbb{Q})$.

(c) $H_j(EG \times_G map_*(S_g, S^n); \mathbb{Q}) \rightarrow H_{j-(n-2)k}(EG \times_G F({S_g}^*, k)/\Sigma_k; \mathbb{Q})$.

(d) $H_j(EG \times_G map(S_g, S^n); \mathbb{Q}) \rightarrow H_{j-(n-2)k}(EG \times_G F(S_g, k)/\Sigma_k; \mathbb{Q})$.

These remarks are used to prove the above theorem as well as the remaining computations in this section, and those given in section 6.

PROOF. The first two parts given in this theorem are clear from the remarks given before Theorem 5.1. Part 3 follows directly from inspection of the Lyndon-Hochschild-Serre spectral sequence for the fibration $EG \times_G map_*(T, S^{2n+1}) \rightarrow BG$ for $G = SL(2, \mathbb{Z})$. This spectral sequence collapses at E_2 as $H^i(SL(2, \mathbb{Z}); \mathbb{Q}[x_1, x_2]) = \{0\}$ for $i \geq 2$.

To prove part 4, notice that $H^*(SL(2, \mathbb{Z}); \mathbb{Q}[x_1, x_2])$ where x_i has degree 2 is precisely $H^*(BDiff^+(T); \mathbb{Q})$ by the remarks in section 3 here or by [**9**]. Furthermore, by Theorem 2.2 here, the homology of the space $EG \times_G map_*(T, S^3)$ gives the homology groups with coefficients in the sign representation $H_*(\Gamma_1^{k,*}; \mathbb{Q}(\pm 1))$ for all k, but with a degree shift.

That degree shift is given as follows where there are isomorphisms for $s > 0$: $H_s(EG \times_G map_*(T, S^3); \mathbb{Q}) \cong \Sigma_{k>0} H_{s-k}(\Gamma_1^{k,*}; \mathbb{Q}(\pm 1))$. Next recall that c has degree 1. Notice that this sum is finite in each degree. Furthermore, by the results in section 4, the groups $H^*(BDiff^+(T); \mathbb{Q})$ are non-zero in degrees 4n+1 for $n > 0$, and in this case the dimension of $H_{4n+1}(EG \times_G map_*(T, S^3_{(1/2)}); \mathbb{Q})$, and $H_{4n+2}(EG \times_G map_*(T, S^3_{(1/2)}); \mathbb{Q})$ is given by $1 + dimension_{\mathbb{R}} M^0_{2n+2}$ where M^0_{2n+2} is the vector space of modular cusp forms based on the standard $SL(2, \mathbb{Z})$ action on the upper 1/2-plane.

Thus the elements of classical polynomial degree k arising from the polynomial algebra $Q[x_1, x_2]$ here corresponds to "configuration space degree" equal to k, and contributes to the homology of the k-th space

$$ESL(2, \mathbb{Z}) \times_{SL(2,\mathbb{Z})} B(T', k) = K(\Gamma_1^{k,*}, 1).$$

The first part here addresses the case of $\Gamma_1^{2k,*}$. Here the monomials $x_1^{2k-i} \cdot x_2^i$ for $2k \geq i \geq 0$ are of classical degree 2k, and the cohomology of $SL(2, \mathbb{Z})$ with coefficients in their linear span contributes $M^0_{2k+2} \oplus \mathbb{R}$. The "configuration space degree" of this contribution is given by 2k, and the homological degree is $1 + degree(x^{2k}) - 2k = 1 + 4k - 2k = 1 + 2k$ as above.

Thus there is a summand for the cohomology of $\Gamma_1^{2k,*}$ with coefficients in $\mathbb{R}(\pm 1)$ given by $M^0_{2k+2} \oplus \mathbb{R}$, and which is concentrated in degree $1 + 2k$.

The monomials $x_1^{2k-i} \cdot x_2^i \cdot c$ for $2k \geq i > 0$ contribute $M^0_{2k+2} \oplus \mathbb{R}$ to the cohomology of $H^*(\Gamma_1^{2k+1,*}; \mathbb{R}(\pm 1\ ))$. The "configuration space degree" here is $1+2k$ while the homological degree is given by $1+degree(x_1^{2k})+1-(1+2k)) = 1+2k$. [The integer 1 + 2(2k) arises from the monomials of weight 2k, and their contribution of modular forms of weight $2k+2$, the "second" "1" corresponds to the degree of c, and the "-(1+2k)" corresponds to shifting down by degree $2k+1$ as the function space in question is $map_*(T, S^3)$.]

The theorem follows. □

Notice that Theorem 2.3 in section 2 is a restatement of the above result.

Next recall that $map(T, X)$ denotes the space of free maps from T to X, the space of all (unpointed) continuous maps, and G = $SL(2, \mathbb{Z})$ below.

THEOREM 5.2. *If $n \geq 1$, then*

1. *The rational cohomology ring of $map(T, S^{2n+1})$ is isomorphic to $\mathbb{Q}[x_1, x_2] \otimes E[c] \otimes E[g]$ where x_i, are of degree 2n, c is of degree 2n-1, and g is of degree 2n+1.*
2. *Furthermore, this ring is an algebra over the integral group ring of $SL(2, \mathbb{Z})$. The action of $SL(2, \mathbb{Z})$ is trivial on both c, and g. The action is specified by setting $x_1 = \binom{1}{0}$, and $x_2 = \binom{0}{1}$.*
3. *There are isomorphisms as follows.*
   (a) $H^k(EG \times_G map(T, S^{2n+1}); \mathbb{Q}) \cong \Sigma_{s+t=k} H^s(G; H^t(map(T, S^{2n+1}); \mathbb{Q}))$
   (b) $\Sigma_{s+t=k} H^s(G; H^t(map(T, S^{2n+1}); \mathbb{Q})) \cong [H^*(G; \mathbb{Q}[x_1, x_2])] \otimes E[c] \otimes E[g]$ *where each x_i has degree 2n.*
4. *If $n = 1$, then $H^*(EG \times_G map(T, S^{2n+1}); \mathbb{Q})$ is isomorphic to*

$$H^*(BDiff^+(T); \mathbb{Q}) \otimes E[c] \otimes E[g]$$

   where c is of degree 1, and g is of degree 3.
5. *Furthermore, there are isomorphisms for $k \geq 2$ given as follows.*

$$H^i(\Gamma_1^k; \mathbb{R}(\pm 1\ )) = \begin{cases} (M^0_{2n} \oplus \mathbb{R}) \oplus (M^0_{2n+2} \oplus \mathbb{R}) & \text{if } k = 2n\ , \text{ and } i = 2n+1, \\ M^0_{2n+2} \oplus \mathbb{R} & \text{if } k = 2n+1, \text{ and } i = 2n+1, \\ M^0_{2n+2} \oplus \mathbb{R} & \text{if } k = 2n+1, \text{ and } i = 2n+3, \\ 0 & \text{otherwise.} \end{cases}$$

PROOF. The proof of this theorem as analogous to the proof given above. The main new point is to work out part 5 of the theorem.

The first part here addresses the case of Γ_1^{2k}. Here the monomials $x_1^{2k-i} \cdot x_2^i$ for $2k \geq i \geq 0$ are of classical degree 2k, and the cohomology of $SL(2, \mathbb{Z})$ with coefficients in their linear span contributes $M^0_{2k+2} \oplus \mathbb{R}$. The homological degree of this contribution is $1 + 4k - 2k = 1 + 2k$ as above. Thus there is a summand for the cohomology of Γ_1^{2k} with coefficients in $\mathbb{R}(\pm 1\ )$ given by $M^0_{2k+2} \oplus \mathbb{R}$, and which is concentrated in degree $1 + 2k$.

The monomials $x_1^{2k-i-2} \cdot x_2^i \cdot c \cdot g$ for $2k-2 \geq i \geq 0$ contribute $M_{2k}^0 \oplus \mathbb{R}$. The "configuration space degree" here is 2k while the homological degree here is $1+2(2k-2)+1+3-2k = 1+2k$.

The monomials $x_1^{2k-i} \cdot x_2^i \cdot c$, and $x_1^{2k-i} \cdot x_2^i \cdot g$ for $2k \geq i \geq 0$ both have "configuration space degree" $2k+1$, and contribute $M_{2k+2}^0 \oplus \mathbb{R}$ in homological degrees $1+4k+1-(2k+1) = 1+2k$, and $1+4k+3-(2k+1) = 3+2k$ respectively for the cohomology of Γ_1^{2k+1} with coefficients in $\mathbb{R}(\pm 1)$.

Theorem 2.4 also follows.

□

6. On $H^*(SL(2,\mathbb{Z}); H^*map_*(T, S^{2n+2}))$

The purpose of this section is to work out the rational cohomology of $SL(2,\mathbb{Z})$ with coefficients in the cohomology of $map_*(T, S^{2n+2})$. The main ingredient here is the exact sequences which arise from the Koszul complex, and their consequences as listed in section 4 here. A cochain complex $(D(2n), d)$ is given in the next paragraph which is useful for keeping track of elements in this computation, and gives the analogue of the Serre spectral sequence for the cohomology of the Heisenberg group.

Consider the differential graded algebra $(D(2n), d)$ specified by the exterior algebra with 3 generators v_1, v_2 of degree $2n+1$, and z of degree $4n+1$ given by $E[v_1, v_2] \otimes E[z]$ where $d(z) = v_1 \cdot v_2$. This algebra is also a differential graded algebra over $SL(2,\mathbb{Z})$ where the action of $SL(2,\mathbb{Z})$ on $E[v_1, v_2]$ is specified by setting $v_1 = \binom{1}{0}$, and $v_2 = \binom{0}{1}$, and the action on z is trivial. As noted in section 2 here, the element $v_1 \cdot v_2$ is fixed by $SL(2,\mathbb{Z})$, and the differential d is $SL(2,\mathbb{Z})$-linear. Recall the $SL(2,\mathbb{Z})$-modules V_{2n+1}, and W_{6n+2} which were specified in section 2, which are given by the fundamental representation of $SL(2,\mathbb{Z})$, $\mathbb{Z} \oplus \mathbb{Z}$ in degrees 2n+1, and 6n+2 respectively. In addition, let $\mathbb{Z}$ denote the integers as a trivial $SL(2,\mathbb{Z})$-module. The proof of the next propostion is by inspection of the definitions, and is omitted.

PROPOSITION 6.1. *For $n \geq 0$, $(D(2n), d)$ is a differential graded algebra over $SL(2,\mathbb{Z})$. In case $n = 1$, the cohomology of $(D(2), d)$ is that of the Heisenberg group, the group of upper triangular matrices in $SL(3,\mathbb{Z})$ with $1'$s along the main diagonal. The cohomology of $(D(2n), d)$ as an $SL(2,\mathbb{Z})$-modules is follows.*

$$H^i(D(2n), d) = \begin{cases} 0 & \text{if } i \neq 0,\ 2n{+}1,\ 6n{+}2,\ 8n{+}3 \\ V_{2n+1} & \text{if } i = 2n{+}1 \\ W_{6n+2} & \text{if } i = 6n{+}2 \\ \mathbb{Z} & \text{if } i = 0,\ 8n{+}3. \end{cases}$$

This proposition is used to work out the cohomology of $map_*(T, S^{2n+2})$ as well as the cohomology of $EG \times_{SL(2,\mathbb{Z})} map_*(T, S^{2n+2})$, and $\Gamma_1^{k,*}$. Additional topological input is given next.

Consider the fibration sequence $map_*(T,X) \to (\Omega X)^2 \to \Omega X$ as given in section 5. In this section, specialize X to an even dimensional sphere of dimension at least 4. This fibration satisfies some additional topological properties if "6" is inverted. That is, there is a morphism of fibrations where the map below $h_2 : \Omega S^{n+1} \to \Omega S^{2n+1}$ denotes the second James-Hopf invariant with homotopy theoretic fibre S^n in case n is odd. Furthermore, there is a morphism of fibrations in which all spaces have been localized at the rational numbers $\mathbb{Q}$:

$$\begin{array}{ccccc}
\Omega S^{2n+1} \times S^{4n+1} & \longrightarrow & \Omega S^{2n+1} \times S^{4n+1} & \longrightarrow & \{*\} \\
\downarrow & & \downarrow & & \downarrow \\
X_1(2n) & \longrightarrow & map_*(T, S^{2n+2}) & \longrightarrow & (\Omega S^{4n+3})^2 \\
\downarrow & & \downarrow & & \downarrow \\
(S^{2n+1})^2 & \longrightarrow & (\Omega S^{2n+2})^2 & \longrightarrow & (\Omega S^{4n+3})^2
\end{array}$$

In addition, part of the next proposition was pointed out in [**1**], while the rest as well as Theorem 6.3 follows by inspection of the methods there using the above morphism of fibrations.

PROPOSITION 6.2. *Assume that* $n > 0$.

1. *There is a fibration* $X_1(2n) \to (S^{2n+1})^2 \to \Omega S^{4n+3}$ *where the map* $(S^{2n+1})^2 \to \Omega S^{4n+3}$ *induces an isomorphism on* $H_{4n+2}(-;\mathbb{Z})$. *Furthermore, the rational cohomology of* $X_1(2n)$ *is isomorphic to the rational cohomology of the cochain complex* $(D(2n), d)$.
2. *Assume that "6" has been inverted. Then there is a homotopy equivalence*
$$map_*(T, S^{2n+2}) \to (\Omega S^{4n+3})^2 \times \Omega S^{2n+1} \times X_1(2n).$$

The action of $SL(2,\mathbb{Z})$ on the cohomology of $map_*(T, S^{2n+2})$ is specified in the next theorem. The proof of this theorem is by inspection and is omitted where $\mathbb{Q}[x_{2n}]$ denotes a polynomial ring in one indeterminate x_{2n} of degree 2n.

THEOREM 6.3. *Assume that* $n > 0$.

1. *The following algebras over* $SL(2,\mathbb{Z})$ *are isomorphic:*
   (a) $H^*(map_*(T, S^{2n+2}); \mathbb{Q})$, *and*
   (b) $\mathbb{Q}[W_{4n+2}] \otimes H^*((D(2n), d) \otimes \mathbb{Q}) \otimes \mathbb{Q}[x_{2n}]$.
2. *In addition*
   (a) $\mathbb{Q}[W_{4n+2}]$ *is given by the polynomial algebra generated by the fundamental (2 dimensional) representation of* $SL(2,\mathbb{Z})$.
   (b) $\mathbb{Q}[x_{2n}]$ *is a trivial* $SL(2,\mathbb{Z})$*-module.*
   (c) *The module* $H^i((D(2n), d) \otimes \mathbb{Q})$ *is*
      (i) *the trivial one dimensional representation for* $i = 0$, *and* $8n+3$.
      (ii) *the fundamental (2 dimensional) representation in degrees* $2n+1$, *and* $6n+2$.
3. *Furthermore, there are isomorphisms of graded abelian groups:*

(a) $H^*[SL(2,\mathbb{Z}); H^*(map_*(T, S^{2n+2}); \mathbb{Q})]$, *and*
(b) $H^*(SL(2,\mathbb{Z}); K)$ *where* K *denotes the coefficient module*

$$\mathbb{Q}[W_{4n+2}] \otimes (H^*(D(2n), d) \otimes \mathbb{Q}) \otimes \mathbb{Q}[x_{2n}].$$

4. *The groups* $\mathbb{Q}[W_{4n+2}] \otimes H^i((D(2n), d)$ *are given as* $SL(2,\mathbb{Z})$*-modules by the following.*
   (a) $\mathbb{Q}[W_{4n+2}] \otimes H^i((D(2n), d) = 0$ *if* $i \neq 0, 2n+1, 6n+2, 8n+3$.
   (b) $\mathbb{Q}[W_{4n+2}] \otimes V_{2n+1}$ *if* $i = 2n+1$.
   (c) $\mathbb{Q}[W_{4n+2}] \otimes W_{6n+2}$ *if* $i = 6n{+}2$.
   (d) $\mathbb{Q}[W_{4n+2}] \otimes \mathbb{Z}$ *if* $i = 0,\ 8n{+}3$.

The next step is to analyze $H^*(SL(2,\mathbb{Z}); H^*(map_*(T, S^{2n+2}); \mathbb{Q}))$ together with the contribution of each summand above to the cohomology of each configuration space, and $\Gamma_1^{k,*}$ for which the language of section 3 will be used. This step breaks up into the following cases.

1. $H^*(SL(2,\mathbb{Z}); \mathbb{Q}[W_{4n+2}])$,
2. $H^*(SL(2,\mathbb{Z}); \mathbb{Q}[W_{4n+2}] \otimes V_{2n+1})$, and
3. $H^*(SL(2,\mathbb{Z}); \mathbb{Q}[W_{4n+2}] \otimes W_{6n+2})$

These group will be enumerated in terms of modular cusp forms, and listed in the following cases where it is convenient to replace $\mathbb{Q}$ by $\mathbb{R}$ in what follows.

Case (1): Notice that $\mathbb{R}[W_{4n+2}]$ is a polynomial ring with generators w_1, and w_2 of "configuration space degree" 2, and homological degree $4n+2$. Write $S_k(2n)$ for the $SL(2,\mathbb{Z})$-module which is the linear span of $w_1^{k-i} \cdot w_2^i$ for $k \geq i \geq 0$.

Recall from section 3 that the group

$$H^i(SL(2,\mathbb{Z}); S_k(2n))$$

satisfies the following properties:

1. It is zero if k is odd, and $i \geq 0$.
2. It is zero if k is even, and $i \geq 2$ or $i = 0$ with $k \geq 1$. (The condition $k \geq 1$ is required in Lemma 2.1 of [**9**].)
3. It is $\mathbb{R}$ if $k = 0$, and $i = 0$.
4. It is isomorphic, as a real vector space, to $M^0_{2t+2} \oplus \mathbb{R}$ for $k = 2t \geq 2$, and $i = 1$. This module is concentrated in $H^1(SL(2,\mathbb{Z}); S_k(2n))$ in total degree $1 + 2t(4n+2) - 4t(2n)$, and thus contributes a direct summand of the homology of the group $H_{1+4t}\Gamma_1^{4t,*}$.
5. $H^1(SL(2,\mathbb{Z}); S_{2k}(2n))$ has "configuration space degree" 4k, and is concentrated in total degree $1 + (2k)(4n+2)$.

The contribution of $H^1(SL(2,\mathbb{Z}); S_{2k}(2n))$ is (a) to the cohomology of $\Gamma_1^{4k,*}$ as the dual homology group has "configuration space degree" $4k$, and (b) has homological degree $1 + (2k)(4n+2) - (4k)(2n) = 1 + 4k$ for $k \geq 1$. This gives $M^0_{2k+2} \oplus \mathbb{R}$ is a direct summand of $H^{4k+1}(\Gamma_1^{4k,*}; \mathbb{R})$. If $k = 0$, there is no contribution.

Since $H^i(SL(2,\mathbb{Z}); S_{1+2k}(2n)) = 0$ by section 3, the modules $S_{1+2k}(2n)$ contribute zero to the cohomology of $\Gamma_1^{4k+2,*}$.

The summand of cohomology corresponding to powers of x_{2n}^q gives that $H^i(\Gamma_1^{k,*};\mathbb{R})$ is a direct summand of $H^i(\Gamma_1^{k+q,*};\mathbb{R})$ for all $q \geq 0$. Thus taking polynomials in x_{2n} then gives that $M_{2k+2}^0 \oplus \mathbb{R}$ is a direct summand of of $H^{4k+1}(\Gamma_1^{4k+q,*};\mathbb{R})$ for all $q \geq 0$, and $k \geq 1$.

Case (2): The contribution of the summand

$$H^*(SL(2,\mathbb{Z}); \mathbb{Z} \otimes \mathbb{R}[W_{4n+2}])$$

where $\mathbb{Z}$ is the trivial representation $H^{8n+3}(D(2n), d)$ is analogous to the above with a shift of "configuration space degree": The representation $H^{8n+3}(D(2n), d)$ occurs in degree 8n+3, and contributes "configuration space degree" 4.

The modules $S_{2k}(2n)$ have "configuration space degree" 4k. The tensor product $S_{2k}(2n) \otimes H^{8n+3}((D(2n), d)$ has "configuration space degree" $4k + 4$, with total degree $(4n+2)(2k) + 8n + 3$, and thus gives classes in homological degree $1 + (4n+2)(2k) + 8n + 3 - (2n)(4k+4) = 4k+4$ corresponding to $H^1(SL(2,\mathbb{Z}); S_{2k}(2n) \otimes H^{8n+3}(D(2n), d))$. Thus $H^{4k+4}(\Gamma_1^{4k+4,*};\mathbb{R})$ has a direct summand as a real vector space given by $M_{2k+2}^0 \oplus \mathbb{R}$ for $k \geq 1$. Notice that the hypothesis that $k \geq 1$ is necessary as required by [**9**].

In case k = 0, then $H^j(SL(2,\mathbb{Z}); H^{8n+3}(D(2n), d))$ is trivial in all degrees $j > 0$, and is $\mathbb{R}$ for $j = 0$. Since $H^{8n+3}((D(2n), d)$ has "configuration space degree" 4, and total degree $8n + 3$, this group contributes $\mathbb{R}$ to $H^3(\Gamma_1^{4,*};\mathbb{R})$.

As before, multiplication by x_{2n}^s gives that $H^{4k+4}(\Gamma_1^{4k+4+s,*};\mathbb{R})$ has $M_{2k+2}^0 \oplus \mathbb{R}$ as a summand for all $s \geq 0$, and $k \geq 1$. Similarly $\mathbb{R}$ is a summand of $H^3(\Gamma_1^{4+q,*};\mathbb{R})$ for all $q \geq 0$.

Case (3): The contribution of the summand

$$H^*(SL(2,\mathbb{Z}); \mathbb{R}[W_{4n+2}] \otimes V_{2n+1})$$

arises as follows.

First, V_{2n+1} has "configuration space degree" 1. By the statement, and proof of proposition 4.1, there is a short exact sequence of coefficient modules for $k \geq 0$ given by

$$0 \to \Lambda_2 \otimes S_{k-1}(2n) \to \Lambda_1 \otimes S_k(2n) \to S_{k+1}(2n) \to 0$$

where $S_{-1}(2n)$ denotes the trivial module $\{0\}$. Furthermore, V_{2n+1} is isomorphic to Λ_1 as an $SL(2,\mathbb{Z})$-module.

Notice that "configuration space degree" is increased by precisely 1 as given next:

1. The "configuration space degree" of $\Lambda_2 \otimes S_{k-1}(2n)$ is $2k$.

2. The "configuration space degree" of $V_{2n+1} \otimes S_k(2n)$ is $1 + 2k$.
3. The "configuration space degree" of $S_{k+1}(2n)$ is $2(k+1) = 2 + 2k$.

In addition, the following properties are satisfied.

1. The groups
$$H^i(SL(2,\mathbb{Z}); V_{2n+1} \otimes S_{2k}(2n))$$
all vanish for $k > 0$, and $i \geq 0$ as the groups $H^1(SL(2,\mathbb{Z}); S_{odd}(2n))$ are zero.
2. By proposition 4.1, there is a direct sum decomposition in case k = 2t, and $t \geq 0$ of
$$H^1(SL(2,\mathbb{Z}); V_{2n+1} \otimes S_{2t+1}(2n))$$
as
$$H^1(SL(2,\mathbb{Z}); \Lambda_2 \otimes S_{2t}(2n)) \oplus H^1(SL(2,\mathbb{Z}); S_{2t+2}(2n) \otimes \mathbb{Z}).$$
3. Notice that $V_{2n+1} \otimes S_{2t+1}(2n)$ has "configuration space degree" given by $1+2(2t+1)$. Thus these summands both have "configuration space degree" equal to $3 + 4t$. Furthermore, the total degree in homology here is $4 + 4t$. The two associated summands in homology are described next.
4. One summand is given by
$$H^1(SL(2,\mathbb{Z}); S_{2t+2}(2n) \otimes \mathbb{Z})$$
for $t \geq 0$. As a vector space, this summand is isomorphic to
$$M^0_{2t+4} \oplus \mathbb{R},$$
and has homological degree given by $4 + 4t$.
5. The second summand is given by
$$H^1(SL(2,\mathbb{Z}); \Lambda_2 \otimes S_{2t}(2n)).$$
As a vector space, this summand is isomorphic to
$$M^0_{2t+2} \oplus \mathbb{R}$$
for $t \geq 1$, and has homological degree $4 + 4t$ as Λ_2 is a trivial $SL(2,\mathbb{Z})$-module.
6. Thus
$$H^{4t+4}(\Gamma_1^{4t+3+q,*}; \mathbb{R})$$
has
$$(M^0_{2t+4} \oplus \mathbb{R}) \oplus (M^0_{2t+2} \oplus \mathbb{R})$$
as a direct summand for $q \geq 0$, and $t \geq 1$.
7. In case t = 0, there is a short exact sequence
$$0 \to \Lambda_2 \to \Lambda_1 \otimes S_1(2n) \to S_2(2n) \to 0,$$
and thus by the natural long exact sequence associated to these coefficients, the groups
$$H^1(SL(2,\mathbb{Z}); \Lambda_1 \otimes S_1(2n)),$$
and
$$H^1(SL(2,\mathbb{Z}); S_2(2n))$$
are isomorphic. Thus there is an isomorphism
$$H^1(SL(2,\mathbb{Z}); \Lambda_1 \otimes S_1(2n)) \to M^0_4 \oplus \mathbb{R}.$$

Furthermore, by a direct computation

$$H^0(SL(2,\mathbb{Z});\Lambda_1 \otimes S_1(2n))$$

is isomorphic to $\mathbb{R}$.

Since $\Lambda_1 \otimes S_1(2n)$ has configuration degree 3, $H^3(\Gamma_1^{3+q,*};\mathbb{R})$ has $\mathbb{R}$ as a direct summand, and $H^4(\Gamma_1^{3+q,*};\mathbb{R})$ has $M_4^0 \oplus \mathbb{R}$ as a direct summand for $q \geq 0$.

Case(4): The contribution of the summand

$$H^*(SL(2,\mathbb{Z});W_{6n+2} \otimes \mathbb{R}[W_{4n+2}])$$

is similar to the previous case and arises as follows:

1. Notice that Λ_1, V_{2n+1}, and W_{6n+2} are isomorphic as $SL(2,\mathbb{Z})$-modules, and so there are isomorphisms
$$H^*(SL(2,\mathbb{Z});W_{6n+2} \otimes S_k(2n)) \to H^*(SL(2,\mathbb{Z});V_{2n+1} \otimes S_k(2n)).$$
2. Thus
   (a) $H^*(SL(2,\mathbb{Z});W_{6n+2} \otimes S_{2k}(2n)) = \{0\}$ for $k \geq 0$.
   (b) The "configuration space degree" of $W_{6n+2} \otimes S_{2t+1}(2n)$ is $3+2(2t+1) = 4t+5$. The group $H^1(SL(2,\mathbb{Z});W_{6n+2}\otimes S_{2t+1}(2n))$ is isomorphic to
   $$(M_{2t+4}^0 \oplus \mathbb{R}) \oplus (M_{2t+2}^0 \oplus \mathbb{R})$$
   for $t \geq 1$, and the total degree of $H^1(SL(2,\mathbb{Z});W_{6n+2} \otimes S_{2t+1}(2n))$ is $2t+1$ for $t \geq 1$.
   (c) If $t = 0$, then $H^1(SL(2,\mathbb{Z});W_{6n+2}\otimes S_1(2n))$ is isomorphic to $M_4^0 \oplus \mathbb{R}$, and $H^0(SL(2,\mathbb{Z});W_{6n+2} \otimes S_1(2n))$ is isomorphic to $\mathbb{R}$.
3. $H^{2t+1}(\Gamma_1^{4t+5+q,*};\mathbb{R})$ has $(M_{2t+4}^0 \oplus \mathbb{R}) \oplus (M_{2t+2}^0 \oplus \mathbb{R})$ as a direct summand for $q \geq 0$, and $t \geq 1$.
4. If $t = 0$, then $H^5(\Gamma_1^{5+q,*};\mathbb{R})$ has $M_4^0 \oplus \mathbb{R}$ as a direct summand, and $H^4(\Gamma_1^{5+q,*};\mathbb{R})$ has $\mathbb{R}$ as a direct summand for $q \geq 0$.

The above exhausts all of the cases (as well as the author), and gives all of the non-zero cohomology classes in $H^*(\Gamma_1^{k,*};\mathbb{R})$. To recapitulate:

1. By Case 1, $H^{4k+1}(\Gamma_1^{4k+q,*};\mathbb{R})$ has $M_{2k+2}^0 \oplus \mathbb{R}$ as a summand for all $q \geq 0$, and $k \geq 1$. If $k = 0$, there is no contribution.
2. By Case 2, $H^{4k}(\Gamma_1^{4k+q,*};\mathbb{R})$ has $M_{2k}^0 \oplus \mathbb{R}$ as a summand for all $q \geq 0$, and $k \geq 1$. In case $k = 0$, then $H^3(\Gamma_1^{4+q,*};\mathbb{R})$ has a summand of $\mathbb{R}$ for all $q \geq 0$.
3. By Case 3, $H^{4k+4}(\Gamma_1^{4k+3+q,*};\mathbb{R})$ has $(M_{2k+4}^0 \oplus \mathbb{R}) \oplus (M_{2k+2}^0 \oplus \mathbb{R})$ as a direct summand for $q \geq 0$, and $k \geq 1$. If $k = 0$, then $H^4(\Gamma_1^{3+q,*};\mathbb{R})$ is isomorphic to $M_4^0 \oplus \mathbb{R}$ for $q \geq 0$. Furthermore, $H^3(\Gamma_1^{3+q,*};\mathbb{R})$ has a summand of $\mathbb{R}$ for all $q \geq 0$ which is distinct from the summand given by case 2.
4. By Case 4, $H^{2k+1}(\Gamma_1^{4k+5+q,*};\mathbb{R})$ has $(M_{2k+4}^0 \oplus \mathbb{R}) \oplus (M_{2k+2}^0 \oplus \mathbb{R})$ as a direct summand for $q \geq 0$ and $k \geq 1$. In case $k = 0$, $H^5(\Gamma_1^{5+q,*};\mathbb{R})$ has $M_4^0 \oplus \mathbb{R}$ as a direct summand, and $H^4(\Gamma_1^{5+q,*};\mathbb{R})$ has $\mathbb{R}$ as a direct summand for $q \geq 0$. There are the following two sub-cases:
   (a) If k = 2j+1, then $H^{4j+3}(\Gamma_1^{8j+9+q,*};\mathbb{R})$ has $(M_{4j+6}^0 \oplus \mathbb{R}) \oplus (M_{4j+4}^0 \oplus \mathbb{R})$ as a direct summand for $q \geq 0$ and $k \geq 1$.

(b) If $k = 2j \geq 1$, then $H^{4j+1}(\Gamma_1^{8j+5+q,*};\mathbb{R})$ has $(M^0_{4j+4}\oplus\mathbb{R})\oplus(M^0_{4j+2}\oplus\mathbb{R})$ as a direct summand for $q \geq 0$. If $k = 0$, then $H^5(\Gamma_1^{5+q,*};\mathbb{R})$ has $M_4^0 \oplus \mathbb{R}$ as a direct summand, and $H^4(\Gamma_1^{5+q,*};\mathbb{R})$ has $\mathbb{R}$ as a direct summand for $q \geq 0$.

5. There are isomorphisms of vector spaces given as follows:
   (a) $H^0(\Gamma_1^{k,*};\mathbb{R})$:
      (i) This group is isomorphic to $\mathbb{R}$.
   (b) $H^{4j}(\Gamma_1^{k,*};\mathbb{R})$ for $j > 0$:
      (i) If k = 1,2, then $H^4(\Gamma_1^{k,*};\mathbb{R})$ is isomorphic to $\{0\}$.
      (ii) If k = 3, then $H^4(\Gamma_1^{k,*};\mathbb{R})$ is isomorphic to $(M_4^0 \oplus \mathbb{R})$.
      (iii) If k = 4, then $H^4(\Gamma_1^{k,*};\mathbb{R})$ is isomorphic to $(M_4^0 \oplus \mathbb{R}) \oplus \mathbb{R}$.
      (iv) If $k \geq 5$, then $H^4(\Gamma_1^{k,*};\mathbb{R})$ is isomorphic to $(M_4^0 \oplus \mathbb{R}) \oplus_2 \mathbb{R}$.
      (v) If $k < 4j - 1$, with $j > 0$, then $H^{4j}(\Gamma_1^{k,*};\mathbb{R})$ is isomorphic to $\{0\}$.
      (vi) If $k = 4j - 1$, with $j > 1$, then $H^{4j}(\Gamma_1^{4j-1,*};\mathbb{R})$ is isomorphic to
      $$(M^0_{2j+2} \oplus \mathbb{R}) \oplus (M^0_{2j} \oplus \mathbb{R}).$$
      (vii) If $k \geq 4j$, with $j > 1$, then $H^{4j}(\Gamma_1^{k,*};\mathbb{R})$ is isomorphic to
      $$H^{4j}(\Gamma_1^{4j-1,*};\mathbb{R}) \oplus (M^0_{2j} \oplus \mathbb{R})$$
      which is isomorphic to
      $$(M^0_{2j+2} \oplus \mathbb{R} \oplus M^0_{2j} \oplus \mathbb{R}) \oplus (M^0_{2j} \oplus \mathbb{R}).$$
   (c) $H^{4j+1}(\Gamma_1^{k,*};\mathbb{R})$:
      (i) If $k \geq 5$, then $H^5(\Gamma_1^{k,*};\mathbb{R})$ is isomorphic to
      $$M_4^0 \oplus \mathbb{R}.$$
      (ii) If $k < 4j - 1$ and $j \geq 1$, then $H^{4j+1}(\Gamma_1^{4j,*};\mathbb{R})$ is isomorphic to $\{0\}$.
      (iii) If $4j \leq k \leq 8j+4$, and $j \geq 1$, then $H^{4j+1}(\Gamma_1^{k,*};\mathbb{R})$ is isomorphic to
      $$M^0_{2j+2} \oplus \mathbb{R}.$$
      (iv) If $k \geq 8j + 5$, and $j \geq 1$, then $H^{4j+1}(\Gamma_1^{k,*};\mathbb{R})$ is isomorphic to
      $$(M^0_{2j+2} \oplus \mathbb{R}) \oplus (M^0_{4j+4} \oplus \mathbb{R} \oplus M^0_{4j+2} \oplus \mathbb{R}).$$
   (d) $H^{4j+2}(\Gamma_1^{k,*};\mathbb{R})$:
      (i) If $j \geq 0$, this group is isomorphic to $\{0\}$.
   (e) $H^{4j+3}(\Gamma_1^{k,*};\mathbb{R})$:
      (i) If $k = 3$, $H^3(\Gamma_1^{k,*};\mathbb{R})$ is isomorphic to $\mathbb{R}$.
      (ii) If $k \geq 4$, $H^3(\Gamma_1^{k,*};\mathbb{R})$ is isomorphic to $\oplus_2\mathbb{R}$.
      (iii) If $k < 8j + 9$, and $j > 0$, then $H^{4j+3}(\Gamma_1^{k,*};\mathbb{R})$ is isomorphic to $\{0\}$.
      (iv) If $k \geq 8j + 9$, and $j > 0$, then $H^{4j+3}(\Gamma_1^{k,*};\mathbb{R})$ is isomorphic to
      $$(M^0_{4j+6} \oplus \mathbb{R}) \oplus (M^0_{4j+4} \oplus \mathbb{R}).$$
   (f) All other groups $H^i(\Gamma_1^{j,*};\mathbb{R})$ vanish.

Remark: The cohomology of the above spaces where $SL(2,\mathbb{Z})$ is replaced by a subgroup Γ can be determined in a similar way. The answers depend on the automorphic cusp forms with respect to Γ as well as the number of equivalence classes of cusps of Γ.

References

[1] C. F. Bödigheimer, F.R. Cohen, and R. J. Milgram, *On configuration spaces* , Math. Zeitschrift, 214 (1993), 179–216.

[2] C. F. Bödigheimer, F.R. Cohen, and L. R. Taylor, *The homology of configuration spaces*, Topology, 28(1989), 111-123.

[3] F.R. Cohen, *On mapping class groups from a homotopy theoretic point of view* , Proceedings of the 1994 Barcelona Algebraic Topology Conference, Birkhäuser, Progress in Mathematics, 136 (1996),119–142.

[4] F.R. Cohen, *On the hyperelliptic mapping class groups, $SO(3)$, and $Spin^c(3)$* , American J. Math., 115 (1993), 389–434.

[5] F.R.Cohen, and J. Pakianathan, *Course notes*, in preparation.

[6] C. Earle, and J. Eells, *A fibre bundle description of Teichmuller theory*, J. Differential Geometry 3(1969), 19-43.

[7] M. Eichler, *Eine Verallgemeinerung der Abelschen Integrale*, Math. Zeit., 67(1957), 267-298.

[8] E. Fadell and L. Neuwirth, *Configuration spaces*, Math. Scand. 10 (1962), 119-126.

[9] M. Furusawa, M. Tezuka, and N. Yagita, *On the cohomology of classifying spaces of torus bundles, and automorphic forms*, J. London Math. Soc., (2)37(1988), 528-543.

[10] E. Getzler, to appear.

[11] W. Magnus, A. Karrass, and D. Solitar, *Combinatorial Group Theory*, Dover Publications, Inc., 1966

[12] J. P. Serre, *A Course in Arithmetic*, Springer-Verlag Graduate Texts in Mathematics, 7, 1970.

[13] J. P. Serre, *Trees.*

[14] G. Shimura, *Introduction to the arithmetic theory of automorphic forms*, Publications of the Mathematical Society of Japan 11, Iwanami Shoten, Tokyo; University Press, Princeton, 1971.

[15] K.K.Tse, *Thesis*, University of Rochester, 1999.

Department of Mathematics, University of Rochester, Rochester, NY 14627
E-mail address: cohf@math.rochester.edu
URL: http://www.math.rochester.edu/u/cohf

Contemporary Mathematics
Volume **279**, 2001

Poincaré duality and deformations of algebras

Bernhard Hanke

ABSTRACT. Let p be a prime number and X a finite dimensional connected $\mathbb{Z}/p$-CW complex. If $H^*(X;\mathbb{F}_p)$ is a Poincaré duality algebra, then by a basic theorem of Smith theory proven by G.E. Bredon, T. Chang and T. Skjelbred, the same is true for $H^*(F;\mathbb{F}_p)$, where $F \subset X^{\mathbb{Z}/p}$ is an arbitrary component of the fixed point set. We give an elementary proof of this fact using deformations of algebras.

The existing proofs of the cited theorem are based on the localization theorem for $\mathbb{Z}/p$-actions that is combined with an inductive argument within the Leray-Serre spectral sequence for the Borel construction $E\mathbb{Z}/p \times_{\mathbb{Z}/p} X$ ([**3, 4**]) or is interpreted as an evaluation theorem ([**2**]). One notorious technical difficulty in these proofs is to show that the obtained Poincaré pairing on each $H^*(F;\mathbb{F}_p)$ is induced by the cup product and an orientation of this cohomology algebra.

In this note we suggest a conceptually more concise treatment of $\mathbb{Z}/p$-actions on Poincaré duality spaces by exploiting the point of view in [**2**]: to regard $H^*(X^{\mathbb{Z}/p};\mathbb{F}_p)$ as an algebraic deformation of an algebra closely related to $H^*(X;\mathbb{F}_p)$. In this context the above mentioned difficulty disappears and makes the discussion of $\mathbb{Z}/p$-actions on Poincaré duality spaces shorter and perhaps more enlightening than in the existing sources.

Throughout the paper, k denotes a field and t and s denote indeterminates. All algebras are assumed to be (graded) commutative.

1. Deformations of algebras, evaluation theorem

If M is a $k[t]$-module or a $k[t]$-algebra and $\alpha \in k$, we set

$$M_\alpha = M/(t-\alpha)M = M \otimes_{k[t]} k_\alpha,$$

where k_α is the ring k equipped with the $k[t]$-multiplicaton induced by $t \mapsto \alpha$. This construction is functorial in M and called *evaluation* of t at α. Let us recall the following definition.

DEFINITION 1. Let A be a k-algebra. The $k[t]$-algebra B is a *one parameter family of deformations* of A, if $B \cong A \otimes_k k[t]$ as $k[t]$-modules and if $B_0 \cong A$ as

2000 *Mathematics Subject Classification.* Primary 57P10; 57S17.
Key words and phrases. finite p-group, group action, Poincaré duality space.

k-algebras. In this situation, the k-algebra B_1 is called a *deformation* of A *along* B.

For later reference we state the following basic facts.

LEMMA 2. *Let M be a finitely generated free $k[t]$-module of rank n and let $\alpha \in k$. Then M_α is a k-vector space of dimension n and the k-vector spaces* $\mathrm{Hom}_{k[t]}(M, M)_\alpha$ *and* $\mathrm{Hom}_k(M_\alpha, M_\alpha)$ *are canonically isomorphic.*

Let X be a finite dimensional $\mathbb{Z}/p$-CW complex, where p is an odd prime number and let g be a generator of $\mathbb{Z}/p$. Additionally, we assume that $H^*(X; \mathbb{F}_p)$ is a finitely generated vector space (this is automatically satisfied in our later applications). As coefficients we will use the field $\mathbb{F}_p$ throughout, if not stated in a different way. Let $(C^*(X), \delta)$ denote the singular cochain complex of X. It has the structure of a differential graded $\mathbb{F}_p[\mathbb{Z}/p]$-module. The equivariant cohomology $H^*_{\mathbb{Z}/p}(X) = H^*(E\mathbb{Z}/p \times_{\mathbb{Z}/p} X)$, where $E\mathbb{Z}/p$ is a free contractible $\mathbb{Z}/p$-CW complex, can be calculated using the cochain complex $\beta^*(X) = \mathrm{Hom}_{\mathbb{Z}[\mathbb{Z}/p]}(\mathcal{E}_*, C^*(X))$, $\mathcal{E}_*$ being a $\mathbb{Z}[\mathbb{Z}/p]$-free resolution of $\mathbb{Z}$ (with the trivial $\mathbb{Z}/p$-operation). Using the special model for $\mathcal{E}_*$ described in [**2**], p. 61, $\beta^*(X)$ is isomorphic to $C^*(X)[s,t]/(s^2)$ as a graded module, where s and t carry gradings 1 and 2, respectively. The $\mathbb{F}_p[t]$-linear differential and the $\mathbb{F}_p[t]$-bilinear product on $C^*(X)[s,t]/(s^2)$ induced by $\beta^*(X)$ are given by

$$\begin{aligned} c &\mapsto \delta c + (-1)^{|c|+1}(1-g)cs\,, \\ cs &\mapsto (\delta c)s + (-1)^{|c|}(1 + \ldots + g^{p-1})ct\,, \end{aligned}$$

and

$$\begin{aligned} (c,d) &\mapsto c \cup d\,, & (c,ds) &\mapsto (c \cup d)s\,, \\ (cs,d) &\mapsto (-1)^{|d|}(c \cup gd)s\,, & (cs,ds) &\mapsto (-1)^{|d|}\sigma(c,d)t\,. \end{aligned}$$

Here, $\cup$ is the product on $C^*(X)$ that is induced by a diagonal and $\sigma(c,d) = \sum_{0 \le i < j \le p-1}(g^i c) \cup (g^j d)$. We obtain an $\mathbb{F}_p[s,t]/(s^2)$-algebra structure on $H^*_{\mathbb{Z}/p}(X)$ that may be identified with the usual $H^*(B\mathbb{Z}/p; \mathbb{F}_p)$-algebra structure on $H^*_{\mathbb{Z}/p}(X)$. The following theorem is well known (cf. [**2**], Theorem 1.4.5).

THEOREM 3 (Evaluation theorem). *The inclusion $i : X^G \hookrightarrow X$ induces an isomorphism of $\mathbb{F}_p[s]/(s^2)$-algebras*

$$H^{(*)}_{\mathbb{Z}/p}(X)_1 \cong H^{(*)}_{\mathbb{Z}/p}(X^{\mathbb{Z}/p})_1 = H^{(*)}(X^{\mathbb{Z}/p})[s]/(s^2).$$

The stars in brackets indicate that degrees are only preserved mod 2. For the last equality, note that $\mathbb{Z}/p$ acts trivially on $X^{\mathbb{Z}/p}$. In the following, let Tor denote $\mathbb{F}_p[t]$-torsion. We set $F^* = H^*_{\mathbb{Z}/p}(X)/\,\mathrm{Tor}$.

COROLLARY 4. $H^{(*)}(X^{\mathbb{Z}/p})[s]/(s^2)$ *is a deformation of* $F_0^{(*)}$ *along* $F^{(*)}$.

PROOF. Note that F^* is a free $\mathbb{F}_p[t]$-module [1], hence $F^* \cong F_0^* \otimes \mathbb{F}_p[t]$ as graded $\mathbb{F}_p[t]$-modules. Furthermore, the map i in Theorem 3 induces an isomorphism $F_1^{(*)} \cong H^{(*)}(X^{\mathbb{Z}/p})[s]/(s^2)$, because $\mathrm{Tor}\, H^*_{\mathbb{Z}/p}(X^{\mathbb{Z}/p}) = 0$. □

[1]Our assumption on $H^*(X)$ ensures that $H^*_{\mathbb{Z}/p}(X)$ is a finitely generated $\mathbb{F}_p[t]$-module. A refined argument shows that Corollary 4 holds without this assumption.

2. Poincaré duality

DEFINITION 5. Let $n \in \mathbb{N}$. We call an $\mathbb{N}$-graded k-algebra A^* a *Poincaré duality algebra* of *formal dimension* n, if $A^0 \cong k$ (i.e. if A^* is *connected*), $A^n \cong k$ and if the map

$$A^* \xrightarrow{\mu_{A^*}} \operatorname{Hom}_k(A^*, A^*) \longrightarrow \operatorname{Hom}_k(A^*, A^n) \cong \operatorname{Hom}(A^*, k)$$

is an isomorphism (μ_{A^*} is adjoint to the multiplication on A^*, i.e. $\mu_{A^*}(a)(x) = a \cdot x$; the second map is induced by the projection $A^* \to A^n$).

In this situation, A^* is a finite dimensional k-vector space and $A^{>n} = 0$. Let $\mu_{A^{>0}} : A^{>0} \to \operatorname{Hom}_k(A^{>0}, A^{>0})$ be adjoint to the multiplication on $A^{>0}$. The following is an immediate consequence of [**2**], Proposition 5.1.4.

LEMMA 6. *Let A^* be a connected $\mathbb{N}$-graded k-algebra with $\dim_k A^* < \infty$. Then the following statements are equivalent.*

i. *A^* is a Poincaré duality algebra.*
ii. $\dim_k \operatorname{Ker} \mu_{A^{>0}} \leq 1$.

The next proposition is the main algebraic ingredient for a deformation theoretic proof of the theorem of Bredon, Chang and Skjelbred in the following section. Roughly speaking, it says that Poincaré duality is preserved under deformations.

PROPOSITION 7. *Let A^*, B^* and E^* be $\mathbb{N}$-graded k-algebras where A^* and B^* are connected. Let $\mathcal{T}$ be a $k[t]$-algebra such that $B \times E$ is a deformation of A along $\mathcal{T}$ (ignoring gradings).*

Suppose we have a $k[t]$-algebra map $\epsilon : \mathcal{T} \to k[t]$ with $\operatorname{Ker} \epsilon_0 = A^{>0}$ and $\operatorname{Ker} \epsilon_1 = B^{>0} \times E^*$. Then, if A^* is a Poincaré duality algebra, B^* is a Poincaré duality algebra.

PROOF. By Lemma 6 is suffices to show that $\dim_k \operatorname{Ker} \mu_{B^{>0}} \leq 1$, if the corresponding statement holds for A^*. We have an exact sequence

$$0 \longrightarrow \operatorname{Ker} \epsilon \longrightarrow \mathcal{T} \xrightarrow{\epsilon} \operatorname{Im} \epsilon \to 0$$

that consists of free $k[t]$-modules and therefore remains exact after applying the functors $(-)_0$ and $(-)_1$. Hence $(\operatorname{Ker} \epsilon)_0 = A^{>0}$ and $(\operatorname{Ker} \epsilon)_1 = B^{>0} \times E^*$ (including the multiplicative structures). Using Lemmas 2 and 6, we get

$$\dim_k \operatorname{Ker} \mu_{B^{>0} \times E^*} = \dim_k \operatorname{Ker} (\mu_{\operatorname{Ker} \epsilon})_1 = \dim_k \operatorname{Ker} (\mu_{\operatorname{Ker} \epsilon})_0 = \dim_k \operatorname{Ker} \mu_{A^{>0}}$$

which is less than or equal to 1 by assumption. It is easy to see that the inequality $\dim_k \operatorname{Ker} \mu_{B^{>0}} \leq \dim_k \operatorname{Ker} \mu_{B^{>0} \times E^*}$ holds, thereby proving our assertion. □

3. Actions on Poincaré duality spaces

Now we can use our techniques to reprove the following important theorem. We will use $\mathbb{F}_p$ as coefficients as before, where p is an odd prime number.

THEOREM 8 ([**3**], [**4**]). *Let X be a finite dimensional connected $\mathbb{Z}/p$-CW complex such that $H^*(X)$ is a Poincaré duality algebra of formal dimension n. Let $F \subset X^G$ be a component of the fixed point set. Then $H^*(F)$ is a Poincaré duality algebra of formal dimension less than or equal to n.*

PROOF. Note that $H^*(S^1 \times_{\mathbb{Z}/p} X) \cong H^*(\beta^*(X)_0)$, where $\mathbb{Z}/p \subset \mathbb{C}$ acts on $S^1 \subset \mathbb{C}$, is a Poincaré duality algebra of formal dimension $n+1$. This follows easily from the above explicit description of the differential on $\beta^*(X)$ which leads to the observation that $H^*(\beta^*(X)_0)$ is the cohomology of the cochain complex $H^*(X)[s]/(s^2)$ with differential $h \mapsto (-1)^{|h|+1}(1-g)hs$, $hs \mapsto 0$, and a bilinear product induced by the explicit formulas from above (a simple spectral sequence argument can be used just as well). We set

$$\begin{aligned} U &= H^*(j)(H^*_{\mathbb{Z}/p}(X)) \subset H^*(S^1 \times_{\mathbb{Z}/p} X), \\ V &= H^*(j)(\operatorname{Tor} H^*_{\mathbb{Z}/p}(X)) \subset U, \end{aligned}$$

where $j : \beta^*(X) \to \beta^*(X)_0$ is the projection. Note that $U \cong H^*_{\mathbb{Z}/p}(X)_0$ and $V \cong (\operatorname{Tor} H^*_{\mathbb{Z}/p}(X))_0$. Using Corollary 4,

$$(H^*(F) \times H^*(X^{\mathbb{Z}/p} \setminus F)) \otimes \mathbb{F}_p[s]/(s^2)$$

is a deformation (not respecting the gradings) of U/V along the one parameter family of deformations $H^{(*)}_{\mathbb{Z}/p}(X)/\operatorname{Tor}$. We want to apply Proposition 7 with $A^* = U/V$, $B^* = H^*(F)[s]/(s^2)$, $E^* = H^*(X^{\mathbb{Z}/p} \setminus F)[s]/(s^2)$, $\mathcal{T} = H^*_{\mathbb{Z}/p}(X)/\operatorname{Tor}$ and $\epsilon : \mathcal{T} \to \mathbb{F}_p[t]$ induced by the inclusion of a point $* \hookrightarrow F \hookrightarrow X$ and the map $H^*_{\mathbb{Z}/p}(*) = \mathbb{F}_p[s,t]/(s^2) \to \mathbb{F}_p[t]$, $s \mapsto 0$. Then the Poincaré duality of $H^*(F)[s]/(s^2)$ implies the Poincaré duality of $H^*(F)$. For Propositio
n 7 being applicable it remains to show that $V = U^\perp$ with respect to the nonsingular bilinear form on $H^*(S^1 \times_{\mathbb{Z}/p} X)$ induced by the cup product and an isomorphism $H^{n+1}(S^1 \times_{\mathbb{Z}/p} X) \cong \mathbb{F}_p$. Note that the singular cochain complex $C^*(X)$ is $\mathbb{F}_p$-cochain homotopy equivalent to $Z^n(X) \oplus C^{<n}(X)$, where $Z^n(X)$ are the cocycles in $C^n(X)$. We consider the following commutative diagram of cochain complexes (with trivial differential in the third column)

$$\begin{array}{ccccc} \beta^*(X) & \xleftarrow{\iota} & (Z^n(X) \oplus C^{<n}(X))[s,t]/(s^2) & \xrightarrow{\pi} & \mathbb{F}_p[s,t]/(s^2) \\ {\scriptstyle t\mapsto 0}\downarrow & & {\scriptstyle t\mapsto 0}\downarrow & & {\scriptstyle t\mapsto 0}\downarrow \\ \beta^*(X)_0 & \xleftarrow{\iota} & (Z^n(X) \oplus C^{<n}(X))[s]/(s^2) & \xrightarrow{\phi} & \mathbb{F}_p[s]/(s^2), \end{array}$$

where π and ϕ are induced by a (fixed) isomorphism $H^n(X) \to \mathbb{F}_p$. The differentials in the second column are induced by the differential on $\beta^*(X)$. Applying the cohomology functor to this diagram, the inclusion maps ι become isomorphisms. Hence $V \subset H^{<n+1}(X \times_{\mathbb{Z}/p} S^1)$, because $\mathbb{F}_p[s,t]/(s^2)$ is $\mathbb{F}_p[t]$-torsion free. This implies $V \subset U^\perp$, as V is an ideal in U. We now consider the universal coefficient sequence for the principal ideal domain $\mathbb{F}_p[t]$

$$0 \to H^*_{\mathbb{Z}/p}(X)_0 \to H^*(\beta^*(X)_0) \to \operatorname{Tor}_{\mathbb{F}_p[t]}(H^*_{\mathbb{Z}/p}(X), (\mathbb{F}_p)_0) \to 0.$$

Using the $\mathbb{F}_p[t]$-free resolution $0 \to \mathbb{F}_p[t] \hookrightarrow \mathbb{F}_p[t] \to (\mathbb{F}_p)_0 \to 0$ for calculating the Tor-term on the right and using the structure theorem for the finitely generated $\mathbb{F}_p[t]$-module $H^*_{\mathbb{Z}/p}(X)$ we obtain

$$\dim_{\mathbb{F}_p} U + \dim_{\mathbb{F}_p} V = \dim_{\mathbb{F}_p} H^*(S^1 \times_{\mathbb{Z}/p} X),$$

hence $V = U^\perp$. The last assertion in the theorem follows, because the deformation parameter t from U/V to $H^{(*)}(X^{\mathbb{Z}/p})[s]/(s^2)$ is positively graded.

□

4. Further remarks

If $p = 2$, the above argument simplifies considerably ([**7**]), because the cohomology ring $H^*(B\mathbb{Z}/2; \mathbb{F}_2)$ does not contain an exterior algebra. In Theorem 8, we did not state the well known fact that (for p odd) the formal dimension of the $\mathbb{F}_p$-cohomology of each component of $X^{\mathbb{Z}/p}$ has the same parity as n. This can be shown using deformation theoretic methods, too, but needs a little more machinery than presented in this note. Because each finite p-group contains a normal subgroup of order p, one can generalize Theorem 8 to actions of finite p-groups on $\mathbb{F}_p$-Poincaré duality spaces. For such actions one can ask about further relations between the algebra structures of $H^*(X; \mathbb{F}_p)$ and $H^*(F; \mathbb{F}_p)$ for the different components $F \subset X^{\mathbb{Z}/p}$. If the formal dimension of $H^*(X; \mathbb{F}_p)$ is even, J.P. Alexander and G.C. Hamrick in [**1**] prove a relation between the Witt classes of $H^{ev.}(X; \mathbb{F}_p)$ and $\sum_{F \subset X^{\mathbb{Z}/p}} H^{ev.}(F; \mathbb{F}_p)$, if X is an integral Poincaré duality space equipped with a $\mathbb{Z}/p$-action. In [**5**] we prove a relation of the Witt classes of total space and fixed point set for $\mathbb{Z}/p$-actions on $\mathbb{F}_p$-Poincaré duality spaces on which the Bockstein operator associated to the exact sequence of coefficients $0 \to \mathbb{Z}/p \to \mathbb{Z}/p^2 \to \mathbb{Z}/p \to 0$ acts as the identity. These theorems do not generalize to actions of finite p-groups in an obvious way, because neither the property of being an integral Poincaré duality space nor having a trivial Bockstein operator need to be preserved by passing from a $\mathbb{Z}/p$-space to its fixed point set components. However, in [**6**] we get a relation between the Witt classes of total space and fixed point set in the case of finite p-group actions on $\mathbb{Z}_{(p)}$-homology manifolds.

The author acknowledges the fruitful remarks of R. Fritsch and V. Puppe during the preparation of this work.

References

1. J.P. Alexander, G.C. Hamrick. Periodic maps on Poincaré duality spaces. *Comment. Math. Helv.*, 53:149-159, 1978.
2. Ch. Allday, V. Puppe. Cohomological methods in transformation groups. Cambridge University Press, 1993.
3. G.E. Bredon. Fixed point sets of actions on Poincaré duality spaces. *Topology*, 12:159-175, 1973.
4. T. Chang, T. Skjelbred. Group actions on Poincaré duality spaces. *Bull. Amer. Math. Soc.* 78:1024-1026, 1972.
5. B. Hanke. Witt classes of inner products and actions of finite p-groups. Dissertation, Universität München, 1999.
6. B. Hanke. Actions of finite p-groups on homology manifolds. Preprint, 1999.
7. M. Raußen. Rational cohomology and homotopy of spaces with circle action. In *Lecture Notes in Mathematics*, 1509:313-325, 1992.

LUDWIG-MAXIMILIANS-UNIVERSITÄT
THERESIENSTRASSE 39
80333 MÜNCHEN, GERMANY
E-mail address: hanke@rz.mathematik.uni-muenchen.de

Contemporary Mathematics
Volume **279**, 2001

An Analog of the May-Milgram Model for Configurations with Multiplicities

Sadok Kallel

To Jim Milgram on his sixtieth birthday

ABSTRACT. We point out a generalization of the May-Milgram and Segal models for iterated loop spaces (and their mapping space analog given by F. Cohen and C.F. Bodigheimer). We phrase our result in terms of labeled configuration spaces of "bounded multiplicity" and in so doing we answer a question of Carlsson and Milgram posed in the handbook. We relate this labeled construction to a theory of Lesh and as a result also obtain a generalization of a theorem of Quillen, Barratt and Priddy. Finally we point out that the stable splittings that occur in the classical case do not persist when higher multiplicities are allowed.

§1. Introduction

Ever since the work of Milgram on iterated loop spaces [M] and its generalization by May [Ma], the May-Milgram configuration space model has come to play a vital role in homotopy theory. Its applications are too numerous to list. Suffices to say that Nishida's nilpotence theorem, Mahowald's infinite family in stable homotopy groups of spheres and most recently Stolz's classification of certain families of constant curvature surfaces [St] make fundamental use of it.

We quickly recall what the model is: Let $F(\mathbb{R}^k, n) \subset (\mathbb{R}^k)^n$ be the subset of n-tuples of disjoint points of $\mathbb{R}^k$, and let X be a connected topological space with basepoint $*$. We assume X has the homotopy type of a CW complex. Observe that both spaces $F(\mathbb{R}^k, n)$ and X^n admit an action by the symmetric group Σ_n given by permuting coordinates. We can then consider the orbit quotients $F(\mathbb{R}^k, n) \times_{\Sigma_n} X^n$ for every $n \geq 1$ (the term

2000 *Mathematics Subject Classification.* 55P47, 55P35, 55P10.
Key words and phrases. Labeled configurations, function spaces, classifying spaces and symmetric products.

corresponding to $n = 0$ is basepoint), and glue them together as follows

1.1 $$C(\mathbb{R}^k, X) = \coprod_{n\geq 0} F(\mathbb{R}^k, n) \times_{\Sigma_n} X^n/\sim$$

where $\sim$ is a (standard) basepoint identification

$$(m_1, \dots, m_n) \times_{\Sigma_n} (x_1, \dots, x_n) \sim$$
$$(m_1, \dots, \hat{m}_i, \dots, m_n) \times_{\Sigma_{n-1}} (x_1, \dots, \hat{x}_i, \dots, x_n) \text{ if } x_i = *.$$

("hat" here means deletion). The theorem of May and Milgram then states that

Theorem 1.2: *There is a map*

$$C(\mathbb{R}^k; X) \longrightarrow \Omega^k\Sigma^k X$$

which is a homotopy equivalence whenever X is connected.

In their paper in the *Handbook of algebraic topology*, G. Carlsson and J. Milgram ([CM], §7) consider the following space: Fix $d \geq 1$ and let $F^d(\mathbb{R}^k, n)$ be the space of ordered n-tuples of vectors in $\mathbb{R}^k$ so that no vector occurs more than d times in the n-tuple (when $d = 1$, $F^1(\mathbb{R}^k, n) = F(\mathbb{R}^k, n)$). Consider as before the "labeled" construction

1.3 $$C^d(\mathbb{R}^k, X) = \coprod_{n\geq 0} F^d(\mathbb{R}^k, n) \times_{\Sigma_n} X^n/\sim$$

Notice we have an inclusion of models $C(\mathbb{R}^k, X) \hookrightarrow C^d(\mathbb{R}^k, X)$. The authors in [CM] raise the question of determining the homotopy type of $C^d(\mathbb{R}^k, X)$ for connected X and $d > 1$. The case $d = 1$ being known (see above), an earlier attempt to answer their question for the case $d = 2$ was carried out by Karagueuzian [Kr]. In this paper we give a general answer to this question for all $d \geq 2$.

Let $SP^d(-)$ be the d-th symmetric product functor. This we recall is defined on spaces X as the quotient $SP^d(X) = X^d/\Sigma_d$ where Σ_d acts by permuting coordinates. We can of course replace $\mathbb{R}^k$ by any (ground) manifold M in 1.3 above and $F^d(\mathbb{R}^k, n)$ by $F^d(M, n)$ in the obvious way (see §2). Our first main result then takes the form.

Theorem 1.4: *Let M be a k-dimensional smooth and parallelizable manifold with non-empty boundary ∂M, and let X be a connected CW complex. Then there is a homotopy equivalence*

$$C^d(M; X) \xrightarrow{\simeq} Map(M, \partial M; SP^d(\Sigma^k X))$$

where the mapping space on the right corresponds to all maps of M into $SP^d(\Sigma^k X)$ sending ∂M to a fixed basepoint.

Remarks and Corollaries 1.5:

• The case $d = 1$ is due to F. Cohen and F. Bodigheimer and reduces to the May-Milgram model when $M = \mathbb{R}^k$ (viewed as the closed disc). There are indications that the full result has been known to them (although this isn't recorded in the literature) and in fact our proof is already largely in [B].

• An analog of theorem 1.4 for $M = \mathbb{R}^n$ based on little cubes of Boardman-Vogt has been obtained by F. Kato [Kt].
• In fact theorem 1.4 admits an extension to stably parallelizable M's. When M is closed, there is also an analog after substituting the space of maps with some appropriate section spaces.
• In the case $d = k = 1$, the space $C^1(\mathbb{R}; X) := C(\mathbb{R}; X)$ (X connected) is homotopy equivalent to the James construction $J(X)$ described as the free monoid on points of X with $*$ as the zero element. The equivalence 1.4 in this case is simply James' equivalence $J(X) \simeq \Omega\Sigma X$.

EXAMPLE: Suppose M is closed, almost parallelizable and $A \subset M$ a closed subset. If $T(A)$ is a tubular neighborhood of A, then $\partial(M - A) \simeq \partial T(A)$. In this case $C^d(M - A; X)$ gets identified with $\mathrm{Map}(M, A; SP^d(\Sigma^k X))$. Choose $A = Q_r$ to be a finite set of points $r \geq 1$. We can identify each point in Q_k with a small closed disk centered at that point. In this case $\partial M - Q_r$ is the boundary of all r disks. The quotient $M/\partial Q_r$ is obtained from M (up to homotopy) by attaching segments between a point $p_0 \in Q_r$ and the remaining $k - 1$ points. This means that $M/\partial Q_k \simeq M \vee \bigvee^{k-1} S^1$ and one gets the following corollary (compare [V])

Corollary 1.6 : *M^k is closed and Q_r a set of r distinct points in M. Then*

$$C^d(M - Q_r; X) \simeq Map^*(M, SP^d(\Sigma^k X)) \times \left(\Omega SP^d(\Sigma^k X)\right)^{r-1}$$

where Map* refers to the space of based maps.

We next discuss the case when X is disconnected. We observe (after Segal) that $C^d(\mathbb{R}^k; X)$ is homotopy equivalent to an associative topological monoid $\bar{C}^d(\mathbb{R}^k; X)$ (cf. §2). This monoid admits a classifying space $B\bar{C}$ and we show (for $k \geq 2$, $d \geq 1$)

Theorem 1.7: $B\bar{C}^d(\mathbb{R}^k; X) \simeq \Omega^{k-1} SP^d(\Sigma^k X)$

Remarks 1.8:
(a) Notice that when X is connected, $\bar{C} = \bar{C}^d(\mathbb{R}^k; X)$ is also connected and hence $\Omega B\bar{C} \simeq \bar{C}$. This yields 1.4 as a corollary. The case $d = 1$ is a theorem of Segal [S].
(b) The case $X = S^0$ has been studied in [K1]. One has $C^d(\mathbb{R}^n, S^0) = \coprod C^d(\mathbb{R}^n, k)$ where $C^d(\mathbb{R}^n, k) = F^d(\mathbb{R}^n, k)/\Sigma_k$. Up to homotopy this is a disconnected topological monoid (see §2). If "+" denotes group completion $\Omega B(-)$, then 1.7 becomes

1.9
$$\left(\coprod_{k\geq 0} C^d(\mathbb{R}^n, k)\right)^+ \simeq \Omega^n SP^d(S^n)$$

For $n = 2$ we recover the equivalence $\left(\coprod_{k\geq 0} \mathrm{Pol}_d^k\right)^+ \simeq \Omega^2 \mathbb{P}^d$ ([GKY], [K2]), where Pol_d^k is the space of complex (monic) degree k polynomials having roots of multiplicity not exceeding d.

In proving the theorems above we avoid entirely the theory of operads ([May]), iterated loop spaces ([CM]) or classifying spaces (as in [S]). Instead we use an interesting shortcut construction given in the form of the "scanning" map originally due to Segal (cf. [K1]). We refer mainly to [B], [K1] for the main ideas/arguments we use in the first part of this note.

PART 2: A CLASSIFYING FAMILY OF SUBGROUPS. The second part of this note ties the above results to work of K. Lesh [L1-2]. First recall that Σ_k acts freely on $F(\mathbb{R}^\infty, k)$ and a model for $B\Sigma_k$ is given by $C(\mathbb{R}^\infty, k)$. The space $\coprod_{k\geq 0} B\Sigma_k$ has the structure of a disconnected monoid with composition being induced from the pairings $\Sigma_n \times \Sigma_m \to \Sigma_{n+m}$. On the other hand we can "deform and add points marching to infinity" (see §2) and hence get stabilizing maps $C(\mathbb{R}^\infty, k) \longrightarrow C(\mathbb{R}^\infty, k+1)$ the limit of which is denoted by $C_\infty(\mathbb{R}^\infty) := B\Sigma_\infty$. The Baratt-Priddy-Quillen theorem (or a version of it) is the special case of 1.9 when $d = 1$ and $n = \infty$; namely it states that

$$1.10 \qquad \left(\coprod B\Sigma_k\right)^+ := \Omega B\left(\coprod_{k\geq 0} B\Sigma_k\right) \simeq QS^0 \simeq_h \mathbb{Z} \times B\Sigma_\infty$$

where $\simeq_h$ means homology equivalence (here $B\Sigma_0$ is basepoint). From this point of view, the space $B\Sigma_\infty = C_\infty(\mathbb{R}^\infty)$ is closely associated with the group completion of the "family" of groups $\Sigma_k, k \geq 1$.

In the exact same way we can stabilize the spaces $C^d(\mathbb{R}^\infty, k)$ to a space $C^d_\infty(\mathbb{R}^\infty)$ and the question is then: what group completion is the space $C^d_\infty(\mathbb{R}^\infty)$ for $d > 1$ trying to describe?

In §4, we recall after T.tom Dieck the notion of a "family" of subgroups and of their classifying spaces. It turns out that given "compatible" families $\mathcal{F}_n$ (each consisting of a collection of subgroups of Σ_n), one can associate to them a topological monoid $\coprod B\mathcal{F}_n$ of which group completion is an infinite loop space ([L1]). As a special case, we consider the collection of subgroups

$$H_{i_1,\dots,i_k} = \Sigma_{i_1} \times \Sigma_{i_2} \times \cdots \times \Sigma_{i_k},\ i_1 + i_2 + \cdots + i_k = n$$

of Σ_n. We define a "family" $\mathcal{F}^d_n$ to consist of all $H_{i_1,\dots,i_k}$ with $i_j \leq d$ and $i_1 + i_2 + \cdots + i_k = n$, together with their subgroups. For every $n \geq 1$, the family $\mathcal{F}^d_n$ affords a classifying space construction $E\mathcal{F}^d_n$ (that is there is a Σ_n space $E\mathcal{F}^d_n$ such that the fixed point set under the action of $H_{i_1,\dots,i_k} \in \mathcal{F}^d_n$ is contractible, and otherwise it is empty). The quotient spaces $B\mathcal{F}^d_n = E\mathcal{F}^d_n/\Sigma_n$, $n \geq 1$, are compatible in the sense that the disjoint union $\coprod B\mathcal{F}^d_n$ has the structure of a topological monoid. We shall prove (cf. §4)

Proposition 1.11: *For all $d \geq 1$, we have the following homology equivalence*

$$\left(\coprod_{n\geq 0} B\mathcal{F}^d_n\right)^+ \simeq_h \mathbb{Z} \times C^d_\infty(\mathbb{R}^\infty)$$

When $d = 1$, $\mathcal{F}_n$ is the trivial family (consisting of the trivial subgroup in Σ_n), $B\mathcal{F}_n = B\Sigma_n$ and one recovers 1.10 this way. We note that theorem 1.11 is closely related to proposition 7.4 of [L2].

PART 3: STABLE SPLITTINGS. In this last part we point out that in general there can be no stable splittings for the labeled configuration space constructions in 1.3 whenever the labels have multiplicity d at least 2. This is in stark contrast with the $d = 1$ case where these splittings (due to Snaith and Kahn) are almost a trademark of the May-Milgram construction.

We owe this section to Fred Cohen who first informed the author of this non-splitting result and to the referee who suggested the line of proof adopted here.

Given the labeled construction $C^d(M, X) = \coprod_{n\geq 0} F^d(M, n) \times_{\Sigma_n} X^n/\sim$, we can consider the successive quotients

$$D_k^d(X) = C_k^d/C_{k-1}^d$$

where $C_k^d = \coprod_{n=0}^k F^d(M, n) \times_{\Sigma_n} X^n/\sim$. We define $D_0^d(X)$ as the basepoint. A standard argument due originally to Dold and Steenrod shows that the space $C^d(M, X)$ splits in homology with direct summands $H_*(D_k^d(X))$. More precisely we shall give in §5 a short proof of the following proposition

Proposition 1.12: *For all $d \geq 1$, there is a Steenrod splitting*

$$H_*(C^d(M, X)) \cong \bigoplus_{k\geq 0} H_*(D_k^d(X))$$

where homology is taken with untwisted coefficients.

The next step is then to check whether such a homology splitting is actually induced from a stable splitting. When X is a connected sphere, this turns out to be the case only when $d = 1$.

Proposition 1.13: *Let $M = \mathbb{R}^k$, $k \geq 1$ and $X = S^j$ a sphere with $j > 1$. Then $C^d(\mathbb{R}^k, S^j)$ stably splits as a bouquet $\bigvee_{k\geq 0} D_k^d$ if and only if $d = 1$.*

Of course the arguments we provide can be applied to other choices of M and X. In fact, this non-stable splitting should be clear in light of the following example. As $d \to \infty$, $C^d(M, X)$ is in homotopy more and more like $SP^\infty(M \ltimes X)$ (where $M \ltimes X$ is the half-smash $M \times X/M \times *$). However SP^∞ is a functor modeled over Eilenberg-MacLane spaces and these abound in cohomology operations which prevent them from splitting. An obvious example: $SP^n(S^2) = \mathbb{P}^n$ the n-th complex projective space, $SP^\infty(S^2) = \mathbb{P}^\infty$ and $H_*(\mathbb{P}^\infty) = \bigoplus H_*(\mathbb{P}^n, \mathbb{P}^{n-1})$ while $\mathbb{P}^\infty$ certainly doesn't split as $\bigvee(\mathbb{P}^n/\mathbb{P}^{n-1})$.

ACKNOWLEDGMENTS: The author is much indebted to Kathryn Lesh, Fred Cohen and the referee for material related to part 2 and part 3 of this paper. The family of subgroups $\mathcal{F}_n^d$ mentioned above and defined in §4 was suggested to the author by K. Lesh and the results of that section follow essentially from a discussion with her. We finally thank Denis Sjerve for his generous support and interest.

§2. Constructions and Notation

NOTATION: A configuration in M is a formal finite sum $\sum n_ix_i$, $n_i \in \mathbf{N}$, $x_i \in M$ $(x_i \neq x_j, i \neq j)$. Such a configuration represents by definition a point of $SP^n(M), n = n_1 + \cdots + n_k$.

Assume M to be a connected manifold (in this paper, M will be either open or compact with non-empty boundary) and define

$$F^d(M,n) = \bigcup \{(\underbrace{x_1,\dots,x_1}_{i_1},\underbrace{x_2,\dots,x_2}_{i_2},\cdots,\underbrace{x_k,\dots,x_k}_{i_k}) \\ \mid x_i \in M, x_i \neq x_j, i \neq j, \text{ and } i_j \leq d,\ i_1 + \cdots + i_k = n\}.$$

We write $C^d(M,n) = F^d(M,n)/\Sigma_n$ for the unoriented configuration space. We now consider the labeled construction 1.3. When $d = 1$ and $M = \mathbb{R}^\infty$, it is customary to write $C^1(\mathbb{R}^\infty; X) = QX$. A stable version of the theorem of May and Milgram asserts that

$$Q(X) \simeq \Omega^\infty\Sigma^\infty X \text{ for connected } X$$

and this of course implies that $\pi_i(Q(X)) \cong \pi_i^s(X)$.

Remark 2.1: One notices that singular homology and stable homotopy sit at opposite ends of the labeled configuration space construction given in the form

$$\coprod_{n\geq 1} F(n) \times_{\Sigma_n} X^n/\sim$$

where $\sim$ is the usual basepoint identification. When $F(n) = *$ and Σ_n acts trivially, we get $SP^\infty(X) = \coprod_n * \times_{\Sigma_n} X^n/_\sim$. The functor $SP^\infty(-)$ gives singular homology by virtue of the well-known theorem of Dold and Thom to the effect that $\pi_*(SP^\infty(X)) \cong \tilde{H}_*(X;\mathbb{Z})$. On the other hand and when $M = \mathbb{R}^\infty$, we get the functor $Q(-)$ and hence $\pi_*^s(X)$. The labeled constructs $C^d(M,-)$ for $d > 1$ provide intermediate functors between stable homotopy and integral homology and this circle of ideas is discussed in [L1-2].

The labeled construction 2.1 defines a bifunctor $C^d(-;-)$ which is a homotopy functor in X and an isotopy functor in M When $X = S^0$, we write $C^d(M) := C^d(M,S^0) = \coprod_{k\geq 0} C^d(M,k)$.

When M has an open end or a boundary, we observe that we can put a monoidal structure on $C^d(M,X)$ up to homotopy. We explain this for $M = \mathbb{R}^k$: Let $\mathbb{R}^k_t$ be given as follows

2.2 $$\mathbb{R}^k_t = \{(x_1,\dots,x_n) \in \mathbb{R}^k, \mid 0 < x_n < t\}$$

and define as in [S] the space

$$\bar{C}^d(\mathbb{R}^k;X) = \{(\zeta,t) \in C^d(\mathbb{R}^k_t;X) \times \mathbb{R}^+\}$$

Again we have that $\bar{C}^d(\mathbb{R}^k,X) \simeq C^d(\mathbb{R}^k,X)$ and this new modified space has now the structure of an associative (*homotopy commutative for* $k \geq 2$) topological monoid with a composition law given by juxtaposition

$$C(\mathbb{R}^k_t;X) \times C(\mathbb{R}^k_{t'};X) \longrightarrow C(\mathbb{R}^k_{t+t'};X),\ (\zeta,\zeta') \mapsto (\zeta + T_t\zeta')$$

where T_t is translation $(0,t'))\longrightarrow(t,t+t')$. In the case $d=k=1$ this is the same up to homotopy as the well-known James construction $J(X)$.

GROUP COMPLETION: Since $\bar{C}^d=\bar{C}^d(\mathbb{R}^k;X)$ is a monoid (possibly disconnected), it admits a "group completion" $\Omega B\bar{C}^d$ (i.e. $\pi_0(\Omega BM)$ is a group completion of $\pi_0(M)$ for M a monoid). A handy description of the homology of this space is given as follows. We shall suppose that X has finitely many components (the countable case follows from a direct limit argument). We then have that $\bar{C}^d(\mathbb{R}^k;X)$ has $\mathbb{N}^m$ components, with $m=|\pi_0(X)|-1$. For each $i\leq|\pi_0(X)|$ choose a point p_i in the i-th component of X (the zero component is the component containing basepoint $*=p_0$). Let $\mathbb{R}^k_t$ be as in 2.2 and choose a point $z_i\in\mathbb{R}^k_{i,i+1}=\mathbb{R}^k_{i+1}-\mathbb{R}^k_i$, for all $i\geq 1$. We can then consider the inclusion

2.3
$$\begin{array}{ccc}\bar{C}^d(\mathbb{R}^k;X) & \xrightarrow{\tau_i} & \bar{C}^d(\mathbb{R}^k;X)\\ (\sum(m_r,x_r),i) & \mapsto & (\sum(m_r,x_r)+(z_i,p_i),i+1)\end{array}$$

The direct limit over these maps is denoted by $\hat{C}^d(\mathbb{R}^k;X)$. When X is connected, we have the equivalence $\hat{C}^d(\mathbb{R}^k;X)\simeq C^d(\mathbb{R}^k;X)$.

EXAMPLE 2.4: We can define maps $C^d(\mathbb{R}^k,i)\xrightarrow{\cong}C^d(\mathbb{R}^k_i,i)\xrightarrow{+z_i}C^d(\mathbb{R}^k,i+1)$ the direct limit of which we write $C^d_\infty(\mathbb{R}^k)$. It is now easy to see that $\hat{C}^d(\mathbb{R}^k;S^0)\simeq\mathbb{Z}\times\lim_i C^d(\mathbb{R}^k;i):=\mathbb{Z}\times C^d_\infty(\mathbb{R}^k)$.

Lemma 2.5: *Let X CW, $k\geq 2$. Then $H_*(\hat{C}^d(\mathbb{R}^k;X))\cong H_*(\Omega B\bar{C}^d(\mathbb{R}^k;X))$.*

PROOF: If we let $\pi=\pi_0(\bar{C}^d)$, then a theorem of Kahn and Priddy (cf. [MS]) states that

2.6
$$H_*(\bar{C}^d)[\pi^{-1}]\cong H_*(\Omega B\bar{C}^d)$$

where the left hand side means localization with respect to the multiplicative set π (provided that π is in the center of $H_*(\bar{C}^d)$ which is the case since $\bar{C}^d$ is homotopy abelian as pointed out earlier). The idea of 2.6 is of course that by inverting π, we are "turning" multiplication by elements of π into isomorphisms (that this is necessary is clear since $\Omega B\bar{C}^d$ is a group and hence the image of π under $M\to\Omega B\bar{C}^d$ must consist of units.) Now notice that the point $(z_i,p_i)\in\bar{C}^d(\mathbb{R}^k,X)$ constructed above represents a point $e_i\in\pi_0(\bar{C}^d)$. The stabilization maps in 2.3 correspond therefore to maps

$$\hat{C}^d(\mathbb{R}^k;X)\simeq\varinjlim_{e_i\in\pi}\left(\bar{C}^d(\mathbb{R}^k;X)\xrightarrow{\cdot e_i}\bar{C}^d(\mathbb{R}^k;X)\right).$$

and this direct limit (by construction) must satisfy $H_*(\hat{C}^d(\mathbb{R}^k;X))\cong H_*(\bar{C}^d(\mathbb{R}^k;X))[\pi^{-1}]$. The claim follows from 2.6. ∎

§3. Proof of Theorem 1.4

The correspondence (known as scanning) which associates to configurations on a parallelizable manifold a mapping space has been described in a few places (cf. [K1], [GKY]). This scanning procedure extends to labeled configuration spaces without modification (and so we refer to [K1] for the details). Let M be parallelizable with non-empty boundary ∂M and $\dim M = k$. Given $x \in M$ we can canonically identify a neighborhood D^k_x of it with the closed disc D^k and hence to every configuration $\zeta \in C^d(M;X)$ we have a map

$$\zeta \times M \longrightarrow C^d(D^k;X)$$

which associates to $x \in M$ part of the configuration ζ lying in D^k_x. We compose with the map $C^d(D^k;X) \longrightarrow C^d(D^k,\partial D^k;X)$ to make this association continuous. Here $C^d(D^k,\partial D^k;X)$ is the quotient of $C^d(D^k;X)$ with the additional identification

$$(m_1,\dots,m_n)\times_{\Sigma_n}(x_1,\dots,x_n)\sim$$
$$(m_1,\dots,\hat{m}_i,\dots,m_n)\times_{\Sigma_n}(x_1,\dots,\hat{x}_i,\dots,x_n),\ \text{if}\ \ m_i\in\partial D^k$$

Equivalently, when points of the ground space D^k tend to the boundary they are discarded together with their labels. To every configuration in $\zeta \in C^d(M;X)$ we then have a map $\zeta \times M \longrightarrow C^d(D^k,\partial D^k;X)$. We can demand that the points making up ζ live away from the boundary (or end) of M and in fact with a bit of care we get a correspondence

$$S' : C^d(M;X) \longrightarrow \mathrm{Map}(M,\partial M; C^d(D^k,\partial D^k;X))$$

where ∂M is sent to basepoint in $C^d(D^k,\partial D^k;X)$. It is not hard to see (by a standard radial retraction argument, cf. [K1], [McD]) that

3.1 $$C^d(D^k,\partial D^k;X) \simeq SP^d((D^k/\partial D^k)\wedge X) = SP^d(\Sigma^k X)$$

and hence S gives rise to the map

$$S : C^d(M;X) \longrightarrow \mathrm{Map}(M,\partial M; SP^d(\Sigma^k X))$$

which extends to the stabilized space $\hat{C}^d(\mathbb{R}^k;X)$ (2.3).

Proposition 3.2: *For X a topological space, scanning induces an (integral) homology equivalence*

$$S_* : H_*(\hat{C}^d(\mathbb{R}^n;X)) \xrightarrow{\cong} H_*(\Omega^n SP^d(\Sigma^n X))$$

S is a homotopy equivalence whenever X is connected.

The arguments that go into verifying 3.2 are by now standard. The idea is to induct on an (ingenious) handle decomposition of the closed unit cube D^n (given for instance by Bodigheimer in [B]) and to use properties of the functor C^d. Define $C^d(M,A)$ for a closed ANR $A \subset M$ to be the quotient of $C^d(M)$ by the identification which requires that points be discarded when they are in A (exactly as in the case $A = \partial D$ above). The functor $C^d(M,A;X)$ is an isotopy functor in M and A. The following easy extension of results in [B], [K1] can be shown

Lemma 3.3: *Let M and N be connected manifolds, $N \subset M$, $M_0 \subset M$ and consider the cofibration $(N, N\cap M_0) \longrightarrow (M,M_0) \longrightarrow (M, N\cup M_0)$. Then*

(a) $C^d(N, N\cap M_0; X)\longrightarrow C^d(M, M_0; X)\longrightarrow C^d(M, N\cup M_0; X)$ *is a quasifibration if* $N\cap M_0\neq\emptyset$ *or* X *connected.*
(b) Assume $N\cap M_0=\emptyset$ *and* N *has an end or a boundary, then*

$$\hat{C}^d(N; X)\longrightarrow\hat{C}^d(M, M_0; X)\longrightarrow C^d(M, N\cup M_0; X)$$

is a quasifibration if X *connected and a homology fibration otherwise.*

SKETCH OF PROOF: Let's consider the case $X=S^0$ and $N\cap M_0=\emptyset$. The point in showing that the sequence of spaces in (b) above is a homology fibration (resp. a quasifibration) boils down to showing that maps

$$\hat{C}^d(N)\xrightarrow{\ \ +\ \ }\hat{C}^d(N)$$

given by adjointing a given set of configurations is a homology equivalence (resp. a weak homotopy equivalence). Because of the very construction of $\hat{C}$, adding configurations simply switches components and since these components are the same, "addition" induces a homology equivalence. This is not (necessarily) a homotopy equivalence since there is no obvious map backwards ("subtraction") which when composed with addition induces the identity on components. When either X is connected or $N\cap M_0\neq\emptyset$ it is possible to "subtract" by moving labels to $*\in X$ or points to $N\cap M_0$ where they get discarded. ∎

SKETCH OF PROOF OF THEOREM 3.2: We let $A^k=S^{k-1}\times D^{n-k}$ denote part of the boundary of the unit cube $D^n=[0,1]^n$ (note that $A^0=\emptyset$). Now retracting then scanning gives a map $C^d(D^n, A^k; X)\longrightarrow\Omega^{n-k}SP^d(\Sigma^nX)$ (cf. [K2]) which in the case $k=n$ (i.e. $A^n=\partial D^n$) is a homotopy equivalence according to 3.1.

Let $I_k\subset D^n$ denote the subset of $(y^1,\dots,y^n)$ such that $y^i=0$ or $y^i=1$ for some $i=k+1,\dots,n$, or $y^k=1$ (that is I_k consists of $D^k\times S^{n-k-1}\subset\partial(D^n=D^k\times D^{n-k})$, for $1\leq k<n$, together with one face of D^k; cf. [B]). Now let $H_k=[0,1]^{k-1}\times[0,\frac{1}{2}]\times[0,1]^{n-k}$. Then there is a cofibration sequence

$$(H_k, H_k\cap I_k)\longrightarrow(D^n, I_k)\longrightarrow(D^n, H_k\cup I_k)$$

The pair $(H_k, H_k\cap I_k)$ can be identified with the pair (D^n, A^{k-1}), while $(D^n, H_k\cup I_k)=(D^n, S^{k-1}\times D^{n-k})$ represents a "handle" (D^n, A^k). Applying the functor $C^d(-;X)$ and then scanning yields the commutative diagram for all $k>1$

$$\begin{array}{ccc}
C^d(D^n, A^{k-1}; X) & \longrightarrow & \Omega^{n-k+1}SP^d(\Sigma^nX)\\
\downarrow & & \downarrow\\
C^d(D^n, I_k; X) & \longrightarrow & \Gamma\\
\downarrow & & \downarrow\\
C^d(D^n, A^k; X) & \longrightarrow & \Omega^{n-k}SP^d(\Sigma^nX)
\end{array}$$

where Γ is an appropriate section space. When $k>1$, $A^{k-1}\neq\emptyset$ and 3.3 asserts that the left vertical sequence is a quasifibration. Note that for $k>0$, D^n retracts onto I_k implying that $C^d(D^n, I_k, X)$ is contractible (similarly so is Γ) and so inductively $C^d(D^n, A^k; X)\simeq\Omega^{n-k}SP^d(\Sigma^nX)$ for $k>1$. When

$k = 1$, $A^{k-1} = A^0 = \emptyset$ and we need to pass to group completed spaces $\hat{C}^d$. In this case we have the diagram

$$\begin{array}{ccc}
\hat{C}^d(D^n;X) & \longrightarrow & \Omega^n SP^d(\Sigma^n X)\\
\downarrow & & \downarrow\\
\hat{C}^d(D^n,*;X) & \xrightarrow{\simeq} & \Gamma\\
\downarrow & & \downarrow\\
C^d(D^n,A^1;X) & \xrightarrow{\simeq} & \Omega^{n-1}SP^d(\Sigma^n X)
\end{array}$$

where the left hand side is now a homology fibration according to 3.3 (or quasifibration if X connected). Since $\Omega^n SP^d(\Sigma^n X)$ must be the homotopy fiber of $\hat{C}^d(D^n,*;X) \to C^d(D^n,A^1;X)$ it follows that $H_*(\hat{C}^d(D^n;X)) \cong H_*(\Omega^n SP^d(\Sigma^n X))$ as asserted. ∎

Proof of Theorem 1.4: A manifold M with non-empty boundary can be obtained from $\mathbb{R}^k$ by a sequence of attachments of handles of index k, $0 \leq k < n$. Theorem 1.2 now follows by repeated use of lemma 3.3 and an inductive argument on a handle decomposition for M (cf. [K1], [McD]). ∎

Proposition 3.4: *For X a CW complex, there is a homotopy equivalence*

$$B\bar{C}^d(\mathbb{R}^k;X) \simeq \Omega^{k-1}SP^d(\Sigma^k X).$$

Proof: First of all there is a homotopy commutative diagram

$$\begin{array}{ccc}
 & \hat{C}^d(\mathbb{R}^k;X) & \\
\swarrow & & \searrow\\
\Omega B\bar{C}^d & \longrightarrow & \Omega^k SP^d(\Sigma^k X)
\end{array}$$

where the right hand map is the "natural fiber" inclusion (see [MS]) and the bottom map is induced by scanning $\bar{C}^d \longrightarrow \Omega^k SP^d(\Sigma^k X)$. By combining 2.5 and 3.2 we see that all maps are homology equivalences. The bottom map is a homology equivalence between two simple spaces (being H-spaces) and also induces an isomorphism at the level of $\pi_1 (= H_1$ being abelian). This is then a homotopy equivalence and the proof is complete (note that X being a CW complex, $\Omega^k SP^d(\Sigma^k X)$ has the homotopy type of a CW complex as well). ∎

§4. Classifying Families of Subgroups of Σ_n and $B\bar{C}^d$

In this section we describe a construction of K. Lesh (cf. [L1-2]) which associates to a (compatible) family of groups an infinite loop space. We then describe how our labeled construction fits in and prove proposition 1.11 of the introduction.

Let G be a group and let $\mathcal{F}$ be a collection of subgroups of G which is closed under conjugation; meaning that

- If $H \in \mathcal{F}$ and $g \in G$, then $g^{-1}Hg \in \mathcal{F}$

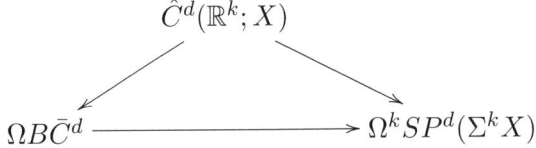

• If $H \in \mathcal{F}$ and K a subgroup of H, then $K \in \mathcal{F}$.
Such a collection is called a *family*.

The prototypical example of a family would be to take all subgroups of a group G (a variant is to consider only the finite subgroups). A less trivial example would be to consider the family of elementary abelian p-subgroups of Σ_n which are generated by disjoint p-cycles together with their subgroups. This family is studied in [L1].

It turns out that to a family $\mathcal{F}$ of subgroups of a group G there is associated a *classifying* space $B\mathcal{F}$ by work of T. tom Dieck. More precisely, tom Dieck constructs a G-space $E\mathcal{F}$ with the property that the fixed point set $E\mathcal{F}^H$ of $H \subset G$ is such that

$$E\mathcal{F}^H \simeq * \text{ for } H \in \mathcal{F}, \text{ and } E\mathcal{F}^H = \emptyset \text{ for } H \notin \mathcal{F}$$

Note that $E\mathcal{F}$ is always contractible since $* \in \mathcal{F}$ for any family. Naturally one then defines the classifying space $B\mathcal{F}$ to be the orbit space of the G action on $E\mathcal{F}$.

EXAMPLE 4.1: Let $\mathcal{F}$ consists only of the trivial subgroup in G. Then $E\mathcal{F} = EG$.

We now specialize to the symmetric groups Σ_n and we suppose that for each n we are given a family $\mathcal{F}_n$ of subgroups for $G = \Sigma_n$. We recall that given two subgroups $H \in \Sigma_n$ and $K \in \Sigma_m$, we have a group $H \times K \in \Sigma_{n+m}$ obtained as the image of the composite

$$H \times K \hookrightarrow \Sigma_n \times \Sigma_m \hookrightarrow \Sigma_{n+m}.$$

DEFINITION 4.2: The families $\{\mathcal{F}_n\}_{n\in\mathbf{Z}^+}$ are *compatible* if whenever $H \in \mathcal{F}_n$ and $K \in \mathcal{F}_m$, then $H \times K \in \mathcal{F}_{n+m}$.

Theorem 4.3 (Lesh): *Let $\{\mathcal{F}_n\}_{n\in\mathbf{Z}^+}$ be a compatible choice of families, then $\coprod B\mathcal{F}_n$ has a monoid structure whose group completion is an infinite loop space $L\mathcal{F}$. Such a space comes equipped with (natural) maps*

$$QS^0 \longrightarrow L\mathcal{F} \longrightarrow \mathbb{Z}.$$

EXAMPLE 4.4: Let $\mathcal{F}_n$ be the family consisting of the trivial subgroup in Σ_n. Then $B\mathcal{F}_n = B\Sigma_n$ and so $L\mathcal{F}$ in this case is the group completion of $\coprod B\Sigma_n$ which is known to correspond by a theorem of Barratt-Priddy and Quillen to the infinite loop space QS^0.

We now relate the above constructions to the spaces $C^d(\mathbb{R}^\infty, n)$ and their stable version $C^d(\mathbb{R}^\infty)$ constructed in §3. Given $n \geq 1$ we consider the following subgroups of Σ_n;

$$H_{i_1,\dots,i_k} = \Sigma_{i_1} \times \Sigma_{i_2} \times \cdots \times \Sigma_{i_k} \subset \Sigma_n,\ i_j \leq n \text{ and } i_1 + i_2 + \cdots + i_k = n.$$

Each such subgroup $H_{i_1,\dots,i_k}$ acts on $F^d(\mathbb{R}^\infty, n)$ by permuting points. Let

$$\mathcal{F}_n^d = \{H_{i_1,\dots,i_k} \mid i_j \leq d,\ i_1+i_2+\cdots+i_k = n, \text{ together with their subgroups}\}.$$

It is not hard to see that $\mathcal{F}_n^d$ satisfies the conditions of a family, and that the newly obtained families $\{\mathcal{F}_n^d\}_{n\in\mathbf{Z}^+}$ form a compatible collection.

Lemma 4.5: $E\mathcal{F}_n^d \simeq F^d(\mathbb{R}^\infty, n)$.

PROOF: Pick $H = H_{i_1,\dots,i_k} \in \mathcal{F}_n^d$ ($i_j \leq d$ and $\sum i_j = n$.) Then

$$(E\mathcal{F}_n^d)^{H_{i_1,\dots,i_k}} = \{(\underbrace{x_1,\dots,x_1}_{i_1},\underbrace{x_2,\dots,x_2}_{i_2},\cdots,\underbrace{x_k,\dots,x_k}_{i_k}) \mid x_i \in \mathbb{R}^\infty\}$$

where the x_i need not be distinct. If we let $X(\mathbb{R}^j)$ be the subset of $(E\mathcal{F}_n^d)^H$ consisting of the $x_i \in \mathbb{R}^j \subset \mathbb{R}^\infty$, then we see that $X(\mathbb{R}^j)$ is open in $\mathbb{R}^{jn}$ and is the complement of hyperplanes of codimension at least j (implying in particular that it is $j-2$ connected). Since $(E\mathcal{F}_n^d)^H$ is the direct limit of $X(\mathbb{R}^j) \hookrightarrow X(\mathbb{R}^{j+1})$ it must be contractible and $(E\mathcal{F}_n^d)^H \simeq *$ as desired. What is left to show is that the fixed point set of $H \notin \mathcal{F}_n^d$ is empty. Observe that any such H must contain a cycle on (at least) $d+1$ letters. The fixed points of such a cycle consists of configurations containing a $d+1$ (or maybe more) repeated point. Such a configuration cannot exist in $F^d(\mathbb{R}^\infty, n)$ (by definition) and $(E\mathcal{F}^d)^H = \emptyset$. ∎

Proposition 4.6: *For all $d \geq 1$, we have the following equivalence*

$$B\left(\coprod_{n\geq 0} B\mathcal{F}_n^d\right) \simeq \Omega^{\infty-1}SP^d(S^\infty).$$

PROOF: By lemma 4.5, $\coprod_{n\geq 0} B\mathcal{F}_n^d = \coprod_{n\geq 0} C^d(\mathbb{R}^\infty, n) = C^d(\mathbb{R}^\infty; S^0) \simeq \bar{C}^d(\mathbb{R}^\infty; S^0)$. On the other hand, it is easy to see that theorem 1.7 still holds true for $k=\infty$ and the claim follows. ∎

Combining now 3.2 with the above proposition yields 1.11 immediately. Note that by construction $\hat{C}^d(\mathbb{R}^\infty, \infty) = \mathbb{Z} \times C_\infty^d(\mathbb{R}^\infty)$ (cf. 2.4).

§5. Stable Splittings

As discussed in the introduction we are concerned with the existence of stable splittings for $C^d(M, X)$ and more generally for the (*filtration n*) subspace

$$C_n^d(M, X) = \coprod_{n=0}^{k} F^d(M, n) \times_{\Sigma_n} X^n / \sim$$

in terms of successive quotients as in

$$C_n^d(M, X) \simeq_s \bigvee_{k=0}^{n} D_k^d(M, X)$$

where $D_0^d(M, X) = *$ and $D_k^d(M, X) = C_k^d / C_{k-1}^d$ for $k \geq 1$. As far as homology is concerned, this splitting always occurs; i.e.

Proposition 5.1: $H_*(C_n^d(M, X)) \cong \bigoplus_{k=0}^n H_*(D_k^d(M, X))$

PROOF:(compare [C]) There are transfer maps

$$\iota_j : C_n^d(M,X) \longrightarrow SP^\infty(C_j^d(M,X)) \longrightarrow SP^\infty(D_j^d)$$

for all $j \leq n$ given in the standard way: the first map associates to each n-tuple in $C_n^d(M,X)$ all choices of j-subtuples and then concatenates them in SP^∞, while the second map is collapsing C_{j-1}^d in C_j^d. The maps ι_j combine to give a map

$$C_n^d(M,X) \longrightarrow \prod_{0\leq j\leq k} SP^\infty(D_j^d) = SP^\infty(\bigvee_{0\leq j\leq k} D_j^d)$$

which in turn can be extended multiplicatively to a map

$$\iota : SP^\infty(C_n^d(M,X)) \longrightarrow SP^\infty(\bigvee_{0\leq j\leq k} D_j^d(M,X))$$

We need then show that ι is a weak equivalence (for then the proposition will follow from the known correspondence: $\pi_*(SP^\infty(X)) \cong \tilde{H}(X;\mathbb{Z})$).

To see that ι is a weak equivalence, we use again the fact that SP^∞ converts cofibrations to fibrations. As in the proof of 3.2 we get the following diagram of fibrations

$$\begin{array}{ccc}
SP^\infty(C_k^d(M,X)) & \longrightarrow & SP^\infty(\bigvee_{j=0}^{k} D_j^d) \\
\downarrow & & \downarrow \\
SP^\infty(C_{k+1}^d(M,X)) & \longrightarrow & SP^\infty(\bigvee_{j=0}^{k+1} D_j^d) \\
\downarrow & & \downarrow \\
SP^\infty(D_{k+1}^d) & \xrightarrow{id} & SP^\infty(D_{k+1}^d)
\end{array}$$

The upper part of the diagram commutes because the transfer map is compatible with the inclusions $C_k^d(M,X) \hookrightarrow C_{k+1}^d(M,X)$ in a natural way. The proof now proceeds by easy induction. ∎

Remark 5.2: A direct limit argument shows that the homology splitting extends to $H_*(C^d(M,X)) \cong \bigoplus_{k\geq 0} H_*(D_k^d(M,X))$.

Remark 5.3: One can also consider the directed system $\iota_d : C^d(M,X) \longrightarrow C^{d+1}(M,X)$ with direct limit $SP^\infty(M \ltimes X)$ where $M \ltimes X$ be the half smash $M \times X/M \times *$ (which can also be written as $M_+ \wedge X$ where M_+ is M with a disjoint point added). The homology splitting for $C^d(M,X)$ is compatible (as is easy to see) with the corresponding homology splitting for $SP^\infty(M \ltimes X)$ (a fact we use in the proof of 5.4 below).

Consider the inclusion $C^d(M,X) \hookrightarrow SP^\infty(M \ltimes X)$. Again X has the homotopy type of CW-complex. We claim that this inclusion is a homotopy equivalence through a range depending on d. To see this, first write

$$D_k^d(M,X) = F^d(M,k)_+ \wedge_{\Sigma_k} X^{(k)}$$

where $X^{(k)}$ is the smash product of $X\wedge\cdots\wedge X$ (k-times) and $F^d(M,k)_+$ is $F^d(M,k)$ with a disjoint basepoint adjointed. From this it follows directly that if M is r-connected, then $D_k^d(M,k)$ is kr-connected (which is the connectivity of $X^{(k)}$).

Proposition 5.4: *Suppose X is r-connected with $r\geq 1$, then the inclusion*

$$\alpha : C^d(M,X)\longrightarrow SP^\infty(M\ltimes X)$$

is a homotopy equivalence up to dimension $r(d+1)$.

PROOF: We first show that α induces an isomorphism in homology up to dimension $r(d+1)$. Remark 5.3 implies that the inclusion $C^d(M,X)\overset{\alpha}{\longrightarrow}$ $SP^\infty(M\ltimes X)$ is stably equivalent to the map

$$\bigvee D_k^d\longrightarrow\bigvee SP^k(M\ltimes X)/SP^{k-1}(M\ltimes X)$$

Clearly $D_k^d = SP^k/SP^{k-1}$ for $k\leq d$ and the map above is the identity on those summands. Homology may then differ starting with the term D_{d+1}^d which has non-trivial homology (possibly) in $r(d+1)+1$ but not before. This implies right away that α_* is an isomorphism in homology up to dimension $r(d+1)$.

To see that this is a homotopy equivalence through that range, it is enough to show $C^d(M,X)$ is simply-connected if X itself is simply connected. To see this, we can apply theorem 1.4 and get that

$$\begin{aligned}\pi_1(C^d(M,X)) &= \pi_1\left(\mathrm{Map}(M,\partial M;SP^d(\Sigma^kX)\right)\\ &= \pi_1\left(\mathrm{Map}^*(M/\partial M,SP^d(\Sigma^kX)\right)\\ &= \Big[\Sigma(M/\partial M),SP^d(\Sigma^kX)\Big]_*\end{aligned}$$

where $k=\dim M$ and $[--]_*$ means based homotopy classes of maps. Now $\Sigma(M/\partial M)$ is a CW complex of dimension $k+1$ whereas $SP^d(\Sigma^kX)$ is $k+1$ connected (which is the connectivity of Σ^kX). The set of based homotopy classes is therefore trivial and the proof follows. ∎

We are now in a position to prove proposition 1.13 of the introduction

Proposition 5.5: *Let $M=\mathbb{R}^k$ and $X=S^j$ with $j>1$. Then $C^d(\mathbb{R}^k,S^j)$ stably splits as a bouquet $\bigvee_{k\geq 0}D_k^d$ if and only if $d=1$.*

PROOF: $C^d(\mathbb{R}^k,S^j)$ is $SP^\infty(\mathbb{R}^k\ltimes S^j)=SP^\infty(S^j)=K(\mathbb{Z},j)$ up to homotopy and through dimension $j(d+1)>2j$ (provided $d>1$). If $\iota_j\in H^j(SP^\infty(S^j))$ denotes the fundamental class, then $\mathrm{Sq}^2(\iota_j)\neq 0\in H^{2j}(SP^2(S^j)) = H^{2j}(SP^\infty(S^j))$ which means that $SP^2(S^j)/S^j = D_2^d(S^j)$ (when $d>1$) cannot split off $SP^\infty(S^j)$ and hence off $C^d(\mathbb{R}^k,S^j)$. ∎

Remark 5.6: When X is not connected, stable splittings can occur for $d>1$. For instance [GKY] verify that $C^d(\mathbb{R}^k;S^0)$ (cf. 1.9) stably splits for all d.

References

[Be] C. Berger, "Opérades cellulaires et espaces de lacets itérés", Annales Institut Fourier, 46 (1996), 1125-1157.

[B] C.F. Bodigheimer, "Stable splittings of mapping spaces", Algebraic topology, Proc. Seattle (1985), Springer lecture notes **1286**, 174–187.

[C] F. Cohen, " The unstable decomposition of $\Omega^2\Sigma^2 X$ and its applications", Math. Z. **182** (1983), no. 4, 553–568.

[CM] G. Carlsson, R.J. Milgram, "Stable homotopy and iterated loop spaces", Handbook of algebraic topology (I.M. James editor), Elsevier 1995.

[GKY] M.A. Guest, A. Koslowski, K. Yamaguchi, "Spaces of polynomials with roots of bounded multiplicity", Fund. Math. 161 (1999), no. 1-2, 93–117.

[K1] S. Kallel, "Particle spaces and generalized Poincaré Dualities", Quart. J. of Math. **52**, part 1 (2001), 45–70.

[K2] S. Kallel, "Bounded multiplicity polynomials and rational maps". To appear in Ann. Inst. Fourier.

[Kt] F. Kato, "On spaces realizing intermediate stages of the Hurewicz map", Master's thesis, Shinshu University 1994.

[Kr] D. Karagueuzian, Thesis, Stanford University 1994.

[L1] K. Lesh, "Infinite loop spaces from group theory", Math. Z. **225** (1997), no.3, 467–483.

[L2] K. Lesh, "A filtration of spectra arising from families of subgroups of symmetric groups", Trans. Amer. Math. Soc. **352** (2000), no. 7, 3211–3237.

[McD] D. McDuff, "Configuration spaces of positive and negative particles", Topology **14**(1975), 91–107.

[MS] D. McDuff, G. Segal, "Homology fibrations and the group completion theorem", Invent. Math. **31** (1976), 279–284.

[M] R.J. Milgram, "Iterated loop spaces", Ann. of Math. **84** (1966), 386–403.

[Ma] J.P. May, "The geometry of iterated loop spaces", Springer lecture notes, **271** (1972).

[Sa] P. Salvatore, "Configuration spaces with summable labels", Math archives AT/9907073.

[S] G. Segal, "Configuration spaces and iterated loop spaces", Invent. Math., **21**(1973), 213–221.

[St] S. Stolz, "Multiplicities of Dupin hypersurfaces", Invent. math. **138** (1999) 2, 253–279.

[V] Vershinin, "Braid groups and loop spaces", Russian Math. surveys **54** (1999), no. 2, 273–350.

LABORATOIRE AGAT, UNIVERSITÉ LILLE I, FRANCE
E-mail address: `sadok.kallel@agat.univ-lille1.fr`

Contemporary Mathematics
Volume **279**, 2001

Configuration Spaces and the Topology of Curves in Projective Space

Sadok Kallel

To Professor Milgram on his 60th birthday

ABSTRACT. We survey and expand on the work of Segal, Milgram and the author on the topology of spaces of maps of positive genus curves into n-th complex projective space, $n \geq 1$ (in both the holomorphic and continuous categories). Both based and unbased maps are studied and in particular we compute the fundamental groups of the spaces in question. The relevant case when $n = 1$ is given by a non-trivial extension which we fully determine.

§1 Introduction and Statement of Results

The topology of spaces of rational maps into various complex manifolds has been extensively studied in the past two decades. Initiated by work of Segal and Brockett in control theory, and then motivated by work of Donaldson in gauge theory, this study has uncovered some beautiful phenomena (cf. [CM], [Hu], [L]) and brought to light interesting relationships between various areas of mathematics and physics (cf. [Hi], [BM],[BHMM]).

Let C_g (or simply C) denote a genus g (compact) Riemann surface ($\mathbb{P}^1$ when $g = 0$) and let V be a complex projective variety. Both C and V come with the choice of preferred basepoints x_0 and $*$ respectively. The main focus of this paper is the study of the geometry of the space

$$\mathrm{Hol}(C, V) = \{f : C \longrightarrow V, f \text{ holomorphic}\}$$

and more particularly its subspace of basepoint preserving maps

$$\mathrm{Hol}^*(C, V) = \{f : C \longrightarrow V, f \text{ holomorphic and } f(x_0) = *\}$$

The space $\mathrm{Hol}^*(C, V)$ doesn't depend up to homeomorphism on the choice of basepoint when V has a transitive group of automorphisms for example. In that situation, the relationship between the based and unbased (or free) mapping spaces is given by the "evaluation"

$$\mathrm{Hol}^*(C, V) \longrightarrow \mathrm{Hol}(C, V) \xrightarrow{ev} V$$

2000 *Mathematics Subject Classification.* 54C35, 55S15, 55R80, 57N65, 55R20.
Key words and phrases. Function spaces, Riemann surfaces, configuration spaces.

which is a holomorphic fibration for V a homogeneous space for example. Here ev evaluates a map at the basepoint of C.

Remark 1.1: By choosing a Riemann surface, we fix the complex structure. We record the link with the theory of "pseudo-holomorphic" curves of Gromov. There one studies the (differential) geometry of spaces of J-holomorphic curves into V, where V is almost complex and J is its almost complex structure. These J-holomorphic curves (of given genus g) correspond when J is integrable to all holomorphic maps of C (with varying complex structure) into V. It is now a theorem of Gromov that for a generic choice of J on V, the space of all such curves (in a given homology class) is a smooth finite dimensional (real) manifold. Let $\mathcal{M}(V,g) = \{(C_g, f) \mid f : C \longrightarrow V \text{ holomorphic}\}$, then there is a projection of $\mathcal{M}(V,g)$, onto the moduli space of curves $\mathcal{M}_g$ sending (C,u) to the isomorphism class of C. The spaces $\mathrm{Hol}(C,V)$ appear then as "fibers" of this projection.

We observe that $\mathrm{Hol}(C,V)$ breaks down into connected components and we write $\mathrm{Hol}_A(C,V)$ for the component of maps f such that $f_*[C] = A$, A a given homology class in $H_2(V)$. When V is simply connected, $[C,V] = [S^2,V] = \pi_2(V) = H_2(V) = \mathbf{Z}^r$ for some rank r and A is completely determined by a multi-degree. The components $\mathrm{Hol}_A(C,V)$ are generally (singular) quasi-projective varieties (see [H2] for example). When $V = \mathbb{P}^n$, we write $\mathrm{Hol}_A(C,\mathbb{P}^n) = \mathrm{Hol}_k(C,\mathbb{P}^n)$ for some $k = \deg A \in \mathbb{Z}$ and these are smooth manifolds as soon as k is big enough (about twice the genus of C). Most of this paper is concerned with the study of these spaces.

Remark 1.2: In physics, spaces of maps between manifolds $\mathrm{Hol}(M,V) \subset \mathrm{Map}(M,V)$ arise in connection with field theory or "sigma models". From the perspective of a physicist, a field on M with values in V is a map $\phi : M \to V$. For example, in the case $M = \mathbf{R}^3$ and $V = \mathbf{S}^3 = \mathbb{R}^3 \cup \{\infty\}$, a map $\phi : \mathbf{R}^3 \to S^3$ could be the field associated to some electrical charge in $\mathbf{R}^3$ (hence vector valued and extending to the point where the charge is located by mapping to ∞). Associated to a field there is an "energy" density or Lagrangian $\mathcal{L}$ (eg. the harmonic measure $\mathcal{L}(\phi) = \frac{1}{2}||d(\phi)||^2$). To an energy density one can in turn associate an "action" which is defined as

$$S[\phi] = \int_M \mathcal{L}(\phi) d\mu(h)$$

where $d\mu(h)$ is the canonical volume measure associated to a metric h on M. Physicists are usually interested in minimizing the action (to determine the dynamics of the system) and hence they are led to study the space of all extrema of this functional. It should be noted that in the case when V is compact, Kahler, a well-known theorem of Eells and Wood asserts that the absolute minima of the energy functional on $\mathrm{Map}(C,V)$ are the holomorphic maps (the critical points here being the harmonic maps).

It has been known since the work of Segal [S] that the topology of holomorphic maps of a given degree $k \in \mathbf{N}$ from C to $\mathbb{P}^n$ compares well with the space of continous maps at least through a range increasing with k. He proved

Theorem 1.3: (Segal) *The inclusion*

$$Hol^*_k(C,\mathbb{P}^n) \hookrightarrow Map^*_k(C,\mathbb{P}^n)$$

induces homology isomorphisms up to dimension $(k-2g)(2n-1)$

A similar statement holds for unbased maps. To simplify notation, we write $\mathrm{Map}_k, \mathrm{Hol}_k, \mathrm{Map}^*_k, \mathrm{Hol}^*_k$ for the corresponding mapping spaces from C into $\mathbb{P}^n$.

When $g=0$ (the rational case) and $n>1$ the homology isomorphism in 1.3 can be upgraded to a homotopy equivalence (cf. [CS]). For $g>0$ however it is not known whether the equivalence in 1.3 holds in homotopy as well; i.e whether the pair $(\mathrm{Map}_k, \mathrm{Hol}_k)$ is actually $(k-2g)(2n-1)$ connected. This is strongly suspected to be true and in this note we give further evidence for this by showing that Hol^*_k and Map^*_k have isomorphic fundamental groups for all n (the relevant case here is $n=1$). In fact we shall show

Theorem 1.4: *Suppose* $k\geq 2g$, *we have isomorphisms*

$$\pi_1(Hol^*_k(C_g,\mathbb{P}^n)) \cong \pi_1(Map^*_0(C_g,\mathbb{P}^n)) \cong \begin{cases} \mathbb{Z}^{2g}, & \textit{when } n>1 \\ G, & \textit{when } n=1 \end{cases}$$

where G *is a cyclic extension of* $\mathbb{Z}^{2g}$ *by* $\mathbb{Z}$ *generated by classes* $e_1,\dots,e_{2g}$ *and* τ *such that the commutators*

$$[e_i, e_{g+i}] = \tau^2$$

and all other commutators are zero.

REMARK[1]: Roughly speaking, the class α can be represented by the one parameter family of maps obtained by rotating roots around poles. Similarly the classes e_i are obtained by rotating roots (or poles) around loops representing the homology generators of C.

Combining this result with a classical result of G. Whitehead (see §2), we deduce

Proposition 1.5: $\pi_1(Map_d(C,S^2))$ *is generated by classes* $e_1,\dots,e_{2g}$ *and* α *such that*

$$\alpha^{2|d|}=1,\ [e_i,e_{g+i}]=\alpha^2$$

and all other commutators are zero. When $d\geq 2g$, $\pi_1(Hol_d(C,\mathbb{P}^1)) \cong \pi_1(Map_d(C,S^2))$.

This description for the continuous mapping space is an earlier result of Larmore-Thomas [LT]. As is clear from 1.5, the components of $\mathrm{Map}(C,\mathbb{P}^n)$ for $n=1$ have different homotopy types. When $n>1$, the fundamental group is however not enough to distinguish between the components. A quick byproduct of our calculations however shows (§8)

[1] J.D.S.Jones has recently informed the author that he has obtained a similar result a few years ago but which he didn't publish.

Proposition 1.6: *$Map_k(C,\mathbb{P}^{2d})$ and $Map_l(C,\mathbb{P}^{2d})$ have different homotopy types whenever l and k have different parity*[2].

The first few sections of this paper are written in a leisurely fashion and are meant in part to survey techniques and ideas in the field most of which carry the deep imprint of Jim Milgram (cf. §4).

In §2 we discuss the rational case (i.e. $g = 0$) and give a short proof of a theorem of Havlicek on the homology of unbased self-morphisms of the sphere. In §3 we introduce a configuration space model (originally given in [K2]) for $\text{Map}^*(C,\mathbb{P}^n)$ and use it to determine the homology of this mapping space. Let $SP^k(M) = M^k/\Sigma_k$ where Σ_k is the cyclic group on k-letters acting by permutations (or equivalently the space of unordered k points on M), and let $SP^k_n(M)$ be the subspace of k points in C no more than n of which are the same ($n \leq k$). When $n = 1$, $SP^k_1(M) = C_k(M)$ is the standard configuration space of k distinct points. When M is a (connected) open manifold or with a (collared) boundary, these spaces can be stabilized (cf. §3) $SP^k_n(M) \longrightarrow SP^{k+1}_n(M) \longrightarrow \cdots$ and we write $SP^\infty_n(M)$ for the (connected) direct limit. The following is discussed in §3

Theorem 1.7 [K2]: *There is a map*

$$SP^\infty_n(C - *) \xrightarrow{\ S\ } Map^*_0(C,\mathbb{P}^n)$$

which is a homotopy equivalence when $n > 1$ and a homology equivalence when $n = 1$.

This model for Map* in terms of 'configurations of bounded multiplicity' turns out to relate quite well with the corresponding model for Hol* and we indicate based on this another proof for Segal's result (§5). Notice that 1.4 shows that the homology equivalence S above cannot be upgraded to a homotopy equivalence when $n = 1$ for already both spaces have different fundamental groups. Indeed the braid group $\pi_1(SP^\infty_1(C_g - *)) = \pi_1(C(C_g - *))$ has a presentation quite different from the central extension given in 1.4.

ACKNOWLEDGEMENTS: We would like to take this occasion to deeply thank Jim Milgram for the many beautiful mathematics we have learned from him. We thank J. F. Barraud, Y. Hantout and J. Nagel for useful discussions we had. Special thanks to the PIms institute in Vancouver for its support while part of this work was carried out.

§2 Preliminaries: The Genus Zero Case

It is customary to write $\text{Rat}(V)$ for the basepoint preserving maps $\text{Hol}^*(S^2,V)$. When $V = \mathbb{P}^n$, these spaces are now very well understood. Let

$$C_k(\mathbb{C},S^{2n-1}) = \coprod_{0\leq i\leq k} F(\mathbb{C},i)\times_{\Sigma_i}(S^{2n-1})^i/\sim$$

[2]it is now verified that all (positive) components of $\text{Map}(C,\mathbb{P}^n)$ have different homotopy type; cf. [KS].

be the standard labeled k-th filtration piece for the May-Milgram model of $\Omega^2 S^{2n+1} \simeq \Omega_0^2(\mathbb{P}^n)$ (cf. [C²M²], [K4], etc). Here $F(\mathbb{C}, i)$ is the set of ordered i-tuples of disctinct points in $\mathbb{C}$ and $\sim$ is a standard basepoint identification identifying an i-tuple with labels to an $i-1$ tuple by discarding a point an its label if the label is at basepoint. Let

$$SP^k(X) = \{\sum k_i x_i, x_i \in X, k_i \in \mathbf{N} \mid x_i \neq x_j \text{ for } i \neq j \text{ and } \sum k_i = k\}$$

be the k-th symmetric product of X (see introduction). Let $SP^k_n(X)$ be the subset of $SP^k(X)$ obtained by restricting to $k_i \leq n$. We then have

Theorem 2.1 ([CS], [K3], [GKY]): *For $k \geq 1$, there are maps and homotopy equivalences*

$$2.1(a) \qquad Rat_k(\mathbb{P}^n) \xleftarrow{\simeq} C_k(\mathbb{C}, S^{2n-1}) \xrightarrow{\simeq} SP_n^{k(n+1)}(\mathbb{C})$$

whenever $n > 1$. When $n = 1$ the spaces are homologous only.

It is interesting to note that no map is known to induce such homology isomorphism between $\mathrm{Rat}_k(\mathbb{P}^n)$ and $SP_n^{k(n+1)}(\mathbb{C})$ $(k > 1)$. When $n > 1$, the left-hand map in 2.1(a) is constructed explicitly in [CS] while the right-hand map is constructed in [K3]. Both configuration space models on either side of 2.1(a) are fairly amenable to calculations and from there the structure of $\mathrm{Rat}_k(\mathbb{P}^n)$ can be made quite explicit (cf. [BM], [C²M²], [K3] and [Ka]).

The space of unbased rational maps is less well understood. The following is due to Havlicek ([H1])

Theorem 2.2: *The Serre spectral sequence for the (holomorphic) fiber bundle*

$$2.2(a) \qquad Rat_k(\mathbf{P}^1) \longrightarrow Hol_k(\mathbf{P}^1) \longrightarrow \mathbf{P}^1$$

has the non-zero differential $d_2(x) = 2k\iota$. The spectral sequence collapses with mod-p coefficients whenever $p = 2$ or p divides k.

We give a short proof for the mod-2 collapse (a general proof and an extension of this result to $\mathrm{Rat}_k(\mathbb{P}^n)$ can be found in [KS]). First of all, to see that 2.2(a) is indeed a fiber bundle, one simply observes that $\mathrm{PSL}_2(\mathbb{C})$ acts transitively on $\mathbb{P}^1$ and if F denotes the stabilizer of a point, then F acts on $\mathrm{Rat}_k(\mathbf{P}^1)$ (by postcomposition). One can see that $\mathrm{Hol}_k(\mathbf{P}^1) = \mathrm{Rat}_k(\mathbf{P}^1) \times_F \mathbb{P}^1$.

Towards the proof of 2.2 (and also in most of §7) we need the following classical result of G. Whitehead. Let X be a based (connected) topological space (with basepoint x_0) and consider the evaluation fibration

$$2.3 \qquad \Omega_f^n X \longrightarrow \mathcal{L}_f^n X \xrightarrow{ev} X$$

where $ev(f) = f(x_0)$, $\mathcal{L}_f^n X = \mathrm{Map}_f(S^n, X)$ is the component of the total mapping space containing a given map $f : S^n \to X$ and $\Omega_f^n X$ the subset of all maps g such that $g(x_0) = f(x_0)$.

Theorem 2.4 [W]: *The homotopy boundary in 2.3 $\partial : \pi_i(X) \to \pi_i(\Omega_f^n X) \cong \pi_{i+n}(X)$ is given (up to sign) by the Whitehead product: $\partial\alpha = [\alpha, f]$.*

Let $\mathcal{L}^2_kS^2 = \mathcal{L}^2_fS^2$ where f is the standard degree k map, and denote by $a \in H^2(S^2)$ the generator. Recall that $\Omega^2S^2 \simeq \mathbb{Z}\times\Omega^2S^3$ and so let e be the generator in $H^1(\Omega^2_kS^2)\cong H^1(\Omega^2S^3)$.

Lemma 2.5: *The Serre spectral sequence for* $\Omega^2_kS^2\longrightarrow\mathcal{L}^2_kS^2\longrightarrow S^2$ *has the (homology) differential* $d_2(a) = 2ke$. *It collapses at* E_2 *with mod-2 coefficients.*

PROOF: For a fibration $F\to E\xrightarrow{p} S^2$ we have the diagram

$$\begin{array}{ccc}
{\scriptstyle \pi_2(E,F) \atop \cong\pi_2(S^2)} & \xrightarrow{\partial} & \pi_1(F)\\
\downarrow{\scriptstyle h} & & \downarrow{\scriptstyle h}\\
H_2(E,F) & \xrightarrow{\partial} & H_1(F)\\
\downarrow{\scriptstyle p_*} & & \downarrow\\
H_2(S^2) & \xrightarrow{\tau} & E^{0,1}
\end{array}$$

where $E^{0,1}\cong H_1(F)$ and τ the transgression. Let $k : S^2\longrightarrow S^2$ denote multiplication by k. From 2.4 and the diagram above, we deduce that

$$\tau : H_2(S^2)\cong\pi_2(S^2)\xrightarrow{k[\iota,-]}\pi_3(S^2)\xrightarrow{ad}\pi_1(\Omega^2_kS^2)\xrightarrow{h}H_1(\Omega^2_kS^2)=\mathbb{Z}$$

$\iota\in\pi_2(S^2)$ is the generator. As is known for even spheres, $h(ad([\iota,\iota])) = 2$ and this yields the differential.

The mod-2 collapse follows from the following very short argument (which we attribute to Fred Cohen): let $E : S^2\longrightarrow\Omega S^3$ be the adjoint map. Then there is a map of (horizontal) fibrations

$$\begin{array}{ccccc}
\Omega^2_kS^2 & \longrightarrow & \mathcal{L}^2_kS^2 & \xrightarrow{ev} & S^2\\
\downarrow{\scriptstyle \Omega^2E} & & \downarrow & & \downarrow{\scriptstyle E}\\
\Omega^3_kS^3 & \longrightarrow & \mathcal{L}^2_k\Omega S^3 & \longrightarrow & \Omega S^3
\end{array}$$

The map E is injective in integral homology while Ω^2E is injective in mod-2 homology. So mod-2, both fiber and base inject in the bottom fibration which is trivial since ΩS^3 is (homotopy equivalent) to a topological group (and translation of basepoint gives the trivialization). The collapse follows in this case. ∎

PROOF OF 2.2: We have inclusions $\mathrm{Rat}(\mathbb{P}^1)\subset\mathrm{Hol}(\mathbb{P}^1)\subset\Omega^2S^2$, where $\mathrm{Rat}(\mathbb{P}^1)$ is the subspace of all holomorphic maps sending the north pole ∞ in $\mathbb{P}^1=\mathbb{C}\cup\infty$ to 1. An element of $\mathrm{Rat}(\mathbb{P}^1)$, say of degree k, is given as a quotient $\frac{p}{q}=\frac{z^k+a_{k-1}z^{k-1}+\cdots+a_0}{z^k+b_{k-1}z^{k-1}+\cdots+b_0}$ where p and q have no roots in common. It is easy to see for example that $\mathrm{Hol}_1(\mathbb{P}^1)$ corresponds to $PSL(2,\mathbb{C})$, the automorphism group of $\mathbb{P}^1$ (which is up to homotopy RP^3), and that

$\mathrm{Rat}_1(\mathbb{P}^1) = \mathbb{C} \times \mathbb{C}^* \simeq S^1$. Consider now the map of fibrations

2.6
$$\begin{array}{ccc} \mathrm{Rat}_k(\mathbb{P}^1) & \hookrightarrow & \Omega^2_k S^2 \\ \downarrow & & \downarrow \\ \mathrm{Hol}_k(\mathbb{P}^1) & \hookrightarrow & \mathcal{L}^2_k S^2 \\ \downarrow & & \downarrow \\ \mathbf{P}^1 & \xrightarrow{=} & S^2 \end{array}$$

According to Segal (theorem 1.3), the top inclusion is an isomorphism in homology up to dimension k. Moreover (cf. [C^{2}M^2], [K1]), $H_*(\mathrm{Rat}_k(\mathbb{P}^1))$ actually injects in $H_*(\Omega^2_0 S^2)$ and it does so in the following nice way. Recall that $\Omega^2_0 S^2 \simeq \Omega^2 S^3 = \Omega^2\Sigma^2 S^1$ and hence it stably splits as an infinite wedge $\bigvee_{j\geq 1} D_j$ where the summands D_j are given in terms of configuration spaces with labels. Stably one finds that

$$\mathrm{Rat}_k(\mathbb{P}^1) \simeq_s \bigvee_{j=1}^{k} D_j.$$

The map of fibers in 2.6 is then an injection and since the bases are same in that diagram, the theorem follows from 2.5 and a spectral sequence comparison argument. ∎

§3 Configuration Space Models and the Higher Genus Case

In this section we describe two configuration space models for each of the mapping spaces $\mathrm{Hol}^*_k(C,\mathbb{P}^n)$ and $\mathrm{Map}^*_k(C,\mathbb{P}^n)$ for $g \geq 1$. A straightforward comparison between both models yields Segal's stability result in §4.

First of all, a map $f \in \mathrm{Hol}_k(C,\mathbb{P}^n)$ can be written locally in the form

$$z \mapsto [p_0(z) : \dots : p_n(z)]$$

where the $p_i(z)$ are polynomials of degree k each and having no roots in common. The p_i's are not global functions on C (otherwise they would be constant) but rather local maps into $\mathbb{C}$ and hence sections of some line bundle. The roots of each p_i give rise to an element $D_i \in SP^k(C)$ (so called *positive divisor*) and conversely D_i only determines p_i up to a non-zero constant (which will be determined if the maps are based). These D_i's (*the root data*) cannot of course have a root in common and if we base our maps so that basepoint $*$ is sent to $[1 : \cdots : 1]$ say, then none of the D_i's contains the basepoint. Define

$$\mathrm{Div}^{n+1}_k(X) = \{(D_0,\dots,D_n) \in SP^k(X)^{n+1} \mid D_0 \cap \cdots \cap D_n = \emptyset\},$$

The previous discussion shows the existence of an inclusion

$$\phi : \mathrm{Hol}^*_k(C,\mathbb{P}^n) \longrightarrow \mathrm{Div}^{n+1}_k(C - *)$$

which associates to a holomorphic map its root data. This correspondence has no inverse since it is not true in general that an (n+1) tuple of degree k-divisors with no roots in common gives rise to a (based) holomorphic map out of C. In fact, this can be understood as follows: the quotients

$\frac{p_i(z)}{p_j(z)}$ are meromorphic maps on C and so for $i \neq j$, $D_i - D_j$ must be the divisor of a function on C (i.e. there is $f : C \to \mathbb{C}$ with roots at D_i and poles at D_j). It is not surprising that this last condition is not satisfied for general pairs (D_i, D_j). We say D_i and D_j are *linearly equivalent* if there is indeed a meromorphic function f on C such that (f) :=zeroes of f-poles of $f = D_i - D_j$. Note that when this is the case D_i and D_j have same degree.

Linear equivalence defines an equivalence relation and one denotes by $J(C)$ the set of all linearly equivalent divisors on C of degree 0. There is a map

3.1 $$\mu : SP^n(C) \longrightarrow J(C),\ D \mapsto [D - nx_0]$$

sending a divisor to the corresponding equivalence class. The above discussion then shows the existence of a homeomorphism

$$\left\{ \begin{array}{c} \text{A (based) holomorphic} \\ X \longrightarrow \mathbb{P}^r \text{ of degree } k \end{array} \right\} \Longleftrightarrow \left\{ \begin{array}{c} \text{An } (n+1) \text{ tuple of positive divisors} \\ D_i \text{ on } C - * \mid D_0 \cap \cdots \cap D_n = \emptyset \\ \deg D_i = k \text{ and } \mu(D_i) = \mu(D_j) \end{array} \right\}$$

To make this correspondence more precise, we need understand the map μ in 3.1. It turns out that:

• $J(C)$ is a g dimensional complex torus (this is a non-trivial fact) and μ is a multiplicative map. In the case $g = 1$, μ is the identity and if one identifies the curve with $\mathbb{C}/L$ for some lattice L, then the Abel-Jacobi condition $\mu(\sum z_i) = \mu(\sum p_j)$, $z_i, p_i \in C$ is equivalent to $\sum z_i = \sum p_j \mod (L)$ (z for zero and p for pole).

• The preimage of a point $[D] = \mu(D) \in J$ is a projective space. To see this let $\mathcal{L}(D)$ be the set of all holomorphic maps on C such that $(f) + D \geq 0$ (i.e. such that $(f) + D \in \bigsqcup_n SP^n(C)$). This is a $\mathbb{C}$-vector space and since $(\alpha f) = (f)$, we have an identification and a map $\mu^{-1}([D]) = \mathbb{P}\mathcal{L}(D) \longrightarrow SP^n(C)$, $[f] \mapsto (f) + D$. This turns out to be a holomorphic embedding ([G],§4, 4.3).

• The complex dimension of $\mu^{-1}([D])$ is denoted by $r(D)$ and a crucial aspect of the classical theory of algebraic curves is the computation of $r(D)$ as D varies in $SP^n(C)$. A very useful interpretation of this dimension is as follows: $r(D) \geq r$ *if and only for any r points of C, there is a divisor D' with $\mu(D') = \mu(D)$ and passing through these r points.*

• If $\deg D = n$, then generically

$$r(D) = max\{0, n - g\}.$$

The exact dimension for every D is determined by Riemann-Roch. This dimension may jump up for certain "special" divisors. However whenever n exceeds $2g - 2$, there are no more jumps and r is uniformally given by $n - g$.

It turns out that for k small (i.e. $k < g$), $\mathrm{Hol}_k^*(C, \mathbf{P}^n)$ is very dependent on the holomorphic structure put on the curve (it can naturally be empty). For example if C is hyperelliptic, then

$$\mathrm{Hol}_{2n+1}(C, \mathbb{P}^1) = \emptyset, 2n + 1 \leq g \text{ (see for instance [FK], Chap3)}$$

Also $\mathrm{Hol}_1(C, \mathbb{P}^n)$ is always empty for all positive genus curves. In the range $k \geq 2g-1$, there is however much better behavior (see [KM])

Lemma 3.2: *Assume $k \geq 2g-1$. Then $Hol_k^*(C, \mathbf{P}^n)$ is a $k(n+1)-ng$ complex manifold.*

REMARK 3.3: $\mathrm{Hol}_k^*(C, \mathbf{P}^n)$ has additionally the structure of a quasiprojective variety. Generally, components of $\mathrm{Hol}(C, V)$ do not have a smooth structure (when V is a smooth projective variety). This already fails for $V = G(n, n+k)$ (the Grassmaniann of n planes in $\mathbb{C}^{n+k}$) when $n > 1$ (see [Ki]). The "expected" dimension of $\mathrm{Hol}_A(C, V)$ (where A is a fixed homology class in $H_2(V)$ and $f_*[C] = A$ for all $f \in \mathrm{Hol}_A$ can however be computed and at generic smooth points it is given by $c_1.A + n(1-g)$. In our case $c_1 = n+1$ ($V = \mathbb{P}^n$) and A is k fold the generator in $H_2(V)$.

§3.1 The configuration space model for $\mathbf{Map}^*(C, \mathbb{P}^n)$

In 1.7 we pointed out to the existence of a model for $\mathrm{Map}(C, \mathbb{P}^n)$ in terms of bounded multiplicity symmetric products $SP_n^k(-)$ defined in §2. More precisely, consider $C - U$ where U is a little open disc around the basepoint $*$ ($C - *$ isotopy retracts onto $C - U$). Let U_k be a nested sequence of neighborhood retracting onto $*$, then one defines the space $SP_n^\infty(C)$ as the direct limit of inclusions $SP_n^k(C - U_k) \longrightarrow SP_n^{k+1}(C - U_{k+1})$, $D \mapsto D + x_k, x_k \in U_k - U_{k+1}$ (here $\{x_k\}$ is a sequence of distinct points converging to $*$).

One can map $SP_n^\infty(C - *)$ to the based mapping space by *scanning*

3.4 $$S : SP_n^\infty(C - *) \longrightarrow \mathrm{Map}_0^*(C, \mathbb{P}^n)$$

This map is given as follows (details in [K2]): identify canonically each neighborhood $D(x)$ of $x \in C - *$ with a closed disc D^2 (this is possible because $C - *$ is parallelizable). It follows that to every configuration in $SP_n^\infty(C - *)$ one can restrict to $D(x)$ and see a configuration in $SP_n^\infty(D(x), \partial D(x)) = SP_n^\infty(D^2, \partial D^2)$ (the relative construction means that configurations are discarded when they get to the boundary). It can be checked that $SP_n^\infty(D^2, \partial D^2) = SP^n(S^2) = \mathbb{P}^n$ and hence 3.4. It turns out that the map S in 3.4. is a homotopy equivalence when $n > 1$ or a homology equivalence when $n = 1$ (cf. [K2]).

Lemma 3.5: *Assume $g \geq 1$, $k \geq n \geq 1$ and $C' = C - *$. Then the collar inclusion $SP_n^k(C') \to SP^\infty(C')$ induces an epimorphism on H_1 (for $n > 1$ it is in fact an isomorphism).*

PROOF: A cohomology class in $H^i(C', \mathbb{Z})$ is represented by (the homotopy class) of a map $C' \longrightarrow K(\mathbb{Z}, i)$ and since the target is an abelian topological group, this map extends multiplicatively to

$$C' \xrightarrow{i_1} SP_n^k(C') \xrightarrow{i_2} SP^k(C') \longrightarrow K(\mathbb{Z}, i)$$

and hence gives rise to a class in $H^i(SP_n^k(C'))$. The "inclusion" $i_1 : C' \to SP_n^k(C')$ (constructed earlier) is then surjective in cohomology and hence injective in homology. Since $\pi_1(SP^k(C')) = H_1(SP^k(C'), \mathbb{Z}) = H_1(C'; \mathbb{Z})$, the composite $i_2 \circ i_1$ must be an isomorphism on H_1. This then implies that $H_1(SP^k_n(C')) \longrightarrow H_1(SP^k(C'))$ is necessarily surjective. It can be checked that as soon as $n \geq 2$, $\pi_1(SP^k_n(C'))$ abelianizes ([K3]) and from there that $H_1(SP^k_n(C')) \cong H_1(SP^k(C')) = \mathbb{Z}^{2g}$. ∎

Lemma 3.6: *Assume $g \geq 1$. The map $Map^*_k(C, \mathbb{P}^n) \xrightarrow{\alpha} Map^*_k(C, \mathbb{P}^\infty) \simeq (S^1)^{2g}$, induced from post-composition with the inclusion $\mathbb{P}^n \hookrightarrow \mathbb{P}^\infty$, is an isomorphism at the level of H_1 when $n > 1$ and a surjection when $n = 1$.*

PROOF: Consider the following homotopy diagram

$$\begin{array}{ccc} SP^\infty_n(C - *) & \xrightarrow{\subset} & SP^\infty(C - *) \\ \downarrow & & \downarrow \simeq \\ \mathrm{Map}^*_0(C, \mathbb{P}^n) & \xrightarrow{\alpha} & \mathrm{Map}^*_0(C, \mathbb{P}^\infty) \end{array}$$

The top map is an isomorphism on H_1 (and only a surjection when $n = 1$) according to the previous lemma. Since both vertical maps are isomorphisms on H_1 as well (by 3.4), the claim follows. ∎

§4 A Spectral Sequence of Milgram

The following spectral sequence appears in special cases in [BCM], [B], [K1] and [KM] and the main ideas trace back to [M1]. Let X be a space with basepoint $*$ and let $E(X) \hookrightarrow SP^\infty(\bigvee^k X)$ be a submonoid of $SP^\infty(\bigvee^k X) = \prod_k SP^\infty(X)$. Given a map $X \longrightarrow E(X)$, it can always be extended to a map $\nu_0 : SP^r(X) \longrightarrow E(X)$ (additively) and then to a map

$$\nu : \coprod_{r \geq 1} SP^r(X) \times E \longrightarrow E, \quad \nu(x, y) = \nu_0(x) + y$$

Our interest is to study the complement $Par(X) = E(X) - \mathrm{Im}(\nu)$ (or *Particle* space). We reserve the notation $C(X)$ for the standard configuration space of distinct points (see 4.1 below). Such generalized families of configuration spaces are studied in [K2].

EXAMPLES 4.1:

• (i) Let $k = 1$ and consider the map $M \longrightarrow E(M) = SP^\infty(M)$, $x \mapsto 2x$. Points in $\mathrm{Im}(\nu)$ are finite sums $\sum n_i x_i$ where $n_i \geq 2$ for at least one index i. In this case

$$Par(M) = C(M) = \{\sum x_i, x_i \neq x_j \text{ for } i \neq j\}$$

Similarly we can map $M \longrightarrow E$ by sending $x \mapsto (n+1)x$ and in this case the discriminant space is the space $SP^\infty_n(M)$ described in 3.4.

• (ii) The divisor space $\mathrm{Div}^n(M)$ introduced in §3 is the particle space obtained as the complement of the diagonal map

$$M \longrightarrow \prod^n SP^\infty(M) = E(M), \; x \mapsto (x, x, \ldots, x)$$

• (iii) If $\mu : M \longrightarrow G$ is a map into an abelian group G, then μ can be extended to $SP^\infty(M)$ and one can define

$$E(M,\mu) = \{(D_1,\dots,D_k) \in \prod_k SP^\infty(M) \mid \mu(D_i) = \mu(D_j) \text{ for all } i,j\}$$

This is a submonoid since by construction $\mu(D+D') = \mu(D) + \mu(D')$. The main example we study in this paper is when $M = C$ is a curve and μ is its Abel-Jacobi map.

Consider the following construction

$$DE(M) = E \times_\nu SP^\infty(cM)$$

where cM is the cone on M and where the twisted product $\times_\nu$ means the identification via ν of the points

$$(\vec{\zeta},(t_1,z_1),\dots,(t_l,z_l)) \sim (\vec{\zeta}+\nu(z_i),(t_1,z_1)\dots\widehat{(t_i,z_i)}\dots(t_l,z_l))$$

whenever $t_i = 0$ is at the base of the cone (here *hat* means deletion). The identification above "cones off" the image of ν in E and so we expect that $H_*(DE) = H_*(E/Im\nu)$. Write

$$E_{n_1,\dots,n_k} = E(M) \cap (SP^{n_1}(M) \times \cdots \times SP^{n_k}(M))$$

and suppose $\nu : M \longrightarrow E(M)$ lands in $E_{l_1,\dots,l_k}, l_j \geq 1$. We can then consider the filtration of DE by subspaces

$$DE_{k_0,k_1,\dots,k_n}(C_g) = \bigcup_{\substack{i_j+ll_j\leq k_j\\ 1\leq j\leq k}} E_{i_0,i_1,\dots,i_n} \times_\nu SP^l(cC_g)$$

There are (well-defined) projection maps

$$p_{k_0,k_1,\dots,k_n} : DE_{k_0,k_1,\dots,k_n} \longrightarrow E_{k_0,k_1,\dots,k_n}/\{\text{Image}(\nu)\}$$

sending $(v_1,\dots,v_s,(t_1,w_1),\dots,(t_r,w_r))$ to $(v_1,\dots,v_s)$. These maps are acyclic and hence induce isomorphisms in homology (the proof in the case of the standard configuration spaces 4.1 (i) is given in [BCM], lemma 4.6 and relies mainly on the fact that $SP^l(cX)$ is contractible for every l).

Let's assume $l_j = d, 1 \leq j \leq k$ and write $(D)E_{i,\dots,i} = (D)E_i$. Suppose $Par_i(M) = E_i - Im(\nu) \cap E_i$ is an oriented manifold of dimension $N(i)$ (it will be for all cases we consider here otherwise we can use $\mathbb{Z}_2$ coefficients). Then by Alexander-Poincaré duality we have an isomorphism $\tilde{H}^{N(i)-*}(Par_i(M);\mathbb{A}) \cong H_*(E_i/Im(\nu);\mathbb{A})$ for commutative rings $\mathbb{A}$. Combining this with the previous paragraph we get

Proposition 4.2: *There are isomorphisms*

$$\begin{aligned}\tilde{H}^{N(i)-*}(Par_i(M);\mathbb{A}) &\cong H_*(DE_i;\mathbb{A})\\ \tilde{H}^{N(i)-*}(Par_i(M-*);\mathbb{A}) &\cong H_*(DE_i/\bigcup_i DE_{i,\dots,i-1,\dots,i};\mathbb{A})\end{aligned}$$

Let LE_i be the quotient $E_{i,\dots,i}/\bigcup E_{i,\dots,i-1,\dots,i}$, $i\geq 1$. The map ν is made out of maps

$$\nu : SP^r(M)\times E_i \longrightarrow E_{i+rd}$$

and so we can consider the quotient model

$$QE_k = DE_{k,\dots,k}/\bigcup DE_{k,\dots,k-1,\dots,k} = \bigcup_{i+dj=k} LE_i\times_\nu (SP^j(cM)/SP^{j-1}(cM))$$

One can filter this complex by j (in which case one obtains an analog of the Eilenberg-Moore spectral sequence with E^2 term a Tor term; see [K1], [BCM]) or one can filter according to i by letting

$$\mathcal{F}_r = \bigcup_{\substack{i\geq r\\ i+dj=r}} LE_i \times_\nu (SP^j(cM)/SP^{j-1}(cM))$$

In this case we get the spectral sequence

Theorem 4.3: *There is a spectral sequence converging to $H^{N(k)-*}Par_k(M-*)$ with E^1 term*

$$E^1 = \coprod_{\substack{i+dj=k\\ i\geq 1}} H_*(LE_i;\mathbb{A})\otimes H_*(SP^j(\Sigma M), SP^{j-1}(\Sigma M);\mathbb{A})$$

and where the d^r differentials are obtained from a chain approximation of the identification maps $\nu : SP^j(M)\times LE_i \longrightarrow LE_{i+d}$, $(x,y)\longrightarrow \nu(x)+y$.

SKETCH OF PROOF: (see [K1] and [KM] for details) Consider $\mathcal{F}_r$ as described above. Then a chain complex for $\mathcal{F}_r$ is given as

$$\bigoplus C_*(LE_i)\otimes_\nu C_*(SP^l(cM)/SP^{l-1}(cM))$$

Here the symmetric product pairing as well as the identification given by t can be chosen to be cellular. We need determine the homology of this complex. An interesting fact is that one can construct chain complexes

$$C_*(SP^\infty(cM)) = C_*(SP^\infty(M))\otimes C_*(SP^\infty(\Sigma M)) \tag{4.4}$$

where the identification above is given as a bigraded differential algebra isomorphism. This follows from the fact (cf.[M1])) that $C_*(SP^\infty(cM))$ can be identified with the acyclic bar construction on $C_*(SP^\infty(M)$ and that $C_*(SP^\infty(\Sigma M))$ can be identified with the reduced bar construction on $C_*(SP^\infty(M)$. Therefore the boundary on cells of $\mathcal{F}_r$ can be made *explicit*. First a cell in $C_*(SP^l(cM)/SP^{l-1}(cM))$ can be written as

$$c_*\otimes |a_1|\cdots|a_l|$$

(according to the decomposition 4.4) and the boundary decomposes into

$$\partial = \partial c_*\otimes |a_1|\cdots|a_l| + \nu_*(c_*\otimes a_1)\otimes |a_2|\cdots|a_l| + c_*\otimes \partial_B(|a_1|\cdots|a_l|) \tag{4.5}$$

where ∂_B is the usual bar differential. The induced boundary on $\mathcal{F}_r/\mathcal{F}_{r+1}$ (which describes d_0) is given by the first and last term and the homology of the complex $\partial : C_*(\mathcal{F}_r/\mathcal{F}_{r+1})\rightarrow C_*(\mathcal{F}_{r+1}/\mathcal{F}_{r+2})$ is given by the expression in 4.3. Remains to identify the d^r differentials and these are deduced by the middle term in 4.5 according to the filtration term in which they land. ∎

§4.1 An application: Configurations with bounded multiplicity

The homology of $SP^{\infty}_n(C'), C' = C - *$ can be calculated from the filtration pieces $SP^k_n(C-*)$ as follows. Start by observing that the configuration space $SP^{\infty}_n(C)$ is the discrimant set in $SP^{\infty}(C-*)$ of the image of $\nu : SP^{\infty}(C) \longrightarrow SP^{\infty}(C)$ which is multiplication by $n+1$ (as described in 4.1 (i)). In this particular case 4.3 takes the form (with field coefficients)
(4.6) *There is a spectral sequence converging to* $H_{2k-*}(SP^k_n(C-*);\mathbb{F})$, $n \geq 1$, *with* $E^1_{i,j}$ *term (to which we refer as* $E^1(k)$ *)*

$$\bigoplus_{\substack{i+(n+1)j=k\\ r+s=*}} H_r(SP^i(C), SP^{i-1}(C);\mathbb{F}) \otimes H_s(SP^j(\Sigma C_g), SP^{j-1}(\Sigma C_g);\mathbb{F}).$$

To see this, one simply observes that $SP^k_n(C-*)$ is open in $SP^k(C-*)$ which is a k dimensional complex manifold and then one applies 4.3. We point out that such a spectral sequence was considered in the case of $n=1$ in ([BCM], theorem 4.1) and for $g=0$ and all n in [K3]. The differentials as explained in the proof of 4.3 are induced from a cellular approximation of the maps

4.7 $$\nu : SP^j(C) \longrightarrow SP^{(n+1)j}(C),\ x \mapsto (n+1)x$$

More explicitly in this case, a chain complex for a Riemann surface C of genus g is given by $2g$ one dimensional classes (which we label $e_1, \dots, e_{2g}$) and a two dimensional orientation class $a = [C]$. Now the homology of $SP^{\infty}(C)$ is generated as a ring by the symmetric products of these classes; i.e.

$$H_*(SP^{\infty}(C)) = E(e_1, \dots, e_{2g}) \otimes \Gamma[a]$$

and $H_*(SP^n(C))$ consists of all n-term products in the complex above (here E is an exterior algebra and Γ is divided power algebra; see [K1] for details). It turns out that this homology **embeds** in $C_*(C)$ and so one can think of the product of these classes as cells as well. We can investigate the boundary term 4.5 on these classes. The map ν is given by the composite $C \xrightarrow{\Delta} C^{\times n} \to SP^n(C)$. The primitive classes e_i map to $\sum 1 \otimes \cdots \otimes e_i \otimes \cdots \otimes 1$ and hence map into $H_*(C) \subset H_*(SP^{n+1}(C))$. For $n > 1$ they clearly vanish in $H_*(SP^{n+1}(C), SP^n(C))$ and so are not seen in the spectral sequence. The class a on the other hand maps via Δ to the class $\sum 1 \otimes \cdots \otimes 1 \otimes C \otimes 1 \otimes \cdots 1 + \sum 1 \otimes \cdots e_i \otimes \cdots \otimes e_j \otimes \cdots \otimes 1$ in $H_2(C^{n+1})$. The projection into $H_*(SP^{n+1}(C), SP^n(C))$ vanishes if $n > 1$ and is non trivial if $n = 1$. More explicitly we have

Lemma 4.8: *When* $n > 1$, *all* d^r *differentials vanish and the spectral sequence above collapses at the* E^1 *term. When* $n = 1$, *there are higher differentials generated by* $d^1(1 \otimes |a|) = 2\sum e_i e_{i+g}$, $1 \leq i \leq g$.

EXAMPLE 4.9 ($H_1(SP^k(C-*))$): There are $2g$ one dimensional (torsion free) classes $\tilde{e}_i$ in $H_1(SP^k_n(C-*))$ corresponding to the classes $e_i a^{k-1} \in E^1_{k,0}$ (which have dimension $2(k-1)+1$). Also and when $n=1$, the class $a^{k-2}|a|$

in E^1 gives rise to a generator in $H_1(SP_1^k(C-*))$ for all $k>1$. But $\partial|a| = 2\sum e_ie_{i+g}$ (according to 4.8) and hence this is a 2-torsion class; i.e.

$$H_1(SP_n^k(C-*);\mathbb{Z}) \cong \begin{cases} \mathbb{Z}^{2g}, & \text{when } n>1, k\geq 1 \\ \mathbb{Z}_2\oplus\mathbb{Z}^{2g}, & \text{when } k>n=1 \end{cases}$$

This also corresponds to $H_1(\mathrm{Map}_0^*(C,\mathbb{P}^n))$ as will be clear shortly (and as is expected according to 1.4!).

EXAMPLE 4.10: There are inclusions $H_*(SP_n^k(C')) \hookrightarrow H_*(SP_n^{k+1}(C'))$ induced by a map of E^1 terms (in 4.6 above)

$$x\otimes y \mapsto xa\otimes y$$

As a corollary one can easily deduce the following standard fact (for another argument, see [K3] this volume): *There are homology splittings (here $SP^0(C') = *$)*

$$H_*(SP_n^k(C')) \cong \bigoplus_{1\leq i\leq k} H_*(SP_n^i(C'), SP_n^{i-1}(C'))$$

§4.2 The homology of $\mathrm{Map}_0^*(C,\mathbb{P}^n)$

We now determine $H_*(\mathrm{Map}^*(C,\mathbb{P}^n);\mathbb{F})$ for $\mathbb{F}=\mathbb{Z}_2,\mathbb{Z}_p$ as a straightforward application of the calculations in §4.1. Recall that when $g>0$, the surface C_g is described topologically by a 2-disc attached to a bouquet of $2g$ circles via the mapping which wraps around as a product of commutators. More precisely one has a cofibration sequence

4.11 $$S^1 \xrightarrow{f} \bigvee^{2g} S^1 \overset{i}{\hookrightarrow} C_g \xrightarrow{\pi} S^2$$

where $i: \bigvee^{2g} S^1 \hookrightarrow C_g$ is the one skeleton inclusion, and the map f is given as a product of commutators $[x_1,x_2][x_3,x_4]\cdots[x_{2g-1},x_{2g}]$ (with the x_i's denoting the generators of $\pi_1(\bigvee_{2g} S^1) = \mathbb{Z}^{*2g}$). Applying the $\mathrm{Map}^*(-,\mathbb{P}^n)$ functor to 4.11 yields the fibration

4.12 $$\Omega^2(\mathbb{P}^n)\longrightarrow \mathrm{Map}^*(C_g,\mathbb{P}^n)\longrightarrow \Omega(\mathbb{P}^n)^{2g}$$

We have $H^*(\Omega(\mathbb{P}^n)^{2g}) = H^*(S^1)^{\otimes 2g}\otimes H^*(\Omega^2 S^{2n+1})$. Note that 4.12 has simple coefficients (since the fiber is a loop space). Write again (cf. §2) $H^*(\Omega^2S^3,\mathbb{Z}_p)$ as an exterior $E(x_1,\dots,x_{2p^{i+1}-1},\dots)$ tensor a truncated algebra $P_T(y_{2p-2},\dots,y_{2p^i-2},\dots)$ where the x's and y's are generators in the stated dimensions.

Theorem 4.13: *Assume $g>1$. Then the Serre spectral sequence for 4.12 collapses at E^2 when $n>1$. When $n=1$, the spectral sequence collapses*

with $\mathbb{F}_2$ coefficients but has the mod-p differentials ($p > 2$)

$$\begin{aligned} d_{p^i}(x_{2p^i-1}) &= \frac{1}{p^i}\left(\sum_1^g f_{2i-1}f_{2i}\right)^{p^i} \\ d_{p^i}(y_{2p^{i+1}-2}) &= \left[\frac{1}{p^i}\left(\sum_1^g f_{2i-1}f_{2i}\right)^{(p-1)p^i}\right] x_{2p^i-1}. \end{aligned}$$

where the f_i are the one dimensional generators of $\Omega(\mathbb{P}^1)^{2g} \simeq (S^1)^{2g} \times (\Omega S^3)^{2g}$.

PROOF: This is a counting argument. The homology splitting in 4.10 holds for $k = \infty$ and combining this with 1.7 (and field coefficients) we see that

$$H_*(\mathrm{Map}_k^*(C, \mathbb{P}^n)) \cong \bigoplus_{k \geq 1} H_*(SP_n^k(C'), SP_n^{k-1}(C'))$$

We simply need identify generators. From 4.6, we read off the correspondence between generators in E^∞ and generators in the serre spectral sequence as follows

$$\begin{aligned} e_i &\mapsto f_i \quad \text{and} \quad |a| \mapsto x_1 \\ |e_i|'s &\mapsto \text{generators of } H^2(\Omega S^3)^{2g} \end{aligned}$$

($|e_i|, |a| \in E(2)$ and $e_i \in E(1)$, but they propagate to $E(k)$ after multiplying by suitable powers of a as indicated in 4.10). Now a generates a divided power algebra in $H_*(SP^\infty(C))$ and hence $|a|$ generates an $H_*(K(\mathbb{Z}, 3))$ in $H_*(SP^\infty(\Sigma C))$ (as is known, the groups $H_*(K(\mathbb{Z}, 3))$ and $H^*(\Omega^2 S^3)$ are formally "dual" to each other). More precisely, we can construct a correspondence between the two as follows (with mod-p coefficients): let γ_i be the divided power generators in $\Gamma(a)$, then

$$|\gamma_{p^i}| \mapsto x_i, \quad |\gamma_{p^i}^{p-1}|\gamma_p| \mapsto y_i$$

The rest of the proof is now straightforward as the non-zero differentials for $p > 2$ are entirely generated by the one in 4.8 (compare [K1]). ∎

§5 Segal's Stability Theorem

In this section we derive Segal's result based on the spectral sequence in 4.3 which in this particular context takes the form

(5.1) *There is a spectral sequence converging to $H^*(Hol_k(C_g, \mathbb{P}^n))$, $k \geq 2g - 1$ with E^1 term*

$${}_hE^1 = \bigoplus_{\substack{i+j=k \\ i \geq 1}} E^1_{i,j} = \bigoplus_{\substack{i+j=k \\ i \geq 1}} H_{*'}(LE_i; \mathbb{A}) \otimes H_{*''}(SP^j(\Sigma C_g). SP^{j-1}(\Sigma C_g); \mathbb{A}).$$

and identifiable d^1 differential. Here $* = 2(k-g)(n+1) + 2g - *' - *''$.

The terms LE_i are constructed out of the Jacobi variety J and its stratifications. Recall that by construction E_i is the pull-back of the diagonal

$$\begin{array}{ccc} E_i & \longrightarrow & SP^i(C_g)\times\cdots\times SP^i(C_g) \\ \downarrow & & \Big\downarrow \underbrace{\mu\times\cdots\times\mu}_{n+1} \\ J & \stackrel{\Delta}{\longrightarrow} & J(C_g)\times\cdots\times J(C_g) \end{array} \tag{5.2}$$

The image under μ of $SP^i(C)$ in $J(C)$ is denoted by W_i. W_i has complex dimensions i and for for $i\geq g$, μ is surjective and $W_i=J$.

We nee recall at this stage that the Abel-Jacobi map $\mu : SP^i(C_g)\rightarrow J(C_g)$ is an analytic fibration in the "stable" range $i\geq 2g-1$ with fiber $\mathbb{P}^{i-g}$ (that the dimension of the fiber stabilizes is a consequence of Riemann-Roch. That μ is actually a fibration is a theorem of Mattuck [Ma]). So from 5.2 it is easy to deduce that for $i\geq 2g$

$$H_*(LE_i,\mathbb{A})\cong H_*(J(C_g),\mathbb{A})\otimes H_*(S^{2(i-g)(n+1)};\mathbb{A}) \tag{5.3}$$

(i.e. LE_i is obtained by first pulling back $SP^i(C)^{n+1}$ over $J(C)$ via Δ and then collapsing the fat wedge in the fiber $(\mathbb{P}^{n-i})^{n+1}$.)

Example 5.4: (the torsion free one-dimensional classes). Note that when $i\geq 2g$, we deduce from 5.3 that there are $2g$ classes in $H_*(LE_k)$ of dimension $2(k-g)(n+1)+2g-1$ which yield $2g$ torsion free generators in $H_1(\mathrm{Hol}_k^*)$.

Stable Classes and Segal's theorem: When $i\geq 2g-1$ (the *stable range*), the terms in ${}_hE^1_{i,j}$ that survive to ${}_hE^\infty$ yield 'stable' classes in $H^*(\mathrm{Hol}^*)$ (and hence in $H_*(\mathrm{Hol}^*)$). The following is shown in ([KM], lemma 8.3)

Proposition 5.5: *The induced map $H_*(Hol_k^*(C_g,\mathbb{P}^n))\longrightarrow H_*(Map_k^*(C_g,\mathbb{P}^n))$ is an isomorphism on the stable classes.*

Proof (sketch of alternate argument): When $i\geq 2g-1$, E_i is an $(i-g)(n+1)+2g$ complex and maps to $SP^{(i-g)(n+1)+2g}(C)$ (via a factoring of the map $E_i\rightarrow SP^i(C)^{n+1}\stackrel{+}{\longrightarrow}SP^{i(n+1)}(C)$ (where $+$ is concatenation). It follows that the stable terms in ${}_hE^1$ (corresponding to $i\geq 2g-1$) map to $E^1(N)$ (see 4.6; here $N=(k-g)(n+1)+2g$)). The main point is that in the stable range the spectral sequence in 5.1 and the one in 4.6 behave identically (same differentials described as in 4.8 above and [KM], lemma 7.8). The stable terms surviving to ${}_hE^\infty$ therefore map isomorphically to $E^\infty(N)$ and so the dual "stable classes" in $H_*(\mathrm{Hol}_k^*)$ have isomorphic image in $H_*(SP^k_n(C-*))$; i.e. in $H_*(SP^\infty_n(C-*))\cong H_*(\mathrm{Map}_0^*(C,\mathbb{P}^n)$. Details omitted. ∎

The unstable terms which appear when $i<2g-1$ in the spectral sequence above form the interesting part and are harder to track down for their existence depends strongly on the geometry of the curve. It follows however that their contribution only starts appearing above a certain range and hence up to that range the homology of $\mathrm{Hol}_k(C_g,\mathbb{P}^n)$ only consists of stable classes and by 5.5 must coincide with that of $\mathrm{Map}_k(C_g, \mathbb{P}^n)$). This is exactly the essence of the stability theorem of Segal and we fill in the details below.

Proposition 5.6: $H_*(LE_i) = 0$ for $* > i(n+1) + 2g$ and for all $1 \leq i \leq 2g$.

PROOF: Define

$$W_i^r = \{x \in \mu(SP^i(C_g)) \mid \mu^{-1}(x) = \mathbb{P}^m, m \geq r\}$$

A well-known theorem of Clifford asserts that in the range $1 \leq i \leq 2g$ the maximum of r is $i/2$; that is $W_i^r = \emptyset$ for $r > \frac{i}{2}$. One can then filter $\mu(SP^i(C)) = W_i$ by the descending filtration

$$W_i^{\frac{i}{2}} \subset \cdots \subset W_i^1 \subset W_i$$

This leads to a spectral sequence converging to $H_*(LE_i)$ with E^1 term

$$\sum_r^{\frac{i}{2}} H_*(W_i^r, W_i^{r+1}) \otimes H_*(S^{2r(n+1)})$$

Since the $W_i^r - W_i^{r+1}$ are algebraic subvarieties of $J(C_g)$ ([ACGH]) we must have that $H_*(W_i^r, W_i^{r+1}) = 0$ for $* > 2g$. This means that each term in the E^1 term above has no homology beyond $2\frac{i}{2}(n+1) + 2g$ and hence

$$H_*(LE_i) = 0 \text{ for } * > 2\frac{i}{2}(n+1) + 2g$$

as asserted. Note that when $i = 2g$, we have that m is uniformly $i - g = g = \frac{i}{2}$ and the proposition is still valid in this case. ∎

Corollary 5.7: (Segal) *The inclusion*

$$Hol_k(C_g, \mathbb{P}^n) \hookrightarrow Map_k(C_g, \mathbb{P}^n)$$

is a homology isomorphism up to dimension $(k - 2g)(2n - 1)$

PROOF: Look in 4.3 at the unstable terms given by

$$H_*(LE_i) \otimes H_{*'}(SP^j(\Sigma C_g), SP^{j-1}(\Sigma C_g)),\ i \leq 2g - 1$$

According to 5.6 these terms vanish for $* > i(n+1) + 2g$. By duality, they contribute therefore no homology to $H_*(\mathrm{Hol}_k(C_g, \mathbb{P}^n); \mathbb{A})$ for

$$* \leq [2(k-g)n + 2k] - [i(n+1) + 2g + 3(k-i)]$$

The expression on the left attains its minimum when $i = 2g$ that is when

$$[2(k-g)n + 2k] - [2g(n+1) + 2g + 3(k-2g)] = (k-2g)(2n-1).$$

So when $* < (k-2g)(2n+1)$ the unstable terms do not contribute to $H_*(\mathrm{Hol}_{k=i+j}(C_g, \mathbb{P}^n); \mathbb{A})$ and the proof now follows from proposition 5.5. ∎

§6 The Fundamental Group

Since the attaching map of the two cell of C_g is given by a product of commutators corresponding to Whitehead products (see 4.11), its suspension must be null-homotopic. This implies that ΣC_g splits as a wedge $\Sigma C_g \simeq \bigvee S^2 \vee S^3$ and one has

6.1 $$\Sigma^i C_g \simeq \bigvee_1^{2g} S^{i+1} \vee S^{i+2},\ i \geq 1.$$

Lemma 6.2: *All components of Map*$^*(C_g, X)$ *are homotopy equivalent, and* $\forall i > 1$, *there is an isomorphism of homotopy groups*

$$\pi_i(Map_0^*(C_g, X)) = \bigoplus^{2g} \pi_{i+1}(X)^{2g} \oplus \pi_{i+2}(X).$$

PROOF: The statement about components is standard and we denote by $\mathrm{Map}_0^*(C_g, X)$ the component of null-homotopic maps. On the other hand and by definition $\pi_i(\mathrm{Map}_0^*(C_g, X)) = [S^i \wedge C_g, X]_*$. The splitting 6.1 yields (*as sets*) the bijection

$$[S^i \wedge C_g, X]_* = [\bigvee_{2g} S^{i+1} \vee S^{i+2}, X]_* = \pi_{i+2}(X) \oplus \bigoplus^{2g} \pi_{i+1}(X)$$

In order for the decomposition above to induce a group homomorphism, it is necessary that 6.1 be a decomposition as "co-H spaces", that is that there is a map $f_i : \Sigma^i C_g \xrightarrow{f_i} \bigvee_1^{2g} S^{i+1} \vee S^{i+2}$ which commutes with the pinch maps up to homotopy. When $i \geq 2$, $f_i = \Sigma f_{i-1}$ is a suspension and hence is automatically a co-H map. The claim follows. ∎

When $i = 1$, the decomposition in 6.2 is not necessarily a group decomposition. The long exact sequence in homotopy associated to $\Omega^2 X \to \mathrm{Map}^*(C_g, X) \to (\Omega X)^{2g}$ together with 6.2 indicate however that there is a short exact sequence of abelian groups

$$0 \longrightarrow \pi_3(X) \longrightarrow \pi_1(\mathrm{Map}_0^*(C_g, X)) \longrightarrow \pi_2(X)^{2g} \longrightarrow 0$$

When $X = \mathbb{P}^n$, $\pi_2(\mathbb{P}^n) \cong \mathbb{Z}$ and $\pi_3(\mathbb{P}^n) \cong \pi_3(S^{2n+1})$ which is zero when $n > 1$. This shows that

6.3(a) $$\pi_1 \mathrm{Map}_0^*(C, \mathbb{P}^n) = \mathbb{Z}^{2g}, \text{ when } n > 1$$

When $n = 1$, there is a short exact sequence

6.3(b) $$0 \longrightarrow \mathbb{Z} \longrightarrow \pi_1(\mathrm{Map}_0^*(C, \mathbb{P}^1)) \longrightarrow \mathbb{Z}^{2g} \longrightarrow 0$$

This is a central extension which turns out to be non-trivial as we now show.

First observe that any Riemann surface has meromorphic functions for sufficiently high degrees k (this is certainly true when $k \geq 2g$ and we will assume this is the case throughout the section). Let f be any such function and construct the map $\mathrm{Rat}_1(\mathbb{P}^1) \longrightarrow \mathrm{Hol}_k^*(C, \mathbb{P}^1)$ by post-composition with

f. $\mathrm{Rat}_1(\mathbb{P}^1)$ is the set of pairs of distinct points (a root and a pole) in $\mathbb{C}$ and has the homotopy type of S^1.

Lemma 6.4: *The inclusion* $f^! : Rat_1(\mathbb{P}^1)\longrightarrow Hol_k^*(C,\mathbb{P}^1)$ *induces an injection* $\mathbb{Z} \hookrightarrow \pi_1(Hol_k^*(C,\mathbb{P}^1))$.

PROOF: Observe that the diagram below commutes strictly

$$\begin{array}{ccc} \mathrm{Rat}_1(\mathbb{P}^1) & \xrightarrow{f^!} & \mathrm{Hol}_k^*(C,\mathbb{P}^1) \\ \downarrow{\scriptstyle i_1} & & \downarrow{\scriptstyle i_2} \\ \Omega_1^2 S^2 & \xrightarrow{g} & \mathrm{Map}_k^*(C,S^2) \end{array} \tag{6.5}$$

The vertical maps are inclusions. The bottom map on π_1 is an injection according to 6.3(b) and hence $g\circ i_1$ is also and injection on π_1 (since i_1 on π_1 is an isomorphism between two copies of $\mathbb{Z}$). The composition $i_2\circ f^!$ is therefore injective on π_1 and hence so is $f^!$ as desired. ∎

Next and from the description of Hol_k^* and $\mathrm{Div}_{k,k}(C-*) = SP^k(C-*)^2-\Delta$ where Δ is the generalized diagonal consisting of pairs of divisors with (at least) one point in common (see §3), we have the pullback diagram

$$\begin{array}{ccc} \mathrm{Hol}_k^*(C,\mathbb{P}^1) & \xrightarrow{\phi} & \mathrm{Div}_{k,k}(C-*) \\ \downarrow{\scriptstyle \pi} & & \downarrow{\scriptstyle \mu^2} \\ J(C) & \xrightarrow{\Delta} & J(C)^2 \end{array} \tag{6.6}$$

where $\pi : \mathrm{Hol}_k^*(C,\mathbb{P}^n)\xrightarrow{p} SP^k(C-*)\xrightarrow{\mu} J(C)$ is the map that sends a holomorphic map to the equivalence class of its divisor of zeros and Δ is the diagonal. The top horizontal map ϕ is an inclusion. We denote by F the homotopy fiber of μ^2 and π.

Lemma 6.7: *The map* $p : Hol_k^*(C,\mathbb{P}^1)\longrightarrow SP^k(C-*)$ *is surjective at the level of fundamental groups.*

PROOF: For $k\geq 2g-1$ the restriction of the Mattuck fibration (cf. §5) to $SP^k(C-*)$ becomes a complex vector bundle

$$\mathbb{C}^{k-g}\longrightarrow SP^k(C-*)\xrightarrow{\mu} J(C)$$

(i.e. the set of linearly equivalent divisors avoiding a point is a hyperplane in $\mathbb{P}^{k-g}$). It follows that μ_* is an isomorphism on $\pi_1(SP^k(C-*))$. Since the fiber of π is connected (see 6.10), it follows that π is surjective at the level of π_1 and hence is p. ∎

Theorem 6.8: *Suppose* $k\geq 2g$. *Then* $\pi_1(Hol_k^*(C,\mathbb{P}^1))$ *is generated (multiplicatively) by classes* $e_1,\dots,e_{2g}$ *and* τ *such that the commutators*

$$[e_i,e_{g+i}]=\tau^2$$

and all other commutators are zero.

We will prove this in a series of lemmas. To begin with, lemmas 6.4 and 6.7 show that there is a sequence

$$0 \longrightarrow \mathbb{Z} \xrightarrow{f_*} \pi_1 \mathrm{Hol}_k^*(C, \mathbb{P}^1) \xrightarrow{p_*} \mathbb{Z}^{2g} \longrightarrow 0 \tag{6.9}$$

which is exact at both ends. In 6.13 we show that this sequence maps to 6.3(b) and hence $\mathrm{Im}(f_*) \subset \mathrm{Ker}\,(p_*)$. That $\mathrm{Ker}(p) \subset \mathrm{Im}(f_*)$ follows from 6.11 and the discussion next.

Lemma 6.10: *$\pi_1(Div_{k,k}(C-*)), k>1$ has torsion free generators $a_i, b_j, 1 \leq i,j \leq 2g$ and τ. The a_i (resp. b_i) are represented by a root (resp. a pole) going around the generators in a symplectic basis of $H_1(C-*)$, while τ is given by a zero (or pole) going around a pole (resp. a zero).*

PROOF: There are a few ways to see that there is a short exact sequence

$$0 \longrightarrow \mathbb{Z} \longrightarrow \pi_1(\mathrm{Div}_{k,k}(C-*)) \longrightarrow \mathbb{Z}^{4g} \longrightarrow 0$$

with generators having the desired properties.

• By considering the map μ^2 in 6.6 and analyzing the fiber F given by the complement of a hyperplane in $\mathbb{C}^{k-g} \times \mathbb{C}^{k-g} = \mathbb{C}^{2(k-g)}$ (the fiber of μ restricted to $SP^k(C-*)$ is $\mathbb{C}^{k-g}$ as indicated in 6.7). Here $\pi_1(F) = \mathbb{Z}$.

• Mimicking the proof of Jones for the case $C = \mathbb{P}^1$. One considers the subspace U of $\mathrm{Div}_{k,k}(C-*))$ consisting of pairs (D_0, D_1) with the roots of D_1 all distinct. This is the complement of a (real) codimension 2 subspace and there is a surjection $\pi_1(U) \longrightarrow \pi_1(\mathrm{Div}_{k,k}(C-*))$. Now there is a fibration $SP^k(C-Q_{k+1}) \longrightarrow \pi_1(U) \longrightarrow \pi_1(F(C-*,k)$ with section (where Q_{k+1} is a set of $k+1$ distinct points). Since $\pi_1(SP^k(C-Q_{k+1})) = H_1(C-Q_{k+1}) = \mathbb{Z}^{2g} \times \mathbb{Z}^k$, there is a semi-direct product of $\mathbb{Z}^{2g} \times \mathbb{Z}^k$ with the ordered braid group $B_k(C-*)$ with quotient at least $\mathbb{Z}^{4g}$. A similar argument as in [S], lemma 6.4 implies the answer. ∎

Lemma 6.11: *$\pi_1(Hol_k^*(C, \mathbb{P}^1))$ has generators τ and $e_i, 1 \leq i \leq g$ which map under $\phi_* : \pi_1(Hol_k^*(C, \mathbb{P}^1)) \hookrightarrow \pi_1(Div_{k,k}(C-*))$ as follows: $\phi_*(\tau) = \tau$ (the generator of the same name) and $\phi_*(e_i) = a_i b_i$. In particular ϕ_* is an injection.*

PROOF: Diagram 6.6 induces a map of fibrations and since $\pi_2(J) = 0$, a map of short exact sequences

$$\begin{array}{ccccccccc}
0 & \longrightarrow & \mathbb{Z} & \longrightarrow & \pi_1(\mathrm{Hol}_k^*(C, \mathbb{P}^1)) & \longrightarrow & \mathbb{Z}^{2g} & \longrightarrow & 0 \\
 & & \downarrow = & & \downarrow \phi_* & & \downarrow \Delta_* & & \\
0 & \longrightarrow & \mathbb{Z} & \longrightarrow & \pi_1(\mathrm{Div}_{k,k}(C-*)) & \longrightarrow & \mathbb{Z}^{4g} & \longrightarrow & 0
\end{array}$$

The map Δ_* is an injection and then so is the middle map. The generator $\tau \in \pi_1(\mathrm{Hol}_k^*)$ constructed in 6.4 corresponds exactly to the generator coming from $\pi_1(F)$ and is described by zeros linking with poles. The rest of the claim follows from the commutativity of the above diagram. ∎

PROOF OF THEOREM 6.8: The generators of π_1 and their images under ϕ having been determined, we next look at the image of the commutators

$$\phi([e_i,e_j]) = a_i\,\underbrace{b_i a_j}\,b_j b_i^{-1}\,\underbrace{a_i^{-1}b_j^{-1}}\,a_j^{-1}$$

Suppose we know that for $k\geq 2$ and $1\leq i,j\leq g$

6.12 $$a_ib_ja_i^{-1}b_j^{-1} = \begin{cases}\tau, & |i-j|=g\\ 1, & \text{otherwise}\end{cases} \in \pi_1(\mathrm{Div}_{k,k}(C-*)),$$

then $\phi[e_i,e_j]$ becomes (after transforming the underbraced terms) and for $|i-j|=g$

$$\begin{aligned}\phi([e_i,e_j]) &= a_i\tau a_jb_ib_jb_i^{-1}b_j^{-1}a_i^{-1}\tau a_j^{-1}\\ &= \tau^2a_ja_i[b_i,b_j]a_i^{-1}a_j^{-1}\\ &= \tau^2[a_i,a_j]=\tau^2\end{aligned}$$

This is the claim in 6.8 and therefore the theorem would follow as soon as we prove 6.12.

Consider the sub-divisor space $\mathrm{Div}^2_{k,1}(C-*))\subset \mathrm{Div}_{k,k}(C-*))$ consisting of pairs $(D_1,q)\in SP^k(C-*)\times(C-*)$ with the property that $q\notin D_1$ (here again $k>1$). There is a projection and a fibration

$$SP^k(C-\{*,q\})\longrightarrow \mathrm{Div}_{k,1}(C-*)\longrightarrow (C-*)$$

The generators $b_i\in\pi_1(C-*)$ act on the fundamental group of the fiber $\pi_1(SP^k(C-\{*,q\})$ and this action describes the commutator map in 6.12. Since $\pi_1(SP^k(C-\{*,q\})$ is abelian, we can look at the action in homology. This has been already determined in ([C^{2}M^2], prop. 10.12) and is exactly given by 6.12 (b_i is τ_{1i} in their notation). This completes the proof. ∎

Proposition 6.13: *For $k\geq 2g$ and all $n\geq 1$, we have an isomorphism*

$$\pi_1(Hol_k^*(C,\mathbb{P}^n))\cong \pi_1(Map_0^*(C,\mathbb{P}^n))$$

PROOF: When $n>1$, one can show that $\pi_1(\mathrm{Hol}_k^*(C,\mathbb{P}^n))$ is abelian (this has been proven in [Ki], lemma 3.5). The equivalence in π_1 when $n>1$ becomes an equivalence at the level of H_1 which we know is true by Segal's theorem. In fact when $n>1$, $\pi_1(\mathrm{Hol}_k^*(C,\mathbb{P}^n))=\pi_1(\mathrm{Map}_k^*(C,\mathbb{P}^n))\cong\mathbb{Z}^{2g}$ by 6.3(a).

Suppose $n=1$. The proof of 6.11 shows that the sequence in 6.9 is actually exact. We would like to see that 6.9 maps to the exact sequence in 6.3(b). The essential point is to show that the following homotopy commutes

$$\begin{array}{ccc}\mathrm{Hol}_k^*(C,\mathbb{P}^1) & \xrightarrow{\quad\quad p\quad\quad} & SP^\infty(C-*)\\ \Big\downarrow i & & \Big\downarrow \simeq\\ \mathrm{Map}_k^*(C,\mathbb{P}^1) & \longrightarrow(\Omega S^2)^{2g}\longrightarrow & (\Omega\mathbb{P}^\infty)^{2g}\end{array}$$

where again p sends a holomorphic map to the divisor of its zeros, the bottom composite $g:\mathrm{Map}_k^*(C,\mathbb{P}^1)\longrightarrow(\Omega\mathbb{P}^\infty)^{2g}$ is restriction to the one skeleton followed by postcomposition with $S^2 \hookrightarrow \mathbb{P}^\infty$. Since both spaces on the far right are Eilenberg-Maclane spaces $(S^1)^g$, one simply need consider the effect of p and $g \circ i$ on cohomology (recall that i is an isomorphism on H^1 by 1.3). Label generators of $H^1((S^1)^{2g})$ by $f_i, 1 \leq i \leq 2g$. As is explicit in theorem 4.13, the $g^*(f_i)$ correspond to the non-trivial torsion free classes in $H^1(\mathrm{Map}_k^*)$ (see 4.13 and 4.9). Similarly the construction of the stable classes in §5 (see example 5.4) implies precisely that $p^*(f_i)$'s are the torsion free generators of $H^1(\mathrm{Hol}_k^*)$. That is $p^* = i^* \circ g^*$ and hence $p \simeq i \circ g$.

Now and upon applying π_1 to the terms of the diagram above as well as to those in 6.5 we get a map from 6.9 down to 6.3(b). Since the end terms of the sequences are isomorphic, it follows by the five lemma that the middle terms are isomorphic; i.e. $\pi_1(\mathrm{Hol}_k^*) \cong \pi_1(\mathrm{Map}_k^*)$. ∎

Corollary 6.14: *$\pi_1(Map_0^*(C, \mathbb{P}^n))$ is generated by classes $e_1, \dots, e_{2g}$ and α such that the commutators $[e_i, e_{g+i}] = \alpha^2$ and all other commutators are zero.*

NOTE: Proposition 6.13 is not sufficient to imply that Segal's homology equivalence 5.7 can be upgraded to a homotopy equivalence. A detailed study of the actions of π_1 on the universal covers is needed.

§7 Spaces of Unbased Maps

There is an important distinction between $\mathrm{Map}^*(C, X)$ and $\mathrm{Map}(C, X)$. For one thing, while the topology of the based mapping space doesn't vary with the degree, this is no longer true in the unbased case. Let $\mathrm{Map}_f(C_g, X)$ denote the component containing a given map f, $x_0 \in C_g$ the basepoint and consider again the "evaluation" fibration

7.1 $$\mathrm{Map}_f^*(C_g, X) \longrightarrow \mathrm{Map}_f(C_g, X) \xrightarrow{ev} X \ , \ ev(g) = g(x_0)$$

Associated to 7.1 is a long exact sequence of homotopy groups

$$\to \pi_i \mathrm{Map}_f^*(C_g, X) \to \pi_i \mathrm{Map}_f(C_g, X) \to \pi_i(X) \xrightarrow{\partial} \pi_{i-1} \mathrm{Map}_f^*(C_g, X) \to$$

Suppose X is simply connected. Notice that given a map $f : C_g \longrightarrow X$, f is null on $\bigvee S^1$ and hence factors (up to homotopy) as in

7.2 $$f : C_g \longrightarrow S^2 \xrightarrow{\bar{f}} X$$

One then has the following diagram of vertical fibrations over X

7.3 $$\begin{array}{ccccc} \Omega_{\bar{f}}^2 X & \longrightarrow & \mathrm{Map}_f^*(C_g, X) & \longrightarrow & (\Omega X)^{2g} \\ \downarrow & & \downarrow & & \downarrow \\ \mathcal{L}_{\bar{f}}^2 X & \longrightarrow & \mathrm{Map}_f(C_g, X) & \longrightarrow & \mathrm{Map}(\bigvee S^1, X) \\ \downarrow{\scriptstyle ev} & & \downarrow{\scriptstyle ev} & & \downarrow{\scriptstyle ev} \\ X & \xrightarrow{=} & X & \xrightarrow{=} & X \end{array}$$

inducing a diagram of homotopy groups and boundary homomorphisms

$$\begin{array}{ccccc} \pi_i(X) & \xrightarrow{=} & \pi_i(X) & \xrightarrow{=} & \pi_i(X) \\ \downarrow \partial_1 & & \downarrow \partial & & \downarrow \partial_2 \\ \pi_{i-1}(\Omega^2(X)) = \pi_{i+1}(X) & \longrightarrow & \pi_{i-1}(\mathrm{Map}^*_f(C_g, X)) & \longrightarrow & \pi_{i-1}(\Omega_{fi}X)^{2g} = \pi_i(X)^{2g} \end{array}$$

Proposition 7.4: *Assume X simply connected, $f : C_g \longrightarrow X$ and $\bar{f}$ such that $C_g \xrightarrow{\pi} S^2 \xrightarrow{\bar{f}} X$ (up to homotopy). Assume $i > 2$. Then the homotopy boundary*

$$\partial : \pi_i(X) \longrightarrow \pi_{i-1}(Map^*_f(C_g, X)) = \pi_i(X)^{2g} \oplus \pi_{i+1}(X)$$

for the fibration $Map^*_f(C_g, X) \longrightarrow Map_f(C_g, X) \xrightarrow{ev} X$ factors through the term $\pi_{i+1}(X)$ and is given (up to sign) by the Whitehead product: $\partial\alpha = [\alpha, \bar{f}]$.

PROOF: The sequence $0 \to \pi_{i+1}(X) \to \pi_{i-1}(\mathrm{Map}^*_f(C_g, X)) \to \pi_i(X)^{2g} \to 0$ splits canonically (according to 6.2) and with respect to that splitting one has $\partial = \partial_1 + \partial_2$. By Whitehead's theorem we have that $\partial_1\alpha = [\alpha, \bar{f}]$ while $\partial_2\alpha = 0$. The proof follows. ∎

REMARK: When $i = 2$ the situation has to be handled differently for $\pi_1(\mathrm{Map}^*_f(C_g, X))$ does not necessarily split as in 7.4.

Lemma 7.5: $\pi_1(\mathrm{Map}_d(C, \mathbb{P}^n)) \cong \mathbb{Z}^{2g}$ when $n > 1$.

PROOF: Consider the Hopf fibration $S^{2n+1} \longrightarrow \mathbb{P}^n \longrightarrow \mathbb{P}^\infty$ and the induced fibration

$$\mathrm{Map}(C, S^{2n+1}) \longrightarrow \mathrm{Map}_d(C, \mathbb{P}^n) \longrightarrow \mathrm{Map}_d(C, \mathbb{P}^\infty) \simeq (S^1)^{2g} \times \mathbb{P}^\infty$$

Since $\pi_1(\mathrm{Map}(C, S^{2n+1})) = 0$ when $n > 1$, the result follows. ∎

We next address the case $n = 1$ and analyze 7.3 when $X = \mathbb{P}^1$. Let $f : S^2 \longrightarrow X$ be a degree d map. The boundary term in the left vertical fibration $\pi_1(\Omega^2_f X) \xrightarrow{\partial} \pi_1(\mathcal{L}^2_f(X))$ is given according to 7.4 by multiplication by $2d$. On the other hand, the right vertical fibration has a section and hence we get the diagram of fundamental groups

$$\begin{array}{ccccccccc} 0 & \longrightarrow & \mathbb{Z} & \longrightarrow & \pi_1(\mathrm{Map}^*_d) & \longrightarrow & \mathbb{Z}^{2g} & \longrightarrow & 0 \\ \downarrow & & \downarrow & & \downarrow & & \downarrow = & & \downarrow \\ 0 & \longrightarrow & \mathbb{Z}_{2d} & \longrightarrow & \pi_1(\mathrm{Map}_d) & \longrightarrow & \mathbb{Z}^{2g} & \longrightarrow & 0 \end{array}$$

Corollary 7.6: (Larmore-Thomas) *$\pi_1(Map_d(C, S^2))$ is generated by classes $e_1, \dots, e_{2g}$ and α such that*

$$\alpha^{2|d|} = 1, \ [e_i, e_{g+i}] = \alpha^2$$

All other commutators are zero. In particular $Map_d(C, \mathbb{P}^1)$ and $Map_{d'}(C, \mathbb{P}^1)$ have different homotopy types whenever $d \neq \pm d'$.

Corollary 7.7: *For $d \geq 2g$, $\pi_1(Hol_d(C, \mathbb{P}^1))$ is generated by $e_1, \dots, e_{2g}$ and α with $\alpha^{2|d|} = 1$, $[e_i, e_{g+i}] = \alpha^2$ and all other commutators are zero.*

PROOF: This is equivalent to showing that $\pi_1(\mathrm{Hol}_d(C,\mathbb{P}^1))$ corresponds to $\pi_1(\mathrm{Map}_d(C,S^2))$ in that range but this follows from a direct comparison of the evaluation fibrations for both mapping spaces and theorem 1.4. ∎

In the case of maps into $\mathbb{P}^n$, $n\geq 2$, the fundamental group is not enough to distinguish between connected components. Proposition 7.4 still implies

Proposition 7.8: $\mathrm{Map}_k(C_g,\mathbb{P}^{2d})$ and $\mathrm{Map}_l(C_g,\mathbb{P}^{2d})$ have different homotopy types whenever $l\not\equiv k(2)$.

PROOF: The long exact sequence for the evaluation fibration gives again

$$\cdots\pi_{2d+1}\mathbb{P}^{2d}\xrightarrow{\partial}\pi_{2d}(\mathrm{Map}_k^*)\to\pi_{2d}(\mathrm{Map}_k)\to\pi_{2d}(\mathbb{P}^{2d})\xrightarrow{\partial}\cdots$$

where Map_k^* stands for $\mathrm{Map}^*(C_g,\mathbb{P}^{2d})$. According to 7.4 the sequence above becomes

$$\longrightarrow\pi_{2d+1}\mathbb{P}^{2d}=\mathbb{Z}\langle\eta\rangle\xrightarrow{k[\iota,\eta]}\pi_{2d+2}(\mathbb{P}^{2d})=\mathbb{Z}_2\to\pi_{2d}(\mathrm{Map}_k)\to 0$$

where ι is the generator of $\pi_2(\mathbb{P}^{2d})$. The Whitehead product pairing for complex projective spaces

$$\pi_{2m+1}(\mathbb{P}^m)\otimes\pi_2(\mathbb{P}^m)\longrightarrow\pi_{2m+2}(\mathbb{P}^m)$$

is worked out in [P]. The result there is that the Whitehead product is zero if m is odd and non-zero if m is even. Since we are in the case $m=2d$, the map $\mathbb{Z}\langle\eta\rangle\xrightarrow{k[\iota,\eta]}\mathbb{Z}_2$ is in fact multiplication by k and the proposition follows right away. ∎

References

[1] [ACGH] E. Arbarello, M. Cornalba, P.A. Griffiths, J. Harris, *Geometry of Algebraic Curves Volume I*, Springer-Verlag, Grundlehren **267** (1985).

[2] [B], Ross Biro, "The structure of cyclic configuration spaces and the homotopy type of Kontsevich's orientation space", Stanford thesis (1994).

[3] [BCM] C.F. Bodigheimer, F.R. Cohen, R.J. Milgram, "Truncated symmetric products and configuration spaces", Math. Zeit., **214** (1993), 179–216.

[4] [BHMM] C. Boyer, J. Hurtubise, B. Mann, R.J. Milgram, "The topology of instanton moduli space: the Atiyah-Jones conjecture", Ann. Math. **137** (1993) 561–609.

[5] [BM] C. Boyer, B. Mann, "Monopoles, non-linear σ models and two fold loop spaces", Comm. Math. Phys. **115** (1988), 571–594.

[6] [C^2M^2] F.R. Cohen, R.L. Cohen, B.M. Mann, R.J. Milgram, "The topology of rational functions and divisors of surfaces", Acta Math., **166**(1991), 163–221.

[7] [CM] R.L. Cohen, R.J. Milgram, "The homotopy types of gauge theoretic moduli spaces", MSRI research publications.

[8] [CS] R.L. Cohen, D.H. Shimamoto, "Rational functions, labelled configurations, and Hilbert schemes", J. London. Math. Soc. **43** (1991) 509–528.

[9] [G] P.A. Griffiths, "Introduction to algebraic curves", Transl. of Math. Monographs, AMS, vol 76 (1989).

[10] [GKY] M.A. Guest, A. Koslowski, K. Yamaguchi, "The space of polynomials with roots of bounded multiplicity", Fund. Math. **161** (1999), no. 1-2, 93–117.

[11] [Ha] V. Hansen, "On the space of maps of a closed surface into the 2-sphere", Math. Scand. **35** (1974), 149–158.

[12] [H1] J.W. Havlicek, "The cohomology of holomorphic self maps of the Riemann sphere", Math. Z. **218** (1995), 179–190.

[13] [H2] J.W. Havlicek, thesis, Stanford University 1994.

[14] [Hi] N.J. Hitchin, "The Geometry and topology of moduli space", Springer Lect. Notes. Math. **1451** 1–49.

[15] [Hu] J.C. Hurtubise, "Moduli spaces and particle spaces", CRM publications, NATO series.

[16] [K1] S. Kallel, "Divisor spaces on punctured Riemann surfaces", Trans. Am. Math. Soc. **350** (1998), 135–164.

[17] [K2] S. Kallel, "Particle spaces on manifolds and generalized Poincaré dualities", Quart. J. of Math. **52**, part 1 (2001), 45–70.

[18] [K3] S. Kallel, "Bounded multiplicity polynomials and rational maps", to appear in Ann. Inst. Fourier.

[19] [K4] S. Kallel, "An analog of the May-Milgram model for configurations with multiplicities", this volume.

[20] [KS] S. Kallel and P. Salvatore, work in progress.

[21] [KM] S. Kallel, R.J. Milgram "The geometry of spaces of holomorphic maps from a Riemann surface into complex projective space", J. Diff. Geometry **47** (1997) 321–375.

[22] [Ki] F.C. Kirwan, "On spaces of maps from Riemann surfaces to Grassmannians and applications to the cohomology of moduli of vector bundles", Ark. Math. **24** 2, (1986), 221–275.

[23] [Ka] Y. Kamiyama, "The modulo 2 homology of the space of rational functions", Osaka J. Math. **28** (1991), 229–242.

[24] [L] B. Lawson, "Spaces of algebraic cycles: levels of holomorphic approximation", Proc. of the Int. Math. Congr. Vol. 1, 2 (Zrich, 1994), 574–584, Birkhuser 1995.

[25] [LM] L.L. Larmore, E. Thomas, "On the fundamental group of a space of sections", Math. Scand. 47 (1980), no. 2, 232–246.

[26] [Ma] A. Mattuck, "Picard bundles", Illinois J. Math., **5**(1961), 550–564.

[27] [M1] R.J. Milgram, "The bar construction and abelian H-spaces", Ill. J. Math. **11** (1967), 242–250.

[28] [M2] R.J. Milgram, "The homology of symmetric products", Trans. Am. Math. Soc. **138** (1969), 251–265.

[29] [P] F.P. Peterson, "Whitehead products and the cohomology structure of principal fibre spaces", Am. J. Math, **82** (1960), 649–652.

[30] [S] G. Segal, "The topology of spaces of rational functions", Acta. Math., **143**(1979), 39–72.

LABORATOIRE AGAT, UNIVERSITÉ LILLE I, FRANCE
E-mail address: `sadok.kallel@agat.univ-lille1.fr`

Contemporary Mathematics
Volume **279**, 2001

Quantum Methods in Algebraic Topology

Max Karoubi

ABSTRACT. This paper offers a new approach of Algebraic Topology through "quantum" differential calculus. We show how this deformation of usual calculus makes Poincaré's lemma true in any characteristic. We are then able to adapt Quillen-Sullivan point of view of homotopy theory to the p-adic case, thanks to the new concept of "neoalgebra". This approach is closely related to the operad method due essentially to M. A. Mandell [**8**].

In this paper, we present a new version of cochains in Algebraic Topology, starting with "*quantum differential forms*". This version provides many examples of modules over the braid group, together with a control of the non commutativity of cup-products on the cochain level. If the quantum parameter q is equal to 1, we recover essentially the commutative differential graded algebra of de Rham-Sullivan forms on a simplicial set [**1**] [**12**]. For topological applications, we may take either $q = 1$ if we are dealing with rational coefficients or $q = 0$ in the general case. In both cases, the quantum formulas are simpler (if $q = 0$ for instance, the quantum exponential $e_q(x)$ is just the function $1/1 - x$).

From this viewpoint, we extract a new structure of "*neo-algebra*"[1]. This structure is detailed in section III of this paper. To a simplicial set X we can associate in a functorial way a neo-algebra $\widehat{\Omega}^*(X)$, which cohomology is canonically isomorphic to the usual one with coefficients in k (k might be an arbitrary commutative ring). As a differential graded algebra, $\widehat{\Omega}^*(X)$ is related to the usual algebra of cochains $C^*(X)$ by a (zigzag) sequence of quasi-isomorphisms.

Using in an essential way some recent results of M. A. Mandell [**8**] [**9**], one may then show that $\widehat{\Omega}^*(X)$ (up to quasi-isomorphisms of neo-algebras) determines the p-adic homotopy type[2] of X (if $k = F_p$). The proof relies on the basic fact that $\widehat{\Omega}^*(X)$ may be provided with an E_∞-algebra structure which is related to the classical one on $C^*(X)$ by a sequence of quasi-isomorphisms.

2000 *Mathematics Subject Classification.* 55, 18.
Key words and phrases. cohomology, quantum differential, neoalgebra, p-adic homotopy type, braid group.

[1]This is closely related to the notion of partial algebra and E_∞-algebra of I. Kriz and P. May [**6**]. As a matter of fact, a neo-algebra is a special case of a partial algebra.

[2]One has to assume that the spaces involved are connected, nilpotent, p-complete and of finite type.

On a more practical level, we can show how to compute Steenrod operations in mod p cohomology, as well as homotopy groups of X from the neo-algebraic data on $\widehat{\Omega}^*(X)$.

Finally in the fourth section of this paper, we see how all the theory can be dualized in the framework of "*neo-coalgebras*".

This paper is mainly expository, although some proofs are sketched. Details will be published elsewhere, as well as applications to homotopy theory (closed model categories, homotopy groups of Moore spaces...). The following URL address:

http://www.math.jussieu.fr/~karoubi/

contains already much complementary informations.

Acknowledgements. I would like to thank M. A. Mandell who brought my attention to his recent work [**8**], [**9**] which is used in section III of this paper. M. Zisman made also some useful comments after a first draft, suggesting many improvements in the presentation.

1. Braided differential graded algebras and q-cohomology.

1.1. Let k be a commutative ring and A be a k-algebra (with unit). A braiding [**5**] on A is given by a k-module *endomorphism* $R : A \otimes_k A \longrightarrow A \otimes_k A$. Let us consider the following properties of R:

α) In the set of endomorphisms of $A^{\otimes 3} = A \otimes_k A \otimes_k A$, the "Yang-Baxter equations" $R_{12}.R_{23}.R_{12} = R_{23}.R_{12}.R_{23}$ are satisfied[3]. If R is an automorphism, this implies that the braid group $\mathcal{B}_n$ acts on $A^{\otimes n}$ (the generators of the braid group are mapped to the automorphisms $R_{i,i+1}$)

β) If 1 is the unit element of A, we have the relations $R(1 \otimes a) = a \otimes 1$ and $R(a \otimes 1) = 1 \otimes a$

γ) If $\mu : A^{\otimes 2} \longrightarrow A$ is the multiplication, we have the following relations among the morphisms from $A^{\otimes 3}$ to $A^{\otimes 2}$

$$R.\mu_{12} = \mu_{23}.R_{12}.R_{23} \quad \text{and} \quad R.\mu_{23} = \mu_{12}.R_{23}.R_{12}$$

δ) The algebra is called R-commutative (or commutative in the quantum sense) if $\mu = \mu.R$ as morphisms from $A^{\otimes 2}$ to A

ε) Finally, if A is a differential graded algebra (DGA in short), R is a morphism of complexes (of degree 0) for the usual differential on $A \otimes_k A$.

If all these properties are satisfied, the differential graded algebra A is called *braided R-commutative* (or simply braided commutative).

1.2. Fundamental example. Let Λ be a commutative k-algebra provided with an *endomorphism* $a \mapsto \overline{a}$. We denote by $\overline{\Omega}^1(\Lambda)$ the cokernel of the morphism $\overline{b} : \Lambda^{\otimes 3} \longrightarrow \Lambda^{\otimes 2}$ defined by $\overline{b}(a_0 \otimes a_1 \otimes a_2) = a_0a_1 \otimes a_2 - a_0 \otimes a_1a_2 + \overline{a}_2a_0 \otimes a_1$ (this is the "twisted" Hochschild boundary). As a left Λ-module, $\overline{\Omega}^1(\Lambda)$ is generated by elements of the type u.dv (class of $u \otimes v$) with the following relation which is a variant of the Leibniz formula:

$$u.d(vw) = uv.dw + u\overline{w}.dv$$

[3]For $i < j$, $R_{ij} = R_{i,j}$ denotes in general the endomorphism of $A^{\otimes n}$ deduced from R carried to the (i,j)-component of the tensor product. In the same way, $\mu_{i,i+1}$ denotes the morphism from $A^{\otimes n}$ to $A^{\otimes (n-1)}$, obtained from the multiplication μ restricted to the $(i, i+1)$-component.

We put now $\overline{\Omega}^0(\Lambda) = \Lambda$ and $\overline{\Omega}^i(\Lambda) = 0$ for $i > 1$. The direct sum $\overline{\Omega}^*(\Lambda) = \overline{\Omega}^0(\Lambda) \oplus \overline{\Omega}^1(\Lambda)$ is obviously a DGA (if we put $u.dv.w = u\overline{w}.dv$), where the following braiding is defined (u, v, w and t being elements of Λ)

$$\begin{aligned}&R(u \otimes v) = v \otimes u\\&R(udv \otimes w) = \overline{w} \otimes udv\\&R(u \otimes vdw) = vdw \otimes u + v(w - \overline{w}) \otimes du\\&R(udv \otimes wdt) = -\overline{w}d\overline{t} \otimes udv\end{aligned}$$

1.3. Theorem. *With the above braiding, the differential graded algebra $A = \overline{\Omega}^*(\Lambda)$ satisfies the axioms $\alpha, \beta, \gamma, \delta$ and ε. Therefore, it is a braided commutative DGA (in the quantum sense).*

Proof. Easy, but tedious.

1.4. An important example of the previous theorem is $\Lambda = k[t]$, the endomorphism $a \mapsto \overline{a}$ being given by $t \mapsto qt$, with $a \in k$. The braided differential graded algebra A, denoted by $\Omega(t)$ or $\Omega^*(t)$, is well known to the experts (see [**7**] for instance). It is generated by the symbols t and dt, with the relations $dt.dt = 0$ and $(t^n dt)t^m = q^m t^{n+m} dt$. If we assume $1 + q + \cdots + q^n$ to be invertible[4] in k for all n ($q = 0$ for instance), Poincaré's lemma is true for $\Omega^*(t)$: the complex

$$0 \longrightarrow \Omega^0(t) \longrightarrow \Omega^1(t) \longrightarrow 0$$

has trivial cohomology, except in degree 0, in which case it is isomorphic to k.

On the other hand, let A and B be two braided DGA's with braidings R and S respectively. The graded tensor product $A\widehat{\otimes}B$ may be provided with the braiding given by the following composition of morphisms:

$$\begin{aligned}A_1\widehat{\otimes}B_1\widehat{\otimes}A_2\widehat{\otimes}B_2 \cong (A_1\widehat{\otimes}A_2)\widehat{\otimes}(B_1\widehat{\otimes}B_2 &\xrightarrow{R\otimes S}\\ &(A_2\widehat{\otimes}A_1)\widehat{\otimes}(B_2\widehat{\otimes}B_1) \cong A_2\widehat{\otimes}B_2\widehat{\otimes}A_1\widehat{\otimes}B_1\end{aligned}$$

(subscripts indicate the selected copy of A or B), where we assume that elements of A and B commute (in the graded sense). It is easy to check the properties listed in 1.1, if R and S satisfy them.

1.5. Of course, these remarks may be applied to an arbitrary number of braided DGA's. In particular, the graded tensor product $\Omega(y, \ldots, y_r) = \Omega(y_1)\widehat{\otimes}\cdots\widehat{\otimes}\Omega(y_r)$ is provided with a structure of braided commutative DGA and Poincaré's lemma is still true (*we always assume the hypothesis of the Note 4*). This last fact can be checked directly — as in the classical case — by introducing an auxiliary parameter t and making the substitution $y_i \mapsto ty_i$; however, the resulting homotopy operator depends on the order of the variables, because the variables dt and t^m do not commute.

After these general preliminaries, we define for all m a *cosimplicial* DGA by the following formula

$$A^{(r)} = \prod_{i_0 < \cdots < i_r} \Omega(x_0, \ldots, \widehat{x}_{i_0}, \ldots, \widehat{x}_{i_r}, \ldots, x_m).$$

[4]From now on, we shall always assume this hypothesis. One should note however that if k is any commutative ring, we may replace k by a suitable localization k' of $k[q]$: it is obtained by inverting the polynomials $1 + q + \cdots + q^n$ for all n. This localization process is faithful on the level of the cohomology (and also on the level of the homotopy type as we shall see later on).

In particular, the two coface operators

$$A^{(0)} = \prod_{1} \Omega(x_0, \dots, \widehat{x}_i, \dots, x_m) \rightrightarrows A^{(1)} = \prod_{i<j} \Omega(x_0, \dots, \widehat{x}_i, \dots, \widehat{x}_j, \dots, x_m)$$

are obvious (we let the variables x_i or $x_j = 0$). The equalizer of these two morphisms defines a commutative braided DGA called $\Omega^*(\Delta_m)$.

1.6. Theorem. *The correspondence $m \mapsto \Omega^*(\Delta_m)$ defines a braided simplicial DGA (for instance, the face operators are defined by the relations $x_i = 1$). In other words, for any non-decreasing map $[s] \longrightarrow [r]$, the following diagram commutes:*

$$\begin{array}{ccc} \Omega^*(\Delta_r) \otimes \Omega^*(\Delta_r) & \xrightarrow{R} & \Omega^*(\Delta_r) \otimes \Omega^*(\Delta_r) \\ \downarrow & & \downarrow \\ \Omega^*(\Delta_s) \otimes \Omega^*(\Delta_s) & \xrightarrow{R} & \Omega^*(\Delta_s) \otimes \Omega^*(\Delta_s) \end{array}$$

1.7. Let $X = X_\natural$ be now a simplicial set. We define the differential graded algebra $\underline{\Omega}^*(X)$ of quantum differential forms on X as the algebra of simplicial maps from $X_\natural$ to $\Omega^*(\Delta_\natural)$.

1.8. Theorem. *If $1 + q + \cdots + q^m$ is invertible in k for all m, the functors $X \mapsto H^n(\underline{\Omega}^*(X))$ are the elements of a (multiplicative) cohomology theory on X which is naturally isomorphic to the usual cohomology with coefficients in k.*

Proof (compare with [**1**]). According to what has been said before, Poincaré's lemma is true for the algebra $\Omega^*(\Delta_r)$: the following complex (where $\Omega^i(\Delta_r) = 0$ for $i > r$)

$$0 \longrightarrow \Omega^0(\Delta_r) \xrightarrow{d_0} \Omega^1(\Delta_r) \xrightarrow{d_1} \dots$$

is acyclic, except in degree 0. The kernel of d_0 is isomorphic to k and is simplicially trivial. On the other hand, for a fixed s, it is easy to see that the homotopy groups of the simplicial abelian group $r \mapsto \Omega^s(\Delta_r)$ are equal to 0. The theorem then follows from a classical result on cohomology theories [**1**].

1.9. *Remark.* This theory is very closely related to the one sketched by my student C. Mouët in [**10**].

1.10. *Remark.* In the above theorem, we may replace the simplicial k-module $r \mapsto \Omega(\Delta_r)$ by a "stabilized" version $r \mapsto \operatorname{colim}_p \Omega^{\otimes p}(\Delta_r)$, which we shall denote by $\widehat{\Omega}(\Delta_r)$ (the inductive system is given by $\omega \mapsto \omega \otimes 1$). This will be necessary in the next sections as we will see (cf. 2.6 for instance).

1.11. *Remark.* It is easy to define a "quantum integral"

$$\int_{\Delta_r} : \Omega^r(\Delta_r) \longrightarrow k$$

starting with the definition of $\Omega^r(\Delta_r)$ given in 1.5. This integral generalizes the well-known "quantum" formula

$$\int_0^1 t^n dt = \frac{1}{1 + q + \cdots + q^n}.$$

In this context, "Stokes' formula" can be written as follows

$$\int_{\Delta_r} d\omega = \sum_{i=0}^{r} (-1)^i \int_{\partial_i \Delta_r} \omega \,.$$

Here ω is of degree $r-1$ and the $\partial_i \Delta_r$ run through all the faces of Δ_r. This quantum integral defines a (non multiplicative) quasi-isomorphism between $\underline{\Omega}^*(X)$ and the complex of classical cochains on X with coefficients in k. In order to define a zigzag sequence of quasi-isomorphisms respecting the multiplicative structure, one has to use the DGA of non commutative differential forms which is detailed in [**3**].

1.12. There is a variant of $\underline{\Omega}^*(X)$ (called $\Omega^*(X)$ in order to avoid any confusion) which is more adapted to infinite complexes, thanks to a well-known notion: the "reduced product" of simplicial and cosimplicial modules. This has been shown to me by M. Zisman and is for instance in a much more general form in the book of A. K. Bousfield and D. M. Kan[5] (with a different terminology). More precisely, let C^* (resp. S_*) be a cosimplicial k-module (resp. a simplicial k-module). Their "reduced product" $C^*\nabla S_*$ is defined as the quotient of the direct sum $\bigoplus_n [C^n \otimes S_n]$ by relations of the form $\Sigma(u^* \otimes 1)(\theta) - \Sigma(1 \otimes u_*)(\theta)$, $\theta \in C^p \otimes S_n$, for any non-decreasing map $u : [p] \longrightarrow [n]$, with the associated morphisms $u_* : S_n \longrightarrow S_p$ and $u^* : C^p \longrightarrow C^n$. On the other hand, we may consider as well the normalized k-module $\overline{S}_*$ (resp. $\overline{C}^*$) regarded as a chain complex (resp. a cochain complex) and take the same type of quotient, also denoted by $\overline{C}^*\nabla\overline{S}_*$. More precisely, in the direct sum of the $\overline{C}^n \otimes \overline{S}_n$ we take the cokernel of $d \otimes 1 - 1 \otimes d'$, where $d : \overline{C}^n \longrightarrow \overline{C}^{n+1}$ (resp. $d' : \overline{S}_n \longrightarrow \overline{S}_{n-1}$) is the differential of the cochain complex (resp. the chain complex).

The general fact about these reduced products is then the following: there exists a canonical isomorphism $C^*\nabla S_* \longrightarrow \overline{C}^*\nabla\overline{S}_*$. This follows simply from the observation that for any k-module M, one has

$$\mathrm{Hom}(C^*\nabla S_*, M) = \mathrm{Hom}_\Delta(C^*, \mathrm{Hom}(S_*, M))$$

and the same type of identity for the Hom functor between cochain complexes

$$\mathrm{Hom}(\overline{C}^*\nabla\overline{S}_*, M) = \mathrm{Hom}(\overline{C}^*, \mathrm{Hom}(\overline{S}_*, M)).$$

Since $\mathrm{Hom}(\overline{C}^*, \mathrm{Hom}(\overline{S}_*, M)) \cong \mathrm{Hom}_\Delta(C^*, \mathrm{Hom}(S_*, M))$ according to the Dold-Kan theorem, the result follows immediately: choose $M = \overline{C}^*\nabla\overline{S}_*$.

1.13. Theorem. *Let us assume that the complex $\overline{S}_*$ has trivial homology. Then an exact sequence of cosimplicial k-modules*

$$0 \longrightarrow C'^* \longrightarrow C^* \longrightarrow C''^* \longrightarrow 0$$

induces an exact sequence of the associated reduced products

$$0 \longrightarrow C'^*\nabla S_* \longrightarrow C^*\nabla S_* \longrightarrow C''^*\nabla S_* \longrightarrow 0$$

if S_* is a flat module or, alternatively, if C'^*, C^* and C''^* are flat modules.

Proof. Since $\overline{C}^*$ is naturally a direct factor in C^* in general, we have also an exact sequence of normalized complexes

$$0 \longrightarrow \overline{C}'^* \longrightarrow \overline{C}^* \longrightarrow \overline{C}''^* \longrightarrow 0.$$

[5]A. K. Bousfield and D. M. Kan. Homotopy limits, completions and localizations, Springer Lecture Notes in Mathematics 304 (1972).

Let us put in general $C_n = C^{-n}$ and consider the total complex associated to the tensor product of *homology* complexes $\overline{C}_{-*} \otimes \overline{S}_*$. The reduced product $\overline{C}^*\nabla\overline{S}_*$ is just the quotient module $\mathrm{Tot}_0/d(\mathrm{Tot}_1)$ in the previous Tot complex. Let us prove first that the homology of this Tot complex is 0. For this, we may assume without loss of generality that $\overline{C}^*$ is bounded (since we start with a cycle lying in a direct sum). We then prove the statement by induction on the size of $\overline{C}^*$, using Künneth's theorem.

This last result shows that $\mathrm{Tot}_0/d(\mathrm{Tot}_1)$ is also $Z_0\mathrm{Tot}$, the k-module of 0-cycles in the Tot complex. On the other hand, according to our flatness assumptions, we have an exact sequence

$$0 \longrightarrow \mathrm{Tot}(\overline{C}'_* \otimes \overline{S}_*) \longrightarrow \mathrm{Tot}(\overline{C}_* \otimes \overline{S}_*) \longrightarrow \mathrm{Tot}(\overline{C}''_* \otimes \overline{S}_*) \longrightarrow 0.$$

The exact sequence required

$$0 \longrightarrow Z_0\mathrm{Tot}(\overline{C}'_* \otimes \overline{S}_*) \longrightarrow Z_0\mathrm{Tot}(\overline{C}_* \otimes \overline{s}_*) \longrightarrow Z_0\mathrm{Tot}(\overline{C}''_* \otimes \overline{S}_*) \longrightarrow 0$$

is then a consequence of the vanishing of $H_1(\mathrm{Tot}(\overline{C}'_* \otimes \overline{S}_*))$.

1.14. Let us apply these general considerations to the case where S_* is the simplicial flat module $\Omega^p(\Delta_*)$. Since $\pi_p(Z^p(\Delta_*)) \cong k$, we can pick a representative $\chi_p \in Z^p(\Delta_p)$ (which vanishes on all the faces). As it is well known (cf. [**4**], section 3 for instance), these forms χ_p may be choosen by induction on p, starting with the obvious choice of χ_0; we write χ_p as the restriction to the last face Δ_p of a form ω_{p+1} belonging to $\Omega^p(\Delta_{p+1})$, vanishing on all the faces except Δ_p (this is in fact the definition of the normalization $\overline{\Omega}(\Delta_{p+1})$). We then choose χ_{p+1} as $d\omega_{p+1}$. We define a morphism

$$\theta_p : \overline{C}^p \longrightarrow \overline{C}^*\nabla\overline{\Omega}^p(\Delta_*)$$

by the formula

$$\theta_p(c) = c \otimes \chi_p + (-1)^{p+1} dc \otimes \omega_{p+1}$$

(we write the elements of $\overline{C}^*\nabla\overline{\Omega}^p(\Delta_*)$ as 0-cycles of the Tot complex; cf. 1.3). Since the diagram

$$\begin{array}{ccc} \overline{C}^p & \longrightarrow & \overline{C}^*\nabla\overline{\Omega}^{[}(\Delta_*) \\ \downarrow & & \downarrow \\ \overline{C}^{p+1} & \longrightarrow & \overline{C}^*\nabla\overline{\Omega}^{p+1}(\Delta_*) \end{array}$$

commutes, the θ_p's define a morphism of cochain complexes.

1.15. Theorem. *The morphism θ above defines a quasi-isomorphism between the complexes $\overline{C}^\natural$ and $\overline{C}^*\nabla\overline{\Omega}^\natural(\Delta_*)$.*

Proof. Without loss of generality, we may assume that the normalized complex $\overline{C}^*$ is bounded. As a direct consequence of 1.13, it is enough to prove the statement when the complex $\overline{C}^*$ is concentrated in a single degree, say n. In this case, it follows from the fact that $\overline{C}^*\nabla\overline{\Omega}^\natural(\Delta_*)$ is the complex $\overline{C}^n \otimes \overline{\overline{\Omega}}^\natural(\Delta_n)$, where $\overline{\overline{\Omega}}^\natural(\Delta_n)$ is the space of differential forms on Δ_n which vanish on *all* the faces. Since $\overline{\overline{\Omega}}^\natural(\Delta_n)$ is flat, its cohomology is $\overline{C}^n \otimes H^n(\Sigma^n)$, where Σ^n is the sphere of dimension n (viewed as the quotient of Δ_n by its boundary).

1.16. Theorem. *Let $C^*(X)$ be the cochain complex associated to a simplicial set X, with coefficients in $k = \mathbb{Z}$ or a field and let us denote by $\Omega^\natural(X)$ the complex of k-modules $C^*(X)\nabla\Omega^\natural(\Delta_*)$. Let us assume as always that $1+q+\cdots+q^m$ is invertible in k for all m. Then we have a natural commutative triangle of quasi-isomorphisms of complexes*

$$\begin{array}{ccccc} \Omega^\natural(X) = & C^*(X)\nabla\Omega^\natural(\Delta_*) & \longrightarrow & \mathrm{Hom}(X_*, \Omega^\natural(\Delta_*)) & = \underline{\Omega}^\natural(X) \\ & \nwarrow & & \nearrow & \\ & & C^\natural(X) & & \end{array}$$

Proof. It follows immediately from the previous considerations. It is also easy to notice that both $\Omega^\natural(X)$ and $\underline{\Omega}^\natural(X)$ are DGA's and that the oblic arrows are NOT morphisms of DGA's. If k contains Q, we may choose the quantum parameter $q = 1$. In this case, $\Omega^\natural(\Delta_*)$ is a *commutative* DGA, as well as $\Omega^\natural(X)$ and $\underline{\Omega}^\natural(X)$.

1.17. *Remark.* More generally, we might consider a sheaf $\mathcal{F}$ of k-modules over a space X. If $\mathcal{F}^p$ denotes the Godement cosimplicial resolution of the sheaf $\mathcal{F}$ by "flasque" sheaves. Then, $\Omega^\natural(X;\mathcal{F}) = \mathcal{F}^*\nabla\Omega^\natural(\Delta_*)$ is an acyclic resolution of $\mathcal{F}$, which we might call the (abstract) de Rham resolution of $\mathcal{F}$. The complex of sections $\Gamma(\mathcal{F}^*\nabla\Omega^\natural(\Delta_*)) = \Gamma(\mathcal{F}^*)\nabla\Omega^\natural(\Delta_*)$ computes the cohomology of X with values in $\mathcal{F}$. The same type of remark applies to the cosimplicial Čech complex $\mathcal{F}(\mathcal{U})$ associated to a covering $\mathcal{U}$ of the space X. Note again that if k contains Q and if choose the quantum parameter q equal to 1, the total complex obtained is a commutative DGA if $\mathcal{F}$ is a sheaf of commutative k-algebras.

1.18. *Remark.* As in 1.10, we may replace $\Omega^\natural(\Delta_*)$ by its "stabilized" version $\widehat{\Omega}^\natural(\Delta_*)$ and define in the same way $\widehat{\Omega}(X)$ or more generally $\widehat{\Omega}(X;\mathcal{F})$ if $\mathcal{F}$ is a sheaf.

2. Symmetric kernel of braided differential graded algebras.

2.1. Let A be a braided DGA with a braiding R. For $i < j$, we recall that $R'_{i,j} = R_{ij}$ is the endomorphism R acting on the (i,j)-components of the tensor product $A^{\otimes n}$ (and the identity on the others). We put $R_{j,i} = \sigma_{i,j}R_{i,j}\sigma_{i,j}$, where $\sigma_{i,j}$ is the obvious transposition (taking into account the signs for the gradation). By definition, the *symmetric kernel* of $A^{\otimes n}$ is the k-submodule of $A^{\otimes n}$ consisting of elements ω such that $R_{u,v}\omega = \sigma_{u,v}\omega$ for all couples[6] (u,v). This symmetric kernel is denoted by $A^{\overline{\otimes} n}$; it is clearly invariant under the action of the symmetric group $\mathfrak{S}_n$.

2.2. Example. Let us suppose that $1 - q^\alpha$ is invertible for all α and consider the braided algebra $A = \Omega(t)$ of 1.4. If we identify $A^{\otimes n}$ with $\Omega(x_1, \ldots, x_n)$, its symmetric kernel is concentrated[7] in degrees 0 and 1: we have ${}^0(A^{\overline{\otimes} n}) = {}^0(A^{\otimes n}) = k[x_1, \ldots, x_n]$ and ${}^1(A^{\overline{\otimes} n}) = d(k[x_1, \ldots, x_n])$. In particular, the inclusion of $A^{\overline{\otimes} n}$ in $A^{\otimes n}$ is a quasi-isomorphism.

2.3. Example. Let us assume moreover that k is a field or the ring of integers $\mathbb{Z}$. Let A and B be two braided commutative DGA's such that the eigenvalues of σR are powers of q in $A^{\otimes 2}$ and $B^{\otimes 2}$. Then, the symmetric kernel of $(A\widehat{\otimes}B)^{\otimes n}$

---

[6] As a matter of fact, the $R_{u,v}$ for $u < v$ are sufficient for the applications we have in mind.

[7] In general, ${}^iC$ denotes the submodule of elements of degree i in the graded module C.

may be identified with $A^{\overline{\otimes} n} \otimes B^{\overline{\otimes} n}$, taking into account the canonical isomorphism $(A\hat{\otimes}B)^{\otimes n} \simeq A^{\otimes n}\hat{\otimes}B^{\otimes n}$.

From these examples, we deduce the following theorem (with the definitions of 1.5):

2.4. Theorem. *Let us suppose that $1 - q^{\alpha}$ is invertible for all α and that k is $\mathbb{Z}$ or a field. Then the inclusion of $\Omega^*(\Delta_r)^{\overline{\otimes} n}$ in $\Omega^*(\Delta_r)^{\otimes n}$ is a quasi-isomorphism.*

2.5. The braided structure of $\Omega^*(\Delta_r)^{\otimes n}$ does not extend to a n-simplicial structure on the k-module of all $\Omega^*(\Delta_{r_1})\otimes\cdots\otimes\Omega^*(\Delta_{r_n})$ for $r_1, \ldots, r_n$ belong to $\mathbb{N}$. However, we can give a n-simplicial meaning to the symmetric kernel if we replace $\Omega^*(\Delta_\bullet)$ by its stabilized version $\widehat{\Omega}^*(\Delta_\bullet)$ with the notations of 1.10. More precisely, let us consider the restriction morphism

$$r : \widehat{\Omega}^*(\Delta_r) \otimes \widehat{\Omega}^*(\Delta_s) \longrightarrow \widehat{\Omega}(\Delta_1) \otimes \widehat{\Omega}^*(\Delta_t)$$

where $t = \mathrm{Inf}(r, s)$. The symmetric kernel of $\widehat{\Omega}^*(\Delta_r) \otimes \widehat{\Omega}^*(\Delta_s)$, denoted by $\widehat{\Omega}^*(\Delta_r)\overline{\otimes}\widehat{\Omega}^*(\Delta_s)$, is defined as the graded k-submodule of $\widehat{\Omega}^*(\Delta_r) \otimes \widehat{\Omega}^*(\Delta_s)$ consisting of the elements ω such that $r(\omega) \in \widehat{\Omega}^*(\Delta_t)^{\overline{\otimes} 2}$. The "symmetric kernel" of $\widehat{\Omega}^*(\Delta_{r_1}) \otimes \cdots \otimes \widehat{\Omega}^*(\Delta_{r_n})$, also denoted by $\widehat{\Omega}^*(\Delta_{r_1})\overline{\otimes}\cdots\overline{\otimes}\widehat{\Omega}^*(\Delta_{r_n})$, is the intersection of the $n(n-1)/2$ partial symmstric kernels obtained by considering all (i, j)-components of the tensor product[8].

2.6. Let us consider now a simplicial set X and the associated differential graded algebra $\Omega^{\natural}(X)$, written simply $\Omega(X)$, defined at the end of §1 as the reduced product $C^*(X)\nabla\Omega(\Delta_*)$. More precisely, we should also consider the "stabilized" version, defined by $\widehat{\Omega}(X) = C^*(X)\nabla\widehat{\Omega}(\Delta_*)$ (cf. 1.10). This notion of reduced product ∇, which we used already many times, can be easily extended to multisimplicial and multicosimplicial modules. In particular, one might consider $[C^*(X) \otimes C^*(X)]\nabla[\Omega(\Delta_*) \otimes \Omega(\Delta_*)]$ as well as $[C^*(X) \otimes C^*(X)]\nabla[\widehat{\Omega}(\Delta_*) \otimes \widehat{\Omega}(\Delta_*)]$. Since $C^*(X)$ is flat (as a $\mathbb{Z}$-module), we can identify these various reduced products as $\Omega(X) \otimes \Omega(X)$ and $\widehat{\Omega}(X) \otimes \widehat{\Omega}(X)$ respectively. By the same method, we can write the nth tensor product as a reduced product of n factors. These identifications enable us to define the symmetric kernel of $\widehat{\Omega}(X)^{\otimes n}$, denoted $\widehat{\Omega}(X)^{\overline{\otimes} n}$, as the reduced product of $C^*(X)^{\otimes n}$ and $\widehat{\Omega}(\Delta_*)^{\overline{\otimes} n}$, where $\widehat{\Omega}(\Delta_*)^{\overline{\otimes} n}$ is defined above. This symmetric kernel $\widehat{\Omega}(X)^{\overline{\otimes} n}$ has two essential properties:

1. The canonical inclusion of $\widehat{\Omega}(X)^{\overline{\otimes} n}$ in $\widehat{\Omega}(X)^{\otimes n}$ is a quasi-isomorphism; it is equivariant for the natural action of the symmetric group $\mathfrak{S}_n$ on both factors.
2. A map α from $\{1, \ldots, n\}$ to $\{1, \ldots, p\}$ induces in a functorial way a morphism of k-modules $\alpha_* : \widehat{\Omega}(X)^{\overline{\otimes} n} \longrightarrow \widehat{\Omega}(X)^{\overline{\otimes} p}$ by the formula

   $$\alpha_*(a_1 \otimes \cdots \otimes a_n) = b_1 \otimes \cdots \otimes b_p$$

   with $b_j = \prod_{\alpha(1)=j} a_i$ (this product is independent of the order). In particular, the product map $\widehat{\Omega}(X)^{\overline{\otimes} n} \longrightarrow \widehat{\Omega}(X)$ is equivariant (with the trivial action of the symmetric group on $\widehat{\Omega}(X)$).

[8]Note that the n-complex associated to the n-simplicial k-module $(r_1, \ldots, r_n) \mapsto \widehat{\Omega}^*(\Delta_{r_1})\overline{\otimes}\cdots\overline{\otimes}\widehat{\Omega}^*(\Delta_{r_n})$ is n-acyclic if $1 - q^{\alpha}$ is invertible for all α.

2.7. Using the previous considerations and some elementary homological algebra, it is easy to define cup i-products and Steenrod operations *on the level of quantum differential forms.* The sequence

$$\widehat{\Omega}(X)^{\otimes n} \hookleftarrow \widehat{\Omega}(X)^{\overline{\otimes} n} \longrightarrow \widehat{\Omega}(X)$$

defines an equivariant morphism from $\widehat{\Omega}(X)^{\otimes n}$ to $\widehat{\Omega}(X)$ in the derived category of $\mathfrak{S}_n$- complexes as we have seen above. From this fact, we deduce a morphism of $k[\mathfrak{S}_n]$-complexes which is well defined up to homotopy[9]

$$B_\natural(\mathfrak{S}_n) \longrightarrow \mathrm{Hom}_\natural(\widehat{\Omega}(X)^{\otimes n}, \widehat{\Omega}(X))$$

where $B_\natural(\mathfrak{S}_n)$ is any projective resolution of k as a $k[\mathfrak{S}_n]$-module. Let us suppose now that $n = p$ is a prime number and let us replace the symmetric group by the cyclic group C_p. We may choose for $B_\natural(C_p)$ the classical acyclic resolution of k by $k[C_p]$-modules of rank 1 (with $k = \mathbb{F}_p$ and the quantum parameter $q = 0$ in order to fix the ideas; other choices are possible). From the previous observations, we deduce morphisms of degree-i, which we might call "cup i-products":

$$\mu_i : \widehat{\Omega}(X)^{\otimes p} \longrightarrow \widehat{\Omega}(X).$$

They are well defined up to homotopy (μ_0 is the usual cup-product map). As it is well known, Steenrod operations can be deduced from the μ_i as morphisms from $H^m(X)$ to $H^{mp-i}(X)$, by taking the composition of μ_i with the pth power operation $P : \widehat{\Omega}(X) \longrightarrow \widehat{\Omega}(X)^{\otimes p}$ which is also equivariant[10]. This can be proved, using for instance the method described in [**4**].

3. Neo-algebras: towards an algebraic description of the homotopy type.

3.1. A "neo-algebra" is given by the following data (1, 2 and 3), subject to the conditions α, β, γ and δ explained below (*this definition will imply that our neo-algebras are just particular cases of partial* DGA's, *as define in* [**6**], p. 40):

1. A differential graded k-module[11] A with a "unit element" $1 \in {}^0A$
2. A differential graded k-submodule A_2 of $A^2 = A^{\otimes 2}$, stable under the action of the group $\mathbb{Z}/2$ acting naturally on A^2 and containing $k.1 \otimes A$ (and therefore $A \otimes k.1$)
3. A "partial multiplication"

$$\mu : A_2 \longrightarrow A$$

which defines a morphism of complexes.

We call μ_{12} (resp. μ_{23}) the partial multiplication on $A_2 \otimes A$ (resp. $A \otimes A_2$) with values in $A^2 = A \otimes A$. On the other hand, for i and j belonging to the set $P = \{1, \dots, n\}$, we denote by $A_{i,j}$ the image of $A_2 \otimes A^{n-2}$ in A^n under the permutation $(1,2) \mapsto (i,j)$ of the factors. If S is a subset of $P \times P$, the k-module A_S is the intersection of all the $A_{i,j}$ where $(i,j) \in S$. In particular, A_n is defined as the module obtained when $S = P \times P$.

---

[9] Hom_0 denotes the k-module of morphisms of degree 0 which are homotopic to the multiplication μ. For $i > 0$, Hom_i is the k-module of *all* morphisms of degree $-i$.

[10] On the first factor, the action of the symmetric group is induced by the signature of the permutations if the differential forms are of odd degree and is the identity otherwise.

[11] We recall that the ${}^rC$ denotes the k-module of elements of degree r in the graded k-module C, and that C^r is the tensor product $C^{\otimes r}$ of r copies of C.

Here are the properties α, β, γ and δ which characterize a neo-algebra (if $A_2 = A^2$, we just recover the definition of a commutative DGA):

α) The inclusion of A_n in A^n is a quasi-isomorphism

β) We have the identity $\mu(1 \otimes a) = \mu(1 \otimes a) = a$ for any element of a of A (*unital axiom*)

γ) The partial multiplication $\mu : A_2 \to A$ is equivariant, the group $\mathbb{Z}/2$ acting trivially on A (*commutativity axiom*)

δ) The k-module $\mu_{12}(A_3)$ is included[12] in A_2. Moreover, we assume that the following diagram commutes (*associativity axiom*):

$$\begin{array}{ccc} A_3 & \xrightarrow{\mu_{12}} & A_2 \\ \mu_{23}\downarrow & & \downarrow \mu \\ A_2 & \xrightarrow{\mu} & A \end{array}$$

As for commutative algebras, these properties imply that any set map from $\{1, \dots, n\}$ to $\{1, \dots, p\}$ induces a functorial morphism $\alpha_* : A_n \longrightarrow A_p$. This property is closely related to the theory of Γ-spaces introduced by G. Segal [**11**].

3.2. Example. The braided differential graded algebra $A = \Omega(\Delta_n)$ defined in 1.5 is a neo-algebra with the symmetric kernel playing the role of A_2 (cf. 2.4 where we assume that $k = \mathbb{Z}$ or a field).

3.3. On the other hand, according to 2.6 $\Omega(X_1) \otimes \cdots \otimes \Omega(X_n)$ may be identified with the reduced product $[C^*(X_1) \otimes \cdot \otimes C^*(X_n)]\nabla[\Omega(\Delta_*) \otimes \cdots \otimes \Omega(\Delta_*)]$. Up to a quasi-isomorphism, we may replace $\Omega(\Delta_*)$ by $\widehat{\Omega}(\Delta_*)$. The "bisimplicial symmetric kernel" $\widehat{\Omega}(\Delta_r)\overline{\otimes}\widehat{\Omega}(\Delta_s)$ is then defined (as in 2.5) to be the k-submodule of $\widehat{\Omega}(\Delta_r) \otimes \widehat{\Omega}(\Delta_s)$ of elements which restrictions to $\widehat{\Omega}(\Delta_t) \otimes \widehat{\Omega}(\Delta_t)$ below to $\widehat{\Omega}(\Delta_t)^{\overline{\otimes}2}$ (with $t = \mathrm{Inf}(r, s)$). If we set $A = \widehat{\Omega}(X)$ and $A_2 =$ the reduced product $[C^*(X) \otimes C^*(X)]\nabla[\widehat{\Omega}(\Delta_*)\overline{\otimes}\widehat{\Omega}(\Delta_*)]$, we can check easily that $\widehat{\Omega}(X)$ is also a neo-algebra.

3.4. If A and B are neo-algebras over $k = \mathbb{Z}$ or a field, it is not difficult to see that $A \otimes B$ is also a neo-algebra (with the usual sign conventions for the tensor product of differential graded k-modules).

3.5. Finally, a morphism between two neo-algebras A and B is defined as a morphism of differential graded k-modules $f : A \longrightarrow B$ such that

1. $(f \otimes f)(A_2) \subset B_2$
2. The following diagram commutes

$$\begin{array}{ccc} A_2 & \longrightarrow & B_2 \\ \downarrow \mu & & \downarrow \mu \\ A & \longrightarrow & B \end{array}$$

[12]By symmetry, this implies the same property for $\mu_{23}(A_3)$.

3.6. Theorem. *Let us consider two connected nilpotent p-complete simplicial sets*[13] *X and Y of finite type. We assume that there is a zigzag sequence of quasi-isomorphisms of neo-algebras*[14]

$$\widehat{\Omega}(X) \longrightarrow A \longleftarrow B \longrightarrow \cdots \longleftarrow \widehat{\Omega}(Y).$$

Then X and Y have the same homotopy type.

Sketch of the proof. Let Z be a simplicial set. According to section II, there is a natural quasi-isomorphism between the differential graded k-modules $\widehat{\Omega}^*(Z)$ and $C^*(Z)$. On the other hand, we can associate to a neo-algebra an E_∞-algebra, using the method in the book of I. Kriz and P. May [**6**]. Let $\mathcal{P}$ (resp. $\mathcal{ESP}$) denote the category of partial algebras (resp. E_∞-algebras, resp. E_∞-simplicial partial algebras). In [**6**] one describes a diagram of categories and functors which is commutative up to isomorphism (φ and ψ being quasi-isomorphisms of underlying differential graded modules)

$$\begin{array}{ccc} & & \mathcal{P} \\ & \mathrm{Id} \nearrow & \\ & & \uparrow \varphi \\ \mathcal{P} & \xrightarrow{V} & \mathcal{ESP} \\ & & \downarrow \psi \\ & \mathrm{W} \searrow & \\ & & \mathcal{E} \end{array}$$

The quasi-isomorphisms in the hypothesis of the theorem

$$\widehat{\Omega}^*(X) \longrightarrow A \longleftarrow B \longrightarrow \cdots \longleftarrow \widehat{\Omega}^*(Y)$$

imply therefore a sequence of quasi-isomorphisms between the associated E_∞-algebras via the functor W.

On the other hand, according to a recent result of M. A. Mandell [**9**], for any finite simplicial set $\mathbb{Z}$, the E_∞-algebras $\widehat{\Omega}^*(Z)$ and $C^*(Z)$ are also related by a sequence of E_∞-algebras quasi-isomorphisms. From the previous conclusion, we deduce that $C^*(X)$ and $C^*(Y)$ are also related by a sequence of E_∞-algebras quasi-isomorphisms.

Since X and Y are nilpotent, a second key result of M. A. Mandell [**8**] implies that X and Y have the same homotopy type, which concludes the proof of our theorem.

A weaker version of the theorem is the following:

3.7. Theorem. *Let us consider two connected finite simplicial sets X and Y such that their homotopy groups are finite p-groups. We assume that there is a zigzag sequence of quasi-isomorphisms of neo-algebras (with the same hypothesis as in the note 14)*

$$\widehat{\Omega}^*(X) \longrightarrow A \longleftarrow B \longrightarrow \cdots \longleftarrow \widehat{\Omega}^*(Y).$$

Then X and Y have the same homotopy type.

[13]This means that its Postnikov tower can be chosen such that each fiber is of type $K(\mathbb{Z}/p, n)$ or $K(\widehat{\mathbb{Z}}p, n)$, where $\widehat{\mathbb{Z}}_p$ denotes the ring of p-adic integers [**8**].

[14]with $k = F_p$ and the quantum parameter q equal to 0.

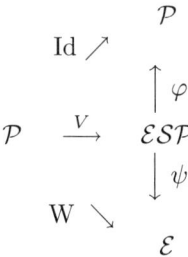

3.8. As a matter of fact, if the homotopy groups of X are finite p-groups, there is a procedure to compute algebraically the homotopy groups of X via a suitable "iteration" of the bar construction [**2**], starting from the neo-algebra $A = \widehat{\Omega}^*(X)$. More precisely, the correspondence $(r_1, \dots, r_n) \mapsto A_{r_1 \cdots r_n}$ defines a n-simplicial graded module (one has to use the base point to define some face maps). The associated total cohomology complex $\mathrm{Tot}(A_{(-r_1)\cdots(-r_n)})$, located in the second quadrant[15], has the cohomology of the nth iterated loop space of X, denoted here $\mathcal{L}^n(X)$. The coalgebra structure on the total complex determines the group structure on $\pi_n(X)$ (note that $H^0(\mathcal{L}^n(X)) = \mathrm{Hom}_{\mathrm{sets}}(\pi_n(X), \mathbb{Z}/p))$.

4. Braided differential graded coalgebras and q-homology. Neo-coalgebras.

4.1. Let A denote the fundamental example of braided DGA defined in 1.2. Its k-dual $\mathrm{Hom}(A,k) = k[[x]] \oplus k[[x]]dx$ is NOT a braided differential coalgebra[16] (the dual of a tensor product is not a tensor product). However, we are going to define a coalgebra of "quantum algebraic currents" $\mathcal{D}(x)$ (in duality with $\Omega(t)$) as a suitable k-submodule of $\mathrm{Hom}(A,k)$, which will be a *covariant algebraic analog of the unit interval* $[0,1]$. Its definition makes use of the quantum exponential[17] $e_q(x)$, considered as an element of $\mathrm{Hom}(A,k)$:

$$e_q(x) = \sum_{n=0}^{\infty} \frac{x^n}{n!_q}.$$

More precisely, let us define the following graded k-submodule $\mathcal{D}(x) = \mathcal{D}_0(x) \oplus \mathcal{D}_1(x)$ of $\mathrm{Hom}(A,k)$: $\mathcal{D}_0(x)$ consists of formal power series $f(x) = U(x)e_q(x)$, where U is a polynomial; the elements of $\mathcal{D}_1(x)$ may be written formally as $T.dx$, where $T \in \mathcal{D}_0(x)$. The duality $\Omega(t) \times \mathcal{D}(x) \to k$ is induced by a "continuous" scalar product on polynomials in t and power series in x, given by the following formulas: $\langle t^n, x^m \rangle = \langle t^n dt, x^m dx \rangle = 0$ if $n \neq m$ and $\langle t^n, x^n \rangle = \langle t^n dt, x^n dx \rangle = n!_q$. Moreover, we assume that elements of different degrees are orthogonal to each other. Another way of looking at the situation is to take as a basis of $\Omega(t)$ the "q-divided powers" $\frac{t^n}{n!_q}$ and $\frac{t^n}{n!_q}dt$. The x^n and $x^n dx$ are then in duality with this basis.

4.2. In order to define the comultiplication of $\mathcal{D}(x)$ in a convenient way, we use a continuous "twisted product" on $\mathrm{Hom}(A,k) \otimes \mathrm{Hom}(A,k)$: we assume that

$$\begin{aligned}
(1 \otimes x^n.(x^m \otimes 1) &= q^{nm}(x^m \otimes 1).(1 \otimes x^n) = q^{nm}(x^m \otimes x^n),\\
(1 \otimes x^n dx).(x^m \otimes 1) &= q^{(n+1)m}(x^m \otimes 1).(1 \otimes x^n dx) = q^{n+1)m}(x^m \otimes x^n dx)\\
(1 \otimes x^n).(x^m dx \otimes 1) &= q^{n(m+1)}(x^m dx \otimes 1).(1 \otimes x^n) = q^{n(m+1)}(x^m dx \otimes x^n)\\
(1 \otimes x^n dx).(x^m dx \otimes 1) &= -q^{(n+1)(n+1)}(x^m dx \otimes 1).(1 \otimes x^n dx)\\
&= -q^{(n+1)(m+1)}(x^m dx \otimes x^n dx)
\end{aligned}$$

[15] which is obtained by considering the *sum* of elements in the diagonals.

[16] Braided DGC in short

[17] We denote by $n!_q$ the product $1_q \times 2_q \times \cdots \times n_q$, where m_q represents the "quantum integer" $\frac{q^m - 1}{q-1}$. Note that if $q = 0$, $n!_q = 1$.

The comultiplication Δ is then deduced from the following formulas

$$\begin{aligned}
\Delta(x^m) &= (x \otimes 1 + 1 \otimes x)^m = \sum_{n=0}^{m} \frac{m!_q}{n!_q (m-n)!_q} x^n \otimes x^{m-n} \\
\Delta(e_q(x)) &= e_q(x) \otimes e_q(x) \\
\Delta(x^m dx) &= (x \otimes 1 + 1 \otimes x)^m (dx \otimes 1 + 1 \otimes dx) \\
\Delta(e_q(x)dx) &= [e_q(x) \otimes e_q(x)](dx \otimes 1 + 1 \otimes dx) \\
&= e_q(x)dx \otimes e_q(qx) + e_q(x)dx
\end{aligned}$$

and in general $\Delta(u.v) = \Delta(u).\Delta(v)$ each time the *product* $u.v$ makes sense in $\mathrm{Hom}(A, k)$ [note that $\overline{e}_q(x) = e_q(qx)$ is equal to $((q-1)x+1).e_q(x)$].

4.3. Finally, the (co)differential $\overline{d} : \mathcal{D}_1(x) \to \mathcal{D}_0(x)$ is defined by

$$\overline{d}[(U(x)e_q(x))dx] = [U(x)e_q(x)].x$$

In order to see that $\overline{d}$ is a differential of coalgebra, it is convenient to introduce formally the variables $X = x \otimes 1$ and $Y = 1 \otimes x$ (with $YX = qXY$). With obvious notations, we then have

$$\overline{d}[\Delta(f(x).dx)] = \overline{d}[f(X+Y).d(X+Y)] = f(X+Y)(X+Y) = \Delta(f(x)x) = \Delta[\overline{d}(f(x).dx)]$$

By definition, $\mathcal{D}(x)$ is the elementary DGC of *quantum algebraic currents on the unit interval* $[0, 1]$. The structure morphisms of $\Omega(t)$ and $\mathcal{D}(x)$ are in duality to each other.

4.4. If $1-q$ is invertible, it is important to notice that $\mathcal{D}(x)$ has two remarkable "group like" elements g (i.e. such that $\Delta(g) = g \otimes g$)). They are $g_1 = e_q(x)$ and $g_0 = e_q(qx) = [(q-1)x+1].e_q(x)$. The two coalgebras-morphisms $\alpha_0 : k \to \mathcal{D}(x)$ and $\alpha_1 : k \to \mathcal{D}(x)$ corresponding to these group-like elements show that $\mathcal{D}(x)$ is the *covariant* algebraic analog of the unit interval, with its two end points (whereas $\Omega(t)$ is the *contravariant* analog).

4.5. On the other hand, *braided* coalgebras are defined dually to braided algebras. For instance, the β axiom in 1.1 for algebras can be translated into the following formula for coalgebras:

$$\mu_{23}R = R_{12}R_{23}\mu_{12} \quad \text{and} \quad \mu_{12}R = R_{23}R_{12}\mu_{23}$$

where μ is the comultiplication and R is the braiding.

With these definitions, taking into account the scalar product defined above, the braiding on $\mathcal{D}(x)$ may be defined by explicit formulas. For $f \in \mathcal{D}_0(x)$, let us put $\overline{f}(x) = f(qx)$. Then we have

$$\begin{aligned}
&R(u \otimes v) = v \otimes u \\
&R(udx \otimes v) = v \otimes u.dx \\
&R(u \otimes v.dx) = v.dx \otimes \overline{u} + (1-q)vx \otimes du = v.dx \otimes \overline{u} + v(x - \overline{x}) \otimes du \\
&R(u.dx \otimes v.dx) = -vdx \otimes \overline{u}.d\overline{x}
\end{aligned}$$

The proof of the following theorem is obvious:

4.6. Theorem. (Poincaré's lemma for $\mathcal{D}(x)$). *If $1-q$ is invertible in k, the homology of the complex*

$$\mathcal{D}_1(x) \xrightarrow{\overline{d}} \mathcal{D}_0(x)$$

is trivial, except in degree 0 where it is isomorphic to k.

4.7. If $x_0, \dots, x_n$ are indeterminates, we define a *simplicial* DGC as follows

$$S_{(r)} = \coprod_{i_0 < \cdots < i_r} \mathcal{D}(x_0, \dots, \widehat{x}_{i_0}, \dots, \widehat{x}_{i_r}, \dots, x_n).$$

In particular, the two face operators

$$S_{(1)} = \coprod_{i<j} \mathcal{D}(x_0, \dots, \widehat{x}_i, \dots, \widehat{x}_j, \dots, x_n) \rightrightarrows S_{(0)} = \coprod_i \mathcal{D}(x_0, \dots, \widehat{x}_i, \dots, x_n)$$

are the obvious inclusions obtained by tensoring the identity with the coalgebra morphism $\alpha_0 : k \to \mathcal{D}(x)$ defined above. The coequalizer of these two morphisms defines a commutative braided DGC which we denote by $\mathcal{D}(\Delta_n)$ or $\mathcal{D}_*(\Delta_n)$. It is easy to see that $\mathcal{D}(\Delta_n)$ is the k-submodule of $\mathcal{D}(x_0, \dots, x_n) - \mathcal{D}(x_0) \otimes \cdots \otimes \mathcal{D}(x_n)$, which consists of sums of tensors of the type $\mathcal{D}(x_0, \dots, \widehat{x}_i, \dots, x_n).e_q(qx_i)$ for various i's. If $q = 0$, $\mathcal{D}(\Delta_n)$ is just the sum of all the $\mathcal{D}(x_0, \dots, \widehat{x}_i, \dots, x_n)$'s inside $\mathcal{D}(x_0, \dots, x_n)$.

4.8. Theorem. *Let us assume $q = 0$. Then the correspondence $n \mapsto \mathcal{D}(\Delta_n)$ defines a cosimplicial braided DGC. In particular, the coface operators are defined by replacing the missing variable x_i by the multiplication with the quantum exponential $e_q(x_i) = \sum_{i=0}^{\infty} (x_i)^t$. Therefore, for any non decreasing set map: $[s] \longrightarrow [r]$, the following diagram commutes*

$$\begin{array}{ccc} \mathcal{D}(\Delta_s) \times \mathcal{D}(\Delta_s) & \xrightarrow{R} & \mathcal{D}(\Delta_s) \times \mathcal{D}(\Delta_s) \\ \downarrow & & \downarrow \\ \mathcal{D}(\Delta_r) \times \mathcal{D}(\Delta_r) & \xrightarrow{R} & \mathcal{D}(\Delta_r) \times \mathcal{D}(\Delta_r) \end{array}$$

Proof. We use the duality between $\mathcal{D}(\Delta_n)$ and $\Omega(\Delta_n)$ made explicit in 4.1. In particular, the algebra morphism $\Omega(t) \to k$ defined by putting $t = 1$ is the transpose of the coalgebra morphism $k \to \mathcal{D}(x)$ which associates to the number 1 the quantum exponential $e_q(x) = \sum_{t=0}^{\infty} x^t$ (since $q = 0$). In the same way, the degeneracy operators are induced by the k-module map of two variables $\phi : \mathcal{D}(x) \otimes \mathcal{D}(y) \to \mathrm{Hom}(A, k) = k[[z]] \oplus k[[z]].dz$ given by the following formula: if $u = \Sigma a_n x^n$, $v = \Sigma b_n y^n$, we have $\phi(u, v) = \Sigma n!_q a_n b_n z^n$, $\phi(u, vdy) = \Sigma(n+1)!_q a_{n+1} b_n z^n dz$, $\phi(udx, v) = \Sigma(n+1)!_q a_n b_{n+1} z^n dz$ and, finally, $\phi(udx, vdy) - 0$. For $q = 0$, these formulas are reduced to $\phi(u, v) = \Sigma a_n b_n z^n$, $\phi(u, vdy) = \Sigma a_{n+1} b_n z^n dz$, $\phi(udx, v) = a_n b_{n+1} z^n dz$ and finally $\phi(udx, vdy) = 0$. We should notice that by choosing u and v functions of the type $x^r e_q(x)$ or $y^s e_q(y)$, the image of ϕ is really included in $\mathcal{D}(z)$.

4.9. The $\mathcal{D}_m(\Delta_*)$ define a *cosimplicial* k-module which is acyclic[18]. This follows from the same argument as in 1.8, where we showed that the homotopy groups of $\Omega^s(\Delta_*)$ are equal to 0. Let know S be a *simplicial* k-module. The reduced product $\mathcal{D}_m(\Delta_*) \nabla S_*$ (cf. 1.12), denoted simply $\mathcal{D}_m(S)$, is by definition the k-module of quantum algebraic currents of degree m associated to the simplicial module S. According to the general considerations of 1.12, $\mathcal{D}_m(S)$ is also isomorphic to $\overline{\mathcal{D}}_m(\Delta_*) \nabla \overline{S}_*$. The differential $\mathcal{D}_m(\Delta_*) \to \mathcal{D}_{m-1}(\Delta_*)$ induces a differential $\mathcal{D}_m(S) \to \mathcal{D}_{m-1}(S)$. An interesting case is when S is the simplicial

[18] Following 1.10, we may also replace $\mathcal{D}_m(\Delta_*)$ by its "stabilization" $\widehat{\mathcal{D}}_n(\Delta_*) = \operatorname{colim}_r \mathcal{D}_m(\Delta_*)^{\otimes r}$, without changing the homology.

chain functor $S(X)$ associated to a simplicial set X: in this case, we note simply $\mathcal{D}_m(X)$ instead of $\mathcal{D}_m(S)$. This notation is coherent with the previous one for $\mathcal{D}_m(\Delta_n)$; as a matter of fact, if X is a finite simplicial set (i.e. with a finite number of non degenerate simplices) and if we put $S = S(X)$ as before, we have $\overline{\mathcal{D}}_m(\Delta_*)\nabla\overline{S}_* \cong \mathrm{Mor}(\overline{S}_*^*, \overline{\mathcal{D}}_m(\Delta_*)) \cong \mathrm{Mor}_\Delta((S)^*, \mathcal{D}_m(\Delta_*))$, where $*$ denotes the dual. When $X = \Delta_n$, we recover $\mathcal{D}_m(\Delta_n)$ as expected.

4.10. The functors $S \mapsto \mathcal{D}_m(S)$ and $X \mapsto \mathcal{D}_m(X)$ satisfy the same formal properties as for the functor $\Omega^m(X)$ defined in 1.12 and 2.6. In particular, from an exact sequence of *flat* simplicial modules

$$0 \to S' \to S \to S'' \to 0$$

we deduce another exact sequence with currents:

$$0 \to \mathcal{D}_m(S') \to \mathcal{D}_m(S) \to \mathcal{D}_m(S'') \to 0.$$

4.11. By induction on m, starting with $m = 0$, we are going to define elements I_m of $\mathcal{D}_m(\Delta_m)$ such that the classes J_m in $\mathcal{D}_m(\Delta_m/\partial\Delta_m)$ generate the reduced homology of the spheres[19]. For this purpose, we look at the following exact sequence of singular complexes

$$0 \to S(\partial\Delta_m) \to S(\Delta_m) \oplus S(*) \to S(\Delta_m/\partial\Delta_m) \to 0.$$

From the previous considerations, we deduce an exact sequence of corresponding reduced modules of algebraic currents

$$0 \to \mathcal{D}_r(\partial\Delta_m) \to \mathcal{D}_r(\Delta_m) \oplus \mathcal{D}_r(*) \to \mathcal{D}_r(\Delta_m/\partial\Delta_m) \to 0.$$

The connecting homomorphism associated to this exact sequence enables us to identify the homology of $\Delta_m/\partial\Delta_m$ with the shifted homology of $\partial\Delta_m$ (in positive degrees). On the other hand, if we write $\partial\Delta_m$ as the union of a cone Λ_{m-1} and a face Δ_{m-1}, we have a Mayer-Vietoris exact sequence

$$0 \to \mathcal{D}_r(\partial\Delta_{m-1}) \to \mathcal{D}_r(\Lambda_{m-1}) \oplus \mathcal{D}_r(\Delta_{m-1}) \to \mathcal{D}_r(\partial\Delta_m) \to 0.$$

Therefore, we can also identify the reduced homology of $\partial\Delta_m$ with the shifted homology of $\partial\Delta_{m-1}$: that's the way the simplicial homology of spheres may be computed. Once the J_m are chosen (in such a way that the cohomology classes are linked by the connecting homomorphisms deduced from the previous exact sequences), we lift them in the currents I_m in $\mathcal{D}_r(\Delta_m)$ and write $\overline{I}_m$ the class in $\overline{\mathcal{D}}_m(\Delta_m)$ and put $\overline{u}_m = \partial_m(\overline{I}_m)$, where $\partial_m : \overline{\mathcal{D}}_m(\Delta_m) \to \overline{\mathcal{D}}_m(\Delta_{m+1})$.

4.12. Theorem. *Let S_* be a flat simplicial complex and $\phi : \overline{S}_m \to \mathcal{D}_m(S)$ defined by associating to a normalized chain c of degree m the sum $\overline{I}_m \otimes c + (-1)^m \overline{u}_m \otimes dc$ in $\overline{\mathcal{D}}_m(\Delta_m) \otimes \overline{S}_m + \overline{\mathcal{D}}_m(\Delta_{m+1}) \otimes \overline{S}_{m-1}$. Then ϕ induces a quasi-isomorphism between the normalized chain complex $\overline{S}_*$ and the complex of quantum algebraic currents $\mathcal{D}_m(S) = \mathcal{D}_m(\Delta_*)\nabla S_*$.*

Proof. As we have seen before, an exact sequence of flat simplicial modules

$$0 \to S'_* \to S_* \to S''_* \to 0$$

induces an exact sequence of complexes

$$0 \to \mathcal{D}(S'_*) \to \mathcal{D}(S_*) \to \mathcal{D}(S''_*) \to 0.$$

[19] This method has been shown to me by M. Zisman.

On the other hand, in order to prove surjectivity, as well as injectivity, we may assume that the normalized complex S_* is bounded. Moreover, by taking inductive limits and reducing the size of the complex $\overline{S}_*$ by induction, we may assume that $\overline{S}_*$ is concentrated in a single degree, say m. The complex $\mathcal{D}(S_*)$ is then isomorphic to $\mathcal{D}_*(\Delta_m/\partial\Delta_m) \otimes \overline{S}_m$. In that case, the theorem becomes clear, since $H_i(\mathcal{D}(S_*)) \cong \overline{S}_m \otimes \overline{H}_i(\Sigma^m)$, where Σ^m is the sphere of dimension m (one has to use the flatness of S_* again).

4.13. *Remark.* If we put $C^* = \mathrm{Hom}(S_*, k)$ and $\Omega^m(S) = C^*\nabla\Omega^m(\Delta_*)$, there is a pairing between $\Omega^m(C)$ and $\mathcal{D}_m(S) = \mathcal{D}_m(\Delta_*)\nabla S_*$. This is induced by the composition

$$\begin{aligned}[C^*\nabla\Omega^m(\Delta_*)] \times [\mathcal{D}_m(\Delta_*)\nabla S_*] &\to C^* \times (\Omega^m(\Delta_*) \otimes \mathcal{D}_m(\Delta_*)) \otimes S_8 \\ &\to C^* \otimes S_* \to k.\end{aligned}$$

4.14. If S_* is a simplicial *coalgebra*, we can provide $\mathcal{D}_*(S)$ with a coalgebra structure. The comultiplication follows from the composition of the following maps

$$\begin{aligned}\mathcal{D}(\Delta_*)\nabla S_* &\to [\mathcal{D}(\Delta_*) \otimes \mathcal{D}(\Delta_*)]\nabla[S_* \otimes S_*)] \\ &\cong [\mathcal{D}(\Delta_*)\nabla S_*] \otimes [\mathcal{D}(\Delta_*)\nabla S_*].\end{aligned}$$

4.15. If S is a commutative *braided* coalgebra, the *symmetric cokernel* of $S^{\otimes n}$ (denoted by $S^{\overline{\otimes} n}$) is the quotient of $S^{\otimes n}$ by the equivalence relation which identifies $\sigma_{u,v}(\omega)$ and $R_{u,v}(\omega)$ for all couples (u, v), with the notations of 2.1. This quotient is stable by the action of the symmetric group. In the case where S is the coalgebra $\mathcal{D}(x)$ defined in 4.1, $C^{\otimes n}$ may be identified with $\mathcal{D}(x_1, \ldots, x_n)$. These currents are linear combinations of elements of the type

$$e_q(x_1) \cdots e_q(x_n).\omega(x_1, \ldots, x_n), \text{ with } \omega(x_1, \ldots, x_n) \in \Omega(x_1, \ldots, x_n).$$

The following theorem may be deduced by duality from the analogous theorem in 2.2.

4.16. Theorem. *Let $S = \mathcal{D}(x)$ be the elementary coalgebra of quantum algebraic currents on the unit interval and let us assume that $q = 0$. Then, all the elements of $S^{\overline{\otimes} n}$ are of degree 0 or 1. In degree 0, we obtained all the elements of degree 0 in $S^{\otimes n}$. The elements of degree 1 in $S^{\overline{\otimes} n}$ are the classes of elements of degree 1 in $S^{\otimes n}$ for the equivalence relation which identifies $f(x_1, \ldots, x_n).dx_i$ and $g(x_1, \ldots, x_n).dx_j$ if the products $f(x_1, \ldots, x_n).x_i$ and $g(x_1, \ldots, x_n).x_j$ coincide. In particular, the quotient map $S^{\otimes n} \to S^{\overline{\otimes} n}$ is a quasi-isomorphism.*

4.17. Let $\widehat{\mathcal{D}}(\Delta_r)$ be the coalgebra of stabilized currents (cf. the footnote 18). Since $\widehat{\mathcal{D}}(\Delta_r)$ is a braided coalgebra, the "symmetric cokernel" $\widehat{\mathcal{D}}(\Delta_r)\overline{\otimes}\widehat{\mathcal{D}}(\Delta_s)$ may be identified with the push-out of $\widehat{\mathcal{D}}(\Delta_r) \otimes \widehat{\mathcal{D}}(\Delta_s)$ in the following diagram (where $t = \mathrm{Inf}(r, s)$)

$$\begin{array}{ccc}\widehat{\mathcal{D}}(\Delta_t) \otimes \widehat{\mathcal{D}}(\delta_t) & \longrightarrow & \widehat{\mathcal{D}}(\Delta_r) \otimes \widehat{\mathcal{D}}(\Delta_s) \\ \downarrow & & \downarrow \\ \widehat{\mathcal{D}}(\Delta_t)\overline{\otimes}\widehat{\mathcal{D}}(\delta_t) & \longrightarrow & \widehat{\mathcal{D}}(\Delta_r)\overline{\otimes}\widehat{\mathcal{D}}(\Delta_s)\end{array}$$

The symmetric cokernel $\widehat{\mathcal{D}}(\Delta_{r_1})\overline{\otimes}\cdots\overline{\otimes}\widehat{\mathcal{D}}(\Delta_{r_n})$ is analogously defined as the quotient of $\widehat{\mathcal{D}}(\Delta_{r_1}) \otimes \cdots \otimes \widehat{\mathcal{D}}(\Delta_{r_n})$ by the sum of all the kernels of the morphisms of $\widehat{\mathcal{D}}(\Delta_{r_i}) \otimes \widehat{\mathcal{D}}(\Delta_{r_j})$ into the various symmetric cokernels $\widehat{\mathcal{D}}(\Delta_{r_i})\overline{\otimes}\widehat{\mathcal{D}}(\Delta_{r_j})$. With these definitions, we can give a n-simplicial meaning to the symmetric cokernel $(r_1, \dots, r_n) \mapsto \widehat{\mathcal{D}}(\Delta_{r_1})\overline{\otimes}\cdots\overline{\otimes}\widehat{\mathcal{D}}(\Delta_{r_n})$, as explained in 2.5 for the dual situation.

4.18. Let us now consider the coalgebra $\widehat{\mathcal{D}}(X)$ of quantum stabilized algebraic currents on the simplicial set X. The "symmetric cokernel" $\widehat{\mathcal{D}}(X)^{\overline{\otimes} n}$ may be defined as the reduced product $(S(X)^{\otimes n})\nabla(\widehat{\mathcal{D}}(\Delta_*)\overline{\otimes}\cdots\overline{\otimes}\widehat{\mathcal{D}}(\Delta_*))$. The quotient morphism $\widehat{\mathcal{D}}(X)^{\otimes n} \to \widehat{\mathcal{D}}(X)^{\overline{\otimes} n}$ is then a quasi-isomorphism.

4.19. With all these definitions, we can say at last what should be a "neo-coalgebra". The axioms are simply dual to the ones for neo-algebras. What is given essentially is a differential graded quotient module C_2 of C^2 and a "partial comultiplication"

$$\mu : C \to C_2$$

with properties dual to α, β, γ and δ in 3.1. In particular, C_n is a quotient module of $C_{n_1} \otimes \cdot \otimes C_{n_p}$ for $n = n_1 + \cdots + n_p$. To any simplicial set X, we can associate the neo-coalgebra $C = \widehat{\mathcal{D}}(X)$ defined above with its symmetric cokernel C_2 defined in 4.17.

4.20. *Remark.* The dual of a coalgebra (resp. a neo-coalgebra) is an algebra (resp. a neo-algebra). Therefore, the dual of $\widehat{\mathcal{D}}(X)$ is a neo-algebra: it is quasi-isomorphic to the neo-algebra $\widehat{\Omega}(X)$ defined in 1.18.

4.21. *Remark.* If the homotopy groups of X are finite p-groups and if C is the neo-coalgebra $\widehat{\mathcal{D}}(X)$, the homology complex[20] $\mathrm{Tot}(C_{(-r_1)\cdots(-r_n)})$ associated to the *n-cosimplicial* graded differential module $(r_1, \dots, r_n) \mapsto C_{r_1\cdots r_n}$ has an homology which is isomorphic to the homology of the n-iterated loop space of X.

References

1. H. Cartan. Théories cohomologiques. *Invent. Math.*, 35:261–271, 1976.
2. W. G. Dwyer. Strong convergence of the Eilenberg-Moore spectral sequence. *Topology*, 13:255–265, 1974.
3. M. Karoubi. Formes différentielles non commutatives et cohomologie à coefficients arbitraires. *Transactions of the AMS*, 347:4277–4299, 1995.
4. M. Karoubi. Formes différentielles non commutatives et opérations de Steenrod. *Topology*, 34:699–715, 1995.
5. C. Kassel. Quantum groups. Graduate Texts in Mathematics. Springer-Verlag, 1995.
6. I. Kriz, P. May. Operads, Algebras Modules and Motives *Astérisque*, 233, 1995.
7. G. Maltsiniotis. Le langage des espaces et des groupes quantiques. *Commun. Math. Phys.*, 151:275–302, 1993.
8. M.-A. Mandell. E_∞-algebras and p-adic homotopy theory; *Topology*, 2000.
9. M.-A. Mandell. Cochain multiplications (to appear).
10. C. Mouët. q-cohomologie non commutative. *C.R. Acad. Sci. Paris, t* 323:849–851, 1996.
11. G. Segal. Categories and cohomology theories. *Topology* 13:293–312, 1974.
12. D. Sullivan. Infinitesimal computations in Topology. *Publ. Math. IHES* 47:269–331, 1977.

Université Paris 7, UFR de Mathématiques, 2, place Jussieu, 75251 Paris Cedex 05
E-mail address: karoubi@math.jussieu.fr

[20]In the Tot homology complex, we should consider the *product* of elements located on the diagonals, in order to get a situation in duality with the cohomological one.

Contemporary Mathematics
Volume **279**, 2001

Adiabatic Limits and Foliations

Kefeng Liu and Weiping Zhang

ABSTRACT. We use adiabatic limits to study foliated manifolds. The Bott connection naturally shows up as the adiabatic limit of Levi-Civita connections. As an application, we then construct certain natural elliptic operators associated to the foliation and present a direct geometric proof of a vanshing theorem of Connes [**4**], which extends the Lichnerowicz vanishing theorem [**6**] to foliated manifolds with spin leaves, for what we call almost Riemannian foliations. Several new vanishing theorems are also proved by using our method.

Introduction

Let M be a compact spin manifold and g^{TM} be a Riemannian metric on TM, the tangent bundle of M. If the scalar curvature k_{TM} of g^{TM} is positive, then a well-known theorem of Lichnerowicz [**6**] states that $\langle \widehat{A}(TM), [M] \rangle = 0$. Here $\widehat{A}(TM)$ is the Hirzebruch $\widehat{A}$-class of TM.

This result is a direct consequence of the Atiyah-Singer index formula [**1**] and the by now standard Lichnerowicz formula, which was also observed by Singer, for Dirac operators on spin manifolds.

Now let (M, F) be a compact foliated manifold. We make the following assumtions: (1) F is spin; (2) there is a metric g^F on F such that the scalar curvature of g^F is positive. Under these two assumptions, A. Connes [**4**] proved that one still has $\langle \widehat{A}(TM), [M] \rangle = 0$. Note that here M needs not be spin and thus $\langle \widehat{A}(TM), [M] \rangle$ may not be an integer. Clearly this extends the Lichnerowicz theorem which corresponds to the special case $F = TM$.

The proof given by Connes in [**4**] is highly noncommutative. It uses, besides an application of the Lichnerowicz formula along the leaves, the longitudinal index theorem of Connes-Skandalis for foliated manifolds [**5**] as well as the cyclic cohomology techniques, see [**4**].

The present paper arose from an attempt to find a direct geometric proof of Connes' vanishing theorem. As we will see, at least for the almost Riemannian foliations studied in [**4**], Connes' vanishing theorem can be directly deduced from a Lichnerowicz type formula and the Atiyah-Singer index theorem.

2000 *Mathematics Subject Classification.* Primary 53C27; Secondary 51H25 .
Key words and phrases. adiabatic limit, foliation.

To make the use of the Lichnerowicz type formula to the foliated manifolds possible, we use the adiabatic limit procedure to blow up the metric along the transversal direction. Some geometric behaviors of the foliation under adiabatic limits are also examined. In particular, we find that the Bott connection [**3**] appears naturally in the limit. Then we construct a class of natural elliptic operators which we called the sub-Dirac operators associated to the foliation. The indices of such operators contain many important geometric information about the foliation. The method we use has more potentials. Actually we can prove certain new vanishing theorems for the foliated manifolds.

This paper is organized as follows. In §1 we discuss the geometry of a foliated manifold under the adiabatic limits. In §2, by constructing the sub-Dirac operators, we give a direct geometric proof of the Connes vanishing theorem for what we call almost Riemannian foliations. In §3, we briefly describe some new vanshing theorems that can be proved by using our method. The final section is an appendix in which we make some remarks on the case of general foliations.

Acknowledgements. Part of this work was done while the first author was visiting the Nankai Institute of Mathematics in Tianjin during the summer of 1995. He would like to thank Prof. S. T. Yau [**8**] who first proposed to use the idea of adiabatic limits to deal with Connes' theorem many years ago. Both of the authors wish to thank the Chinese National Science Foundation for its financial support. The first author was also partially supported by the NSF and the Sloan Foundation, and the second author by the Ministry of Education of China and the Qiu Shi Foundation.

§1. Adiabatic limits and foliations

The purpose of this section is to establish some formulas concerning the adiabatic limit of foliations.

Let (M, F) be a closed foliation, that is, F is an integrable sub-bundle of TM. Let g^F be a metric on F. Let g^{TM} be a metric on TM which restricted to g^F on F. Let $F^\perp$ be the orthogonal complement of F in TM with respect to g^{TM}. Then we have the following orthogonal splitting,

$$\begin{aligned} TM &= F \oplus F^\perp, \\ g^{TM} &= g^F \oplus g^{F^\perp}, \end{aligned} \tag{1.1}$$

where $g^{F^\perp}$ is the restriction of g^{TM} to $F^\perp$.

It is clear that we can and we will make the identification that

$$TM/F = F^\perp. \tag{1.2}$$

Let p, $p^\perp$ be the orthogonal projection from TM to F, $F^\perp$ respectively. Let ∇^{TM} be the Levi-Civita connection of g^{TM} and ∇^F (resp. $\nabla^{F^\perp}$) be the restriction of ∇^{TM} to F (resp. $F^\perp$). That is,

$$\begin{aligned} \nabla^F &= p \nabla^{TM} p, \\ \nabla^{F^\perp} &= p^\perp \nabla^{TM} p^\perp. \end{aligned} \tag{1.3}$$

Now for any $\varepsilon > 0$, let g_ε^{TM} be the metric

$$g^{TM,\varepsilon} = g^F \oplus \frac{1}{\varepsilon} g^{F^\perp}. \tag{1.4}$$

Let $\nabla^{TM,\varepsilon}$ be the Levi-Civita connection of $g^{TM,\varepsilon}$. Let $\nabla^{F,\varepsilon}$ (resp. $\nabla^{F^\perp,\varepsilon}$) be the restriction of $\nabla^{TM,\varepsilon}$ to F (resp. $F^\perp$). We will examine the behavior of $\nabla^{TM,\varepsilon}$ as $\varepsilon \to 0$.

The process of taking the limit $\varepsilon \to 0$ is called adiabatic limit.

The standard formula for Levi-Civita connection gives us the following sets of formulas,

$$\begin{aligned} \nabla^{F,\varepsilon} &= \nabla^F, \\ p\nabla_X^{TM,\varepsilon}p^\perp &= p\nabla_X^{TM}p^\perp, \text{ for } X \in \Gamma(F), \end{aligned} \tag{1.5}$$

and

$$\begin{aligned} \left\langle \nabla_V^{TM,\varepsilon}U, X\right\rangle = \langle \nabla_V^{TM}U, X\rangle &- \frac{1}{2}\left\langle X, \nabla_V^{TM}U + \nabla_U^{TM}V\right\rangle \\ &+\frac{1}{2\varepsilon}\left\langle X, \nabla_V^{TM}U + \nabla_U^{TM}V\right\rangle, \end{aligned} \tag{1.6}$$

for $X \in \Gamma(F)$, $U, V \in \Gamma(F^\perp)$. Furthermore,

$$\begin{aligned} p^\perp\nabla_X^{TM,\varepsilon}p &= \varepsilon p^\perp\nabla_X^{TM}p, \text{ for } X \in \Gamma(F), \\ \left\langle \nabla_V^{TM,\varepsilon}Y, U\right\rangle &= -\frac{1}{2}\left\langle Y, \nabla_V^{TM}U + \nabla_U^{TM}V\right\rangle + \frac{\varepsilon}{2}\langle Y, [U,V]\rangle \end{aligned} \tag{1.7}$$

and

$$\begin{aligned} \nabla_V^{F^\perp,\varepsilon} &= \nabla_V^{F^\perp}, \\ \left\langle \nabla_X^{F^\perp,\varepsilon}U, V\right\rangle &= \langle [X,U], V\rangle - \frac{1}{2}\left\langle X, \nabla_V^{TM}U + \nabla_U^{TM}V\right\rangle - \frac{\varepsilon}{2}\langle X, [U,V]\rangle \end{aligned} \tag{1.8}$$

for $X \in \Gamma(F)$, $U, V \in \Gamma(F^\perp)$.

Let L be a leave of F, then $F^\perp$ is a flat bundle along L, carrying with the canonical Bott connection [**3**]

$$\dot{\nabla}_X^L = p^\perp[X,U], \ X \in \Gamma(TL), \ U \in \Gamma(F^\perp). \tag{1.9}$$

Following [**2**], set

$$\omega^L = \left(g^{F^\perp}\right)^{-1}\dot{\nabla}^L g^{F^\perp} = \dot{\nabla}^{L*} - \dot{\nabla}^L, \tag{1.10}$$

where $\dot{\nabla}^{L*}$ is the dual of $\dot{\nabla}^L$ with respect to $g^{F^\perp}$. Then

$$\widetilde{\nabla}^L = \dot{\nabla}^L + \frac{1}{2}\omega^L \tag{1.11}$$

is the natural unitary connection on $F^\perp|_L$ associated to $\dot{\nabla}^L$.

We can now state the main result of this section as follows.

Theorem 1.1. *Along each leave L, the following identity holds,*

$$\lim_{\varepsilon\to 0} \nabla^{F^\perp,\varepsilon}\Big|_L = \widetilde{\nabla}^L. \tag{1.12}$$

Proof. For any $X \in \Gamma(TL)$, $U, V \in \Gamma(F^\perp)$, one has

$$\begin{aligned} \omega^L(X)(U,V) &= \left\langle \dot{\nabla}_X^{L*}U, V\right\rangle - \left\langle \dot{\nabla}_X^L U, V\right\rangle \\ &= -\left\langle U, \dot{\nabla}_X^L V\right\rangle - \left\langle \dot{\nabla}_X^L U, V\right\rangle + X\langle U, V\rangle \\ &= -\langle U, [X,V]\rangle - \langle [X,U], V\rangle + X\langle U,V\rangle \\ &= -\left\langle U, \nabla_X^{TM}V - \nabla_V^{TM}X\right\rangle - \left\langle \nabla_X^{TM}U - \nabla_U^{TM}X, V\right\rangle + X\langle U,V\rangle \\ &= -\left\langle \nabla_U^{TM}V, X\right\rangle - \left\langle \nabla_V^{TM}U, X\right\rangle - \left\langle U, \nabla_X^{TM}V\right\rangle \end{aligned}$$

$$-\langle\nabla_X^{TM}U,V\rangle+X\langle U,V\rangle. \tag{1.13}$$

Note that the last three terms cancel. So (1.12) follows directly from (1.8), (1.9), (1.11) and (1.13). □

Remark 1.2. Conversely, one sees from Theorem 1.1 that the Bott connection shows up naturally from the above adiabatic limit procedure.

§2. The Connes vanishing theorem for almost Riemannian foliations

In this section we present a direct geometric proof of the Connes vanishing theorem for a special class of foliated manifolds which we call almost Riemannian foliations.

This section is organized as follows. In a), we define what we call almost Riemannian foliations. In b) we construct a class of elliptic operators specially defined for our purpose and prove the corresponding Lichnerowicz formula. In c) we prove the Connes vanishing theorem for almost Riemannian foliations.

a). Almost Riemannian foliations

We use the same assumptions and notations as in §1. First note that the 1-form ω^L defined in (1.10) does not depend on the metric g^F. Also recall that, if $\omega=0$, then (M,F,g^{TM}) is a Riemannian foliation. This motivates the following definition.

Definition 2.1. Let g^F be a metric on F. If there is a series of metrics g_ε^{TM}, $\varepsilon>0$, such that as $\varepsilon\to 0$, the corresponding one form defined in (1.10) verifies that

$$\omega_\varepsilon\to 0, \tag{2.1}$$

then we say that (M,F,g^F) admits an almost Riemannian structure.

Remark 2.2. The equation (2.1) should be interpreted as follows: as $\varepsilon\to 0$, for any $X\in\Gamma(F)$, one has

$$|\omega_\varepsilon(X)|_{g_\varepsilon^{TM}}\to 0. \tag{2.2}$$

b). A class of elliptic operators associated to Spin subbundles of the tangent bundle

From now on we make the special assumption that F is oriented, spin and carries a fixed spin structure. We also assume that $F^\perp$ is oriented and that both $p=\dim F$ and $q=\dim F^\perp$ are even.

Let $S(F)$ be the bundle of spinors associated to (F,g^F). For any $X\in\Gamma(F)$, denote by $c(X)$ the Clifford action of X on $S(F)$. Since $p=\dim F$ is even, we have the splitting

$$S(F)=S_+(F)\oplus S_-(F) \tag{2.3}$$

and $c(X)$ exchanges $S_\pm(F)$.

Let $\Lambda(F^{\perp,*})$ be the exterior algebra bundle of $F^\perp$. Then $\Lambda(F^{\perp,*})$ carries a canonically induced metric $g^{\Lambda(F^{\perp,*})}$ from $g^{F^\perp}$. For any $U\in\Gamma(F^\perp)$, let $U^*\in\Gamma(F^{\perp,*})$ be the corresponding dual of U with respect to $g^{F^\perp}$.

Now for $U \in \Gamma(F^{\perp})$, set

$$c(U) = U^* \wedge -i_U, \ \widehat{c}(U) = U^* \wedge +i_U, \tag{2.4}$$

where $U^*\wedge$ and i_U are the exterior and inner multiplications by U^* and U on $\Lambda(F^{\perp,*})$ respectively. One has the following obvious identities,

$$\begin{aligned} c(U)c(V) + c(V)c(U) &= -2\langle U, V\rangle_{g^{F^\perp}}, \\ \widehat{c}(U)\widehat{c}(V) + \widehat{c}(V)\widehat{c}(U) &= 2\langle U, V\rangle_{g^{F^\perp}}, \\ c(U)\widehat{c}(V) + \widehat{c}(V)c(U) &= 0 \end{aligned} \tag{2.5}$$

for $U,\ V \in \Gamma(F^{\perp})$.

Let $h_1, \cdots, h_q$ be an oriented local orthonormal basis of $F^{\perp}$. Set

$$\tau\left(F^{\perp}, g^{F^{\perp}}\right) = \left(\frac{1}{\sqrt{-1}}\right)^{\frac{q(q+1)}{2}} c(h_1)\cdots c(h_q). \tag{2.6}$$

Then

$$\tau\left(F^{\perp}, g^{F^{\perp}}\right)^2 = \mathrm{Id}_{\Lambda(F^{\perp,*})}. \tag{2.7}$$

Denote

$$\Lambda_{\pm}\left(F^{\perp,*}\right) = \left\{h \in \Lambda\left(F^{\perp,*}\right) : \ \tau\left(F^{\perp}, g^{F^{\perp}}\right)h = \pm h\right\}. \tag{2.8}$$

Then $\Lambda_{\pm}(F^{\perp,*})$ are sub-bundles of $\Lambda(F^{\perp,*})$. Also, since q is even, one verifies that for any $h \in \Gamma(F^{\perp})$, $c(h)$ anti-commutes with τ, while $\widehat{c}(h)$ commutes with τ. Thus $c(h)$ exchanges $\Lambda_{\pm}(F^{\perp,*})$.

We will view both vector bundles

$$S(F) = S_+(F) \oplus S_-(F) \tag{2.9}$$

and

$$\Lambda\left(F^{\perp,*}\right) = \Lambda_+\left(F^{\perp,*}\right) \oplus \Lambda_-\left(F^{\perp,*}\right) \tag{2.10}$$

as super vector bundles. Their $\mathbf{Z}_2$ graded tensor product is given by

$$\begin{aligned} S(F)\widehat{\otimes}\Lambda(F^{\perp,*}) = &\left[S_+(F)\otimes\Lambda_+\left(F^{\perp,*}\right) \oplus S_-(F)\otimes\Lambda_-\left(F^{\perp,*}\right)\right] \\ &\bigoplus \left[S_+(F)\otimes\Lambda_-\left(F^{\perp,*}\right) \oplus S_-(F)\otimes\Lambda_+\left(F^{\perp,*}\right)\right]. \end{aligned} \tag{2.11}$$

For $X \in \Gamma(F)$, $U \in \Gamma(F^{\perp})$, the operators $c(X)$, $c(U)$ and $\widehat{c}(U)$ extends naturally to $S(F)\widehat{\otimes}\Lambda(F^{\perp,*})$.

The connections ∇^F, $\nabla^{F^{\perp}}$ lift to $S(F)$ and $\Lambda(F^{\perp,*})$ naturally , and preserve the splitting (2.9) and (2.10). We write them as

$$\begin{aligned} \nabla^{S(F)} &= \nabla^{S_+(F)} \oplus \nabla^{S_-(F)}, \\ \nabla^{\Lambda(F^{\perp,*})} &= \nabla^{\Lambda_+(F^{\perp,*})} \oplus \nabla^{\Lambda_-(F^{\perp,*})}. \end{aligned} \tag{2.12}$$

Then $S(F)\widehat{\otimes}\Lambda(F^{\perp,*})$ carries the induced tensor product connection

$$\nabla^{S(F)\widehat{\otimes}\Lambda(F^{\perp,*})} = \nabla^{S(F)} \otimes \mathrm{Id}_{\Lambda(F^{\perp,*})} + \mathrm{Id}_{S(F)} \otimes \nabla^{\Lambda(F^{\perp,*})}. \tag{2.13}$$

And similarly for $S_{\pm}(F)\widehat{\otimes}\Lambda_{\pm}(F^{\perp,*})$.

Let $S \in \Omega^1(T^*M)\otimes\Gamma(\mathrm{End}(TM))$ be defined by

$$\nabla^{TM} = \nabla^F + \nabla^{F^{\perp}} + S. \tag{2.14}$$

Then for any $X \in \Gamma(TM)$, $S(X)$ exchanges $\Gamma(F)$ and $\Gamma(F^{\perp})$ and is skew-adjoint with respect to g^{TM}.

For any vector bundle E over M. By an integral polynomial of E we will mean a vector bundle $\phi(E)$ which is a polynomial in the exterior and symmetric powers of E with integral coefficients.

Let $\phi(F^\perp)$ be an integral polynomial of $F^\perp$, then $\phi(F^\perp)$ carries a naturally induced metric $g^{\phi(F^\perp)}$ from $g^{F^\perp}$ and also a naturally induced Hermitian connection $\nabla^{\phi(F^\perp)}$ induced from $\nabla^{F^\perp}$.

Our main concern will be on the Z_2-graded vector bundle

$$\left(S(F)\widehat{\otimes}\Lambda\left(F^{\perp,*}\right)\right)\otimes\phi(F^\perp) \tag{2.15}$$

which is

$$\left[S_+(F)\otimes\Lambda_+\left(F^{\perp,*}\right)\otimes\phi\left(F^\perp\right)\oplus S_-(F)\otimes\Lambda_-\left(F^{\perp,*}\right)\otimes\phi\left(F^\perp\right)\right]$$
$$\bigoplus\left[S_+(F)\otimes\Lambda_-\left(F^{\perp,*}\right)\otimes\phi\left(F^\perp\right)\oplus S_+(F)\otimes\Lambda_-\left(F^{\perp,*}\right)\otimes\phi\left(F^\perp\right)\right].$$

The Clifford actions $c(X)$, $c(U)$ and $\widehat{c}(U)$ for $X\in\Gamma(F)$, $U\in\Gamma(F^\perp)$ extends further to these bundles by acting as identity on $\phi(F^\perp)$.

We can also form the tensor product metric on the new bundles as well as the tensor product connection

$$\nabla^{(S(F)\widehat{\otimes}\Lambda(F^{\perp,*}))\otimes\phi(F^\perp)}=\nabla^{S(F)\widehat{\otimes}\Lambda(F^{\perp,*})}\otimes\mathrm{Id}_{\phi(F^\perp)}+\mathrm{Id}_{S(F)\widehat{\otimes}\Lambda(F^{\perp,*})}\otimes\nabla^{\phi(F^\perp)}, \tag{2.16}$$

and similarly for the $\pm$ sub-bundles.

Now let $\{f_i\}_{i=1}^p$ be an oriented orthonormal basis of F. Recall that $\{h_s\}_{s=1}^q$ is an orthonormal basis of $F^\perp$. The elliptic operators which are the main concern of this subsection can be defined as follows. It is introduced mainly for the reason that the vector bundle $F^\perp$ might well be non-spin.

Definittion 2.2. Let $D_{F,\phi(F^\perp)}$ be the operator mapping from

$$\Gamma(S(F)\widehat{\otimes}\Lambda(F^{\perp,*})\otimes\phi(F^\perp))$$

to itself defined by

$$\begin{aligned}D_{F,\phi(F^\perp)}=&\sum_{i=1}^p c(f_i)\nabla_{f_i}^{(S(F)\widehat{\otimes}\Lambda(F^{\perp,*}))\otimes\phi(F^\perp)}\\&+\sum_{s=1}^q c(h_s)\nabla_{h_s}^{(S(F)\widehat{\otimes}\Lambda(F^{\perp,*}))\otimes\phi(F^\perp)}\\&+\frac{1}{2}\sum_{i,j=1}^p\sum_{s=1}^q\langle S(f_i)f_j,h_s\rangle c(f_i)c(f_j)c(h_s)\\&+\frac{1}{2}\sum_{s,t=1}^q\sum_{i=1}^p\langle S(h_s)h_t,f_i\rangle c(h_s)c(h_t)c(f_i).\end{aligned} \tag{2.17}$$

It is easy to verify that $D_{F,\phi(F^\perp)}$ is a first order formally self adjoint elliptic differential operator. Furthermore it anti-commutes with the $\mathbf{Z}_2$ grading operator of the super vector bundle $(S(F)\widehat{\otimes}\Lambda(F^{\perp,*}))\otimes\phi(F^\perp)$.

Let $D_{+,F,\phi(F^\perp)}$ (resp. $D_{-,F,\phi(F^\perp)}$) be the restriction of $D_{F,\phi(F^\perp)}$ to the even (resp. odd) sub-bundle of $(S(F)\widehat{\otimes}\Lambda(F^{\perp,*}))\otimes\phi(F^\perp)$. Then one has

$$D_{+,F,\phi(F^\perp)}^*=D_{-,F,\phi(F^\perp)}. \tag{2.18}$$

Let $\Delta^{F,\phi(F^\perp)}$ be the Bochner Laplacian defined by

$$\Delta^{F,\phi(F^\perp)} =$$

$$\sum_{i=1}^{p}\left(\nabla_{f_i}^{(S(F)\widehat{\otimes}\Lambda(F^{\perp,*}))\otimes\phi(F^\perp)} + \frac{1}{2}\sum_{j=1}^{p}\sum_{s=1}^{q}\langle S(f_i)f_j, h_s\rangle\, c(f_j)c(h_s)\right)^2$$

$$+\sum_{s=1}^{q}\left(\nabla_{h_s}^{(S(F)\widehat{\otimes}\Lambda(F^{\perp,*}))\otimes\phi(F^\perp)} + \frac{1}{2}\sum_{t=1}^{q}\sum_{j=1}^{p}\langle S(h_s)h_t, f_j\rangle c(h_t)c(f_j)\right)^2$$

$$-\left(\nabla_{\Sigma_{i=1}^{p}\nabla_{f_i}^{TM}f_i}^{(S(F)\widehat{\otimes}\Lambda(F^{\perp,*}))\otimes\phi(F^\perp)} + \frac{1}{2}\sum_{j=1}^{p}\sum_{s=1}^{q}\left\langle S\left(\sum_{i=1}^{p}\nabla_{f_i}^{TM}f_i\right)f_j, h_s\right\rangle c(f_j)c(h_s)\right)$$

$$-\left(\nabla_{\Sigma_{s=1}^{q}\nabla_{h_s}^{TM}h_s}^{(S(F)\widehat{\otimes}\Lambda(F^{\perp,*}))\otimes\phi(F^\perp)} + \frac{1}{2}\sum_{t=1}^{q}\sum_{j=1}^{p}\left\langle S\left(\sum_{s=1}^{q}\nabla_{h_s}^{TM}h_s\right)h_t, f_j\right\rangle c(h_t)c(f_j)\right). \tag{2.19}$$

Let k_{TM} be the scalar curvature of the metric g^{TM}, let $R^{F^\perp}$ be the curvature tensor of $\nabla^{F^\perp}$. Let $R^{\phi(F^\perp)}$ be the curvature of $\nabla^{\phi(F^\perp)}$. Then we can state the corresponding Lichnerowicz type formula for $D_{F,\phi(F^\perp)}$ as follows.

Theorem 2.3. *The following identity holds,*

$$D_{F,\phi(F^\perp)}^2 = -\Delta^{F,\phi(F^\perp)} + \frac{1}{2}\sum_{i,j=1}^{p} c(f_i)c(f_j)R^{\phi(F^\perp)}(f_i,f_j)$$

$$+\sum_{i=1}^{p}\sum_{s=1}^{q}c(f_i)c(h_s)R^{\phi(F^\perp)}(f_i,h_s) + \frac{1}{2}\sum_{s,t=1}^{q}c(h_s)c(h_t)R^{\phi(F^\perp)}(h_s,h_t)$$

$$+\frac{k_{TM}}{4} + \frac{1}{4}\sum_{i=1}^{p}\sum_{r,s,t=1}^{q}\left\langle R^{F^\perp}(f_i,h_r)h_t, h_s\right\rangle c(f_i)c(h_r)\widehat{c}(h_s)\widehat{c}(h_t)$$

$$+\frac{1}{8}\sum_{i,j=1}^{p}\sum_{s,t=1}^{q}\left\langle R^{F^\perp}(f_i,f_j)h_t, h_s\right\rangle c(f_i)c(f_j)\widehat{c}(h_s)\widehat{c}(h_t)$$

$$+\frac{1}{8}\sum_{s,t,r,l=1}^{q}\left\langle R^{F^\perp}(h_r,h_l)h_t, h_s\right\rangle c(h_r)c(h_l)\widehat{c}(h_s)\widehat{c}(h_t). \tag{2.20}$$

Proof. We first assume that $F^\perp$, and thus TM also, is spin. Then $D_{F,\phi(F^\perp)}$ is just the standard Dirac operator on M twisted by $S(F^\perp)\otimes\phi(F^\perp)$ which carries the canonical connections induced from $\nabla^{F^\perp}$,

$$D_{F,\phi(F^\perp)} : \Gamma\left(S(TM)\otimes S\left(F^\perp\right)\otimes\phi\left(F^\perp\right)\right) \longrightarrow \Gamma\left(S(TM)\otimes S\left(F^\perp\right)\otimes\phi\left(F^\perp\right)\right). \tag{2.21}$$

With this observation, (2,20) is a direct corollary of the usual Lichnerowicz formula [**6**] for twisted Dirac operators.

Now observe that (2.20) is a local formula, and that locally we can always assume TM is spin. Thus the formula is proved. The interested reader may also proceed to verify (2.20) by a direct computation. □

Remark 2.4. The definition of $D_{F,\phi(F^\perp)}$ does not use the condition that F is integrable.

Remark 2.5. While locally $D_{F,\phi(F^\perp)}$ can be seen as a twisted Dirac operator, the key point here is that its definition relies *only* on the spin structure of F.

c). Almost Riemannian foliations and the Connes vanishing theorem

Recall that (M,F) is a compact foliated manifold and we assume that F is oriented, spin and both $p=\dim F$ and $q=\dim F^\perp$ are even, and that TM/F is oriented.

For any $x\in M$, let L_x be the leave passing through x. Then g^{TF} restricts to a Riemannian metric on L_x. So it determines a scalar curvature $k_F(x)$.

We now make the basic assumption that k_F is positive on M. Since M is compact, there is then a positive number $\delta>0$, such that

$$k_F(x)>\delta \quad \text{for any} \quad x\in M. \tag{2.22}$$

The second important assumption for this subsection is that (M,F,g^F) is an almost Riemannian foliated manifold in the sense of Definition 2.1.

We fix temporarily a splitting

$$TM=F\oplus F^\perp. \tag{2.23}$$

Given a metric $g^{F^\perp}$ on $F^\perp$, recall that we have defined

$$\omega=\left(g^{F^\perp}\right)^{-1}\dot\nabla g^{F^\perp} \tag{2.24}$$

along the leaves of F, or we can say, along F.

By [**2**], one sees that

$$\dot\nabla^2=-\frac{1}{4}\omega^2 \tag{2.25}$$

along F.

Let $\phi(F^\perp)$ be an integral polynomial of $F^\perp$. Let $\dot\nabla_\phi$ be the lift of $\dot\nabla$ to $\phi(F^\perp)$ along F. Then $\dot\nabla_\phi$ is also a flat connection on $\phi(F^\perp)$ along F.

Let ω_ϕ be the lift of ω to $\phi(F^\perp)$. It is easy to see that ω_ϕ can be represented as linear combinations of the powers of ω. In particular, if

$$||\omega||:=\sup_{x\in M}\{|\omega(X)|:X\in\Gamma(F),|X|\le 1\}<1, \tag{2.26}$$

where $|\cdot|$ is the norm with respect to the metric g^{TM}, then there exists a constant $C_\phi>0$ such that for any $X\in\Gamma(F)$ with $|X|\le 1$,

$$|\omega_\phi(X)|\le C_\phi||\omega||. \tag{2.27}$$

Also, similar to (2.25), we have

$$\left(\dot\nabla_\phi\right)^2=-\frac{1}{4}\omega_\phi^2 \tag{2.28}$$

along F.

Now for any $\varepsilon>0$, let $g^{TM,\varepsilon}$ be the metric

$$g^{TM,\varepsilon}=g^F\oplus\frac{1}{\varepsilon}g^{F^\perp}. \tag{2.29}$$

Let $D_{F,\phi(F^\perp),\varepsilon}$ be the elliptic operator as constructed in last subsection but for the metric $g^{TM,\varepsilon}$. Let $\Delta^{F,\phi(F^\perp),\varepsilon}$ be the corresponding Bochner Laplacian. We will use suitable subscript or superscript on the corresponding geometric quantities to indicate that they are with respect to $g^{TM,\varepsilon}$.

We now examine the behaviour of $D^2_{F,\phi(F^\perp),\varepsilon}$ when $\varepsilon \to 0$. Recall that $\{f_i\}_{i=1}^p$, $\{h_s\}_{s=1}^q$ constitute an orthonormal basis of $g^{TM} = g^F \oplus g^{F^\perp}$. Thus $\{f_i\}_{i=1}^p$, $\{\sqrt{\varepsilon}h_s\}_{s=1}^q$ is an orthonormal basis for $g^{TM,\varepsilon}$.

By applying Theorem 2.3 to this situation, we get

$$\begin{aligned}
D^2_{F,\phi(F^\perp),\varepsilon} = -\Delta^{F,\phi(F^\perp),\varepsilon} &+ \frac{1}{2}\sum_{i,j=1}^{p} c(f_i)c(f_j)R^{\phi(F^\perp),\varepsilon}(f_i,f_j) \\
&+\sqrt{\varepsilon}\sum_{i=1}^{p}\sum_{s=1}^{q} c(f_i)c(h_s)R^{\phi(F^\perp),\varepsilon}(f_i,h_s) \\
&+\frac{\varepsilon}{2}\sum_{s,t=1}^{q} c(h_s)c(h_t)R^{\phi(F^\perp),\varepsilon}(h_s,h_t) + \frac{k_{TM,\varepsilon}}{4} \\
+\frac{\sqrt{\varepsilon}}{4}&\sum_{i=1}^{p}\sum_{r,s,t=1}^{q} \left\langle R^{F^\perp,\varepsilon}(f_i,h_r)h_t,h_s\right\rangle c(f_i)c(h_r)\widehat{c}(h_s)\widehat{c}(h_t) \\
+\frac{1}{8}&\sum_{i,j=1}^{p}\sum_{s,t=1}^{q} \left\langle R^{F^\perp,\varepsilon}(f_i,f_j)h_t,h_s\right\rangle c(f_i)c(f_j)\widehat{c}(h_s)\widehat{c}(h_t) \\
+\frac{\varepsilon}{8}&\sum_{s,t,r,l=1}^{q} \left\langle R^{F^\perp,\varepsilon}(h_r,h_l)h_t,h_s\right\rangle c(h_r)c(h_l)\widehat{c}(h_s)\widehat{c}(h_t). \qquad (2.30)
\end{aligned}$$

We first examine the behaviour of $k_{TM,\varepsilon}$ as $\varepsilon \to 0$. By definition, one has

$$\begin{aligned}
k_{TM,\varepsilon} = \sum_{i,j=1}^{p}\left\langle R^{TM,\varepsilon}(f_i,f_j)f_i,f_j\right\rangle &+ \sum_{s,t=1}^{q}\varepsilon\left\langle R^{TM,\varepsilon}(h_s,h_t)h_s,h_t\right\rangle \\
&+\sum_{i=1}^{p}\sum_{s=1}^{q}\left\langle R^{TM,\varepsilon}(f_i,h_s)f_i,h_s\right\rangle, \qquad (2.31)
\end{aligned}$$

where $R^{TM,\varepsilon}$ is the curvature of $\nabla^{TM,\varepsilon}$, the Levi-Civita connection of $g^{TM,\varepsilon}$.

Recall that p, $p^\perp$ are the orthogonal projections, with respect to g^{TM}, from TM to $F, F^\perp$ respectively. From (1.5) to (1.8), we deduce that

$$\begin{aligned}
\left\langle R^{TM,\varepsilon}(f_i,f_j)f_i,f_j\right\rangle &= \left\langle R^{F}(f_i,f_j)f_i,f_j\right\rangle \\
+\varepsilon\left\langle \nabla^{TM}_{f_j}f_j, p^\perp\nabla^{TM}_{f_i}f_i\right\rangle &- \varepsilon\left\langle \nabla^{TM}_{f_i}f_j, p^\perp\nabla^{TM}_{f_j}f_i\right\rangle, \qquad (2.32)
\end{aligned}$$

and, for $X \in \Gamma(F)$, $U \in \gamma(F^\perp)$ that

$$\begin{aligned}
&\left\langle R^{TM,\varepsilon}(X,U)X,U\right\rangle = \\
&\varepsilon\left\langle \nabla^{TM}_{X}p\nabla^{TM}_{U}X,U\right\rangle - \left\langle \nabla^{TM}_{U}\left(p\nabla^{TM,\varepsilon}_{X}X\right),U\right\rangle \\
&-\varepsilon\left\langle \nabla^{TM}_{p[X,U]}X,U\right\rangle - \left\langle \nabla^{TM}_{p^\perp[X,U]}X,U\right\rangle
\end{aligned}$$

$$+\frac{1}{2}\left\langle X,\left[U,p^{\perp}[X,U]\right]\right\rangle-\frac{\varepsilon}{2}\left\langle X,\left[U,p^{\perp}[X,U]\right]\right\rangle$$

$$-\frac{1}{2}\left\langle X,\nabla^{TM}_{p^{\perp}\nabla_U^{TM,\varepsilon}X}U+\nabla_U^{TM}\left(p^{\perp}\nabla_U^{TM,\varepsilon}X\right)\right\rangle$$

$$+\left\langle\left[X,p^{\perp}\nabla_U^{TM,\varepsilon}X\right],U\right\rangle+\frac{\varepsilon}{2}\left\langle X,\left[U,p^{\perp}\nabla_U^{TM,\varepsilon}X\right]\right\rangle$$

$$-\left\langle\nabla_U^{TM}\left(p^{\perp}\nabla_X^{TM,\varepsilon}X\right),U\right\rangle$$

$$=-\varepsilon\left\langle p\nabla_U^{TM}X,\nabla_X^{TM}U\right\rangle-\varepsilon\left\langle\nabla^{TM}_{p[X,U]}X,U\right\rangle$$

$$-\frac{\varepsilon}{2}\left\langle X,\left[U,p^{\perp}[X,U]\right]\right\rangle+\frac{\varepsilon}{2}\left\langle X,p^{\perp}\left(\nabla_U^{TM,\varepsilon}X\right)\right\rangle$$

$$-\varepsilon\left\langle\nabla_U^{TM}\left(p^{\perp}\nabla_X^{TM}X\right),U\right\rangle+\left\langle p\nabla_X^{TM}X,\nabla_U^{TM}U\right\rangle$$

$$+\frac{1}{2}\left\langle X,\nabla_U^{TM}\left(p^{\perp}[X,U]\right)+\nabla^{TM}_{p^{\perp}[X,U]}U\right\rangle+\left\langle\left[X,p^{\perp}\nabla_U^{TM,\varepsilon}X\right],U\right\rangle$$

$$-\frac{1}{2}\left\langle X,\nabla^{TM}_{p^{\perp}\nabla_U^{TM,\varepsilon}X}U+\nabla_U^{TM}\left(p^{\perp}\nabla_U^{TM,\varepsilon}X\right)\right\rangle,\tag{2.33}$$

while for $U,\ V\in\Gamma(F^{\perp})$, we have

$$\left\langle R^{TM,\varepsilon}(U,V)U,V\right\rangle$$

$$=-\frac{1}{2}\left\langle p\nabla_V^{TM,\varepsilon}U,\nabla_U^{TM}V+\nabla_V^{TM}U\right\rangle+\left\langle\nabla^{TM}_{p[U,V]}U,V\right\rangle$$

$$+\frac{\varepsilon}{2}\left\langle p\nabla_V^{TM,\varepsilon}U,[V,U]\right\rangle+\frac{1}{2}\left\langle p\nabla_V^{TM,\varepsilon}U,\nabla_U^{TM}V+\nabla_V^{TM}U\right\rangle$$

$$-\frac{\varepsilon}{2}\left\langle p\nabla_U^{TM,\varepsilon}V,[V,U]\right\rangle+\left\langle R^{F^{\perp}}(U,V)U,V\right\rangle-\left\langle\nabla^{TM,\varepsilon}_{p[U,V]}U,V\right\rangle$$

$$=\left\langle R^{F^{\perp}}(U,V)U,V\right\rangle+\frac{1}{2}\left\langle[U,V],p\left(\nabla_U^{TM}V+\nabla_V^{TM}U\right)\right\rangle$$

$$-\left(\frac{1}{2}-\varepsilon\right)\left\langle p[U,V],[U,V]\right\rangle.\tag{2.34}$$

On the other hand, by (1.7), one has

$$p^{\perp}\nabla_U^{TM,\varepsilon}X=\sum_{s=1}^{q}\left\langle\nabla_U^{TM,\varepsilon}X,h_s\right\rangle h_s$$

$$=-\sum_{s=1}^{q}\left\{\frac{1}{2}\left\langle X,\nabla_U^{TM}h_s+\nabla_{h_s}^{TM}U\right\rangle h_s+\frac{\varepsilon}{2}\left\langle X,[U,h_s]\right\rangle h_s\right\}$$

$$=\frac{1}{2}\sum_{s=1}^{q}\omega(X)(U,h_s)h_s-\frac{\varepsilon}{2}\sum_{s=1}^{q}\left\langle X,[U,h_s]\right\rangle h_s.\tag{2.35}$$

Here we have used the notation of §1 and also formula (1.13). Also since the Bott connection (1.9) is flat, by (1.10) one verifies directly that (cf. [**2**])

$$\dot{\nabla}\omega=-\frac{1}{4}\omega^2,\tag{2.36}$$

along F.

Equations (2.35) and (2.36) can be used to control the term

$$\left\langle \left[X, p^{\perp}\nabla_U^{TM,\varepsilon} X\right], U\right\rangle \tag{2.37}$$

in (2.33). Combining with (1.13), (2.22) and (2.32)–(2.36), one finds that there are positive constants $\varepsilon_0 > 0$, $C > 0$ such that if $0 < \varepsilon < \varepsilon_0$, then

$$k_{TM,\varepsilon} > \delta - C\|\omega\|. \tag{2.38}$$

Now examining the other curvature terms appearing in (2.30). This part is easier, as we have the convergence formula (1.12).

Note that $\dot{\nabla}$ as well as $\dot{\nabla}_\phi$ are flat along F. By (1.11), (2.27) and (2.28), we know that whenever $\|\omega\| < 1$, there exist positive constants C', ε_0', such that if $0 < \varepsilon < \varepsilon_0'$, then

$$\begin{aligned}\Bigg| \frac{1}{8}\sum_{i,j=1}^{p}\sum_{s,t=1}^{q} & \left\langle R^{F^{\perp},\varepsilon}(f_i,f_j)h_t,h_s\right\rangle c(f_i)c(f_j)\widehat{c}(h_s)\widehat{c}(h_t)\\
+\frac{\sqrt{\varepsilon}}{4}\sum_{i=1}^{p}\sum_{r,s,t=1}^{q} & \left\langle R^{F^{\perp},\varepsilon}(f_i,h_r)h_t,h_s\right\rangle c(f_i)c(h_r)\widehat{c}(h_s)\widehat{c}(h_t)\\
+\frac{\varepsilon}{8}\sum_{s,t,r,l=1}^{q} & \left\langle R^{F^{\perp},\varepsilon}(h_r,h_l)h_t,h_s\right\rangle c(h_r)c(h_l)\widehat{c}(h_s)\widehat{c}(h_t)\\
& +\frac{1}{2}\sum_{i,j=1}^{p} c(f_i)c(f_j)R^{\phi(F^{\perp}),\varepsilon}(f_i,f_j)\\
& +\sqrt{\varepsilon}\sum_{i=1}^{p}\sum_{s=1}^{q} c(f_i)c(h_s)R^{\phi(F^{\perp}),\varepsilon}(f_i,h_s)\\
& +\frac{\varepsilon}{2}\sum_{s,t=1}^{q} c(h_s)c(h_t)R^{\phi(F^{\perp}),\varepsilon}(h_s,h_t)\Bigg| \leq C'\|\omega\|.\end{aligned} \tag{2.39}$$

Now since we have assumed that (M, F, g^F) is almost Riemannian, we can and we will choose g^{TM} such that

$$\|\omega\| < 1, \quad (C + C')\|\omega\| < \frac{\delta}{8}. \tag{2.40}$$

From (2.30), (2.38), (2.39) and (2.40), we see that there is a positive constant ε_0'' such that when $0 < \varepsilon < \varepsilon_0''$, we have

$$D^2_{F,\phi(F^{\perp}),\varepsilon} > -\Delta^{F,\phi(F^{\perp}),\varepsilon} + \frac{\delta}{8} > 0, \tag{2.41}$$

since $-\Delta^{F,\phi(F^{\perp}),\varepsilon}$ is clearly nonnegative.

Now we can prove the following result which is the Connes vanishing theorem for almost Riemannian foliations.

Theorem 2.6. *Let (M, F, g^F) be an almost Riemannian foliation, assume M is compact and transversally oriented. If F is spin and $k_F > 0$, then*

$$\left\langle \widehat{A}(F)p\left(F^{\perp}\right), [M]\right\rangle = 0, \tag{2.42}$$

where $p(F^{\perp})$ is any Pontrjagin class of $F^{\perp}$.

Proof: By (2.18), (2.41), one deduces that

$$\operatorname{ind} D_{+,F,\phi(F^{\perp}),\varepsilon} = 0. \tag{2.43}$$

Thus by applying the Atiyah-Singer index theorem [**1**], or by a direct heat kernel evaluation, we get

$$\operatorname{ind} D_{+,F,\phi(F^{\perp})} = \int_M \widehat{A}(F)L\left(F^{\perp}\right)\operatorname{ch}\left(\phi\left(F^{\perp}\right)\right) = \operatorname{ind} D_{+,F,\phi(F^{\perp}),\varepsilon} = 0, \tag{2.44}$$

where $L(F^{\perp})$ is the Hirzebruch L-class of $F^{\perp}$.

Now it is a standard fact in topology that any rational Pontrjagin class of $F^{\perp}$ can be expressed as a rational linear combination of the classes of the form $L(F^{\perp})\operatorname{ch}(\phi(F^{\perp}))$. So the theorem follows from (2.44). □

Corollary 2.7. *Under the hyperthesis of Theorem 2.6, one has* $\langle \widehat{A}(TM), [M]\rangle = 0$.

§3. New vanishing theorems

By slightly modifying the construction of the operator defined in §2, one can also prove the following new vanishing results.

Theorem 3.1. *Let (M, F) be an oriented almost Riemannian foliation with F also oriented. If M, instead of F, is spin, then we still have $\langle \widehat{A}(F)p(TM/F), [M]\rangle = 0$, under the same condition that F admits a metric of positive scalar curvature over M.*

Theorem 3.2. *Under the same assumptions as in Theorem 2.6, we have $\langle \widehat{A}(F)e(TM/F), [M]\rangle = 0$, where $e(TM/F)$ is the Euler class of TM/F.*

Appendix. Remarks on the general case

We first show that the almost isometric foliations studied in [**4**] are almost Riemannian in the sense of §2.

Let (M, F) be a foliation, let TM/F be the transversal bundle. Let G be the holonomy groupoid of (M, F) (cf. [**4**], [**7**]), then G acts on TM/F. Let us assume E is a proper subbundle of TM/F. We choose a splitting of TM/F as

$$TM/F = E \oplus (TM/F)/E. \tag{A.1}$$

Let q_1, q_2 be the dimensions of E and $(TM/F)/E$ respectively.

Definition A.1. If there is a metric $g^{TM/F}$ on TM/F with its restriction to E and $(TM/F)/E$ such that the action of G on TM/F takes the form

$$\begin{pmatrix} O(q_1) & A \\ 0 & O(q_2) \end{pmatrix}, \tag{A.2}$$

where $O(q_1)$, $O(q_2)$ are orthogonal matrices of rank q_1 and q_2 respectively, and A is a $q_1 \times q_2$ matrix, then we say that (M, F) carries an almost isometric structure.

Clearly the existence of the almost isometric structure does not depend on the splitting (A.1). Let g^{TF} be a metric on F. Choose a splitting

$$TM = F \oplus F^{\perp}. \tag{A.3}$$

We can and we will identify TM/F with $F^{\perp}$. Thus E and $(TM/F)/E$ are identified with sub-bundles of $F^{\perp}$ as $F_1^{\perp}$ and $F_2^{\perp}$ respectively. Let $g^{F^{\perp}}$ be the metric on $F^{\perp}$ corresponding to the metric on $g^{TM/F}$, let $g^{F_1^{\perp}}$ and $g^{F_2^{\perp}}$ be the restrictions of $g^{F^{\perp}}$ to $F_1^{\perp}$ and $F_2^{\perp}$ respectively. Then we have the orthogonal splitting

$$\begin{aligned} TM &= F \oplus F_1^{\perp} \oplus F_2^{\perp}, \\ g^{TM} &= g^F \oplus g^{F_1^{\perp}} \oplus g^{F_2^{\perp}}. \end{aligned} \tag{A.4}$$

Recall that the action of the holonomy groupoid G can be defined with respect to the canonical flat connection on F, the Bott connection defined in (1.9). Then the almost isometric condition can be rewritten locally as

$$\begin{aligned} \langle [X, U_i], V_i \rangle + \langle U_i, [X, V_i] \rangle &= X \langle U_i, V_i \rangle, \\ \langle [X, U_1], U_2 \rangle &= 0, \end{aligned} \tag{A.5}$$

where $X \in \Gamma(F)$, U_i, $V_i \in \Gamma(F_i^{\perp})$, $i = 1,\ 2$.

Equation (A.5) can be rewritten as

$$\begin{aligned} \langle X, \nabla_{U_i} V_i + \nabla_{V_i} U_i \rangle &= 0, \\ \langle \nabla_X U_1, U_2 \rangle + \langle X, \nabla_{U_1} U_2 \rangle &= 0. \end{aligned} \tag{A.6}$$

Thus for any $X \in \Gamma(F)$, $U = U_1 + U_2 \in \Gamma(F_1^{\perp}) \oplus \Gamma(F_2^{\perp}) = \Gamma(F^{\perp})$, one has, in view of (1.13), that

$$\begin{aligned} \omega(X)(U,U) &= -\frac{1}{2} \left\langle X, \nabla_U^{TM} U \right\rangle = -\frac{1}{2} \left\langle X, \nabla_{U_1}^{TM} U_2 + \nabla_{U_2}^{TM} U_1 \right\rangle \\ &= -\frac{1}{2} \left\langle U_1, \nabla_X^{TM} U_2 \right\rangle + \frac{1}{2} \left\langle U_1, \nabla_{U_2}^{TM} X \right\rangle = -\frac{1}{2} \left\langle [X, U_2], U_1 \right\rangle. \end{aligned} \tag{A.7}$$

Now for any $\gamma > 0$, set

$$g_\gamma^{TM} = g^F \oplus g^{F_1^{\perp}} \oplus \frac{1}{\gamma} g^{F_2^{\perp}}.$$

Clearly, if $U = U_1 + U_2 \in \Gamma(F^{\perp})$ is of norm one in g^{TM}, then $U(\gamma) = U_1 + \sqrt{\gamma} U_2 \in \Gamma(F^{\perp})$ also has norm one in g_γ^{TM}. Let ω_γ be the form as constructed in §1, corresponding to g_γ^{TM}. Then by (A.7), we have

$$\omega_\gamma(X)(U(\gamma), U(\gamma)) = -\frac{1}{2} \sqrt{\gamma} \left\langle [X, U_2], U_1 \right\rangle \tag{A.8}$$

from which one sees that

$$||\omega_\gamma||_{g_\gamma^{TM}} = \sqrt{\gamma} ||\omega||_{g^{TM}}. \tag{A.9}$$

Taking γ to be as small as possible, we arrive at the following

Proposition A.2. *Any almost isometric foliation carries an almost Riemannian structure.*

Now recall that in [**4**], Connes first proved his vanishing theorem for compact almost isometric foliations by using the techniques of cyclic cohomology, and then pass to non-compact manifolds to prove the general case. Thus, what we have done in §2 may be thought of as direct geometric approach of the first step of Connes' proof. As for the general case, it seems one needs a full geometric understanding of the Connes fibration constructed in [**4**], which is non-compact. We leave this for further studies.

References

1. M. F. Atiyah and I. M. Singer. The index of elliptic operators I. *Annals of Math.*, 87 (1968), 484-530.
2. J.-M. Bismut and W. Zhang. An extension of a theorem by Cheeger and Müller. *Astérisque,* Tom. 205, (1992) Paris.
3. R. Bott. On a topological obstruction to integrability. *Global Analysis: Proc. Symp. Pure Math.*, vol.16, (1970), 127-131.
4. A. Connes. Cyclic cohomology and the transverse fundamental class of a foliation. *Geometric Methods in Operator Algebras*, H. Araki eds., pp. 52-144, Pitman Research Notes in Math. Series, vol. 123, 1986.
5. A. Connes and G. Skandalis. The longitudinal index theorem for foliations. *Publ. Res. Inst. Math. Sci. Kyoto*, 20 (1984), 1139-1183.
6. A. Lichnerowicz. Spineurs harmoniques. *C. R. Acad. Sci. Paris, Série A,* 257 (1963), 7-9.
7. E. Winkelnkemper The graph of a foliation. *Ann. Global Anal. Geom.*, 1 (1983), 51-75.
8. S. T. Yau Private communications. 1992.

DEPARTMENT OF MATHEMATICS, ULCA, LOS ANGELES, CA 90095
E-mail address: liu@math.ucla.edu

NANKAI INSTITUTE OF MATHEMATICS, TIANJIN 300071, P. R. CHINA.
E-mail address: weiping@nankai.edu.cn

Contemporary Mathematics
Volume **279**, 2001

Legendrian Links of Topological Unknots

Klaus Mohnke

ABSTRACT. We use an estimate on the Thurston–Bennequin invariant of a Legendrian link in terms of its Kauffman–polynomial to show that links of topological unknots, e.g. the Borromean rings or the Whithead link, may not be represented by Legendrian links of Legendrian unknots.

All Legendrian knots representing the unknot are classified by their Thurston–Bennequin number tb, and their rotation r (see [**2**]).

Thus the Legendre knot given by the wavefront of the 'eye' which has $tb = -1, r = 0$ is referred to as the *trivial Legendrian knot.*

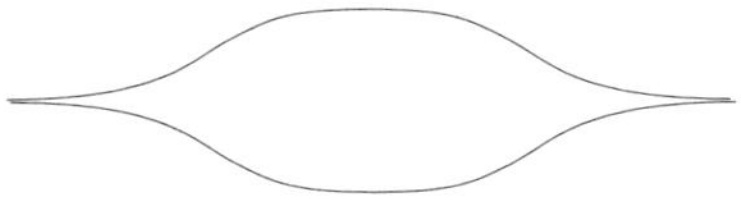

FIGURE 1. A front projection of the Legendrian unknot

Back then there was no other obvious obstruction for Legendrian *links* consisting of (topological) unknots and Eliashberg asked the question: "Given a link of topological unknots, can it be realized as a link of [...] Legendrian unknots?"

The answer is negative in general and the new obstructions are given by a sharper inequality on the Thurston–Bennequin number governed by the Kauffman polynomial $K(x,t)$ which was found by Lee Rudolph in [**4**] (for further reference see e.g. [**1**] and [**3**]). It simply states that the Thurston–Bennequin number is not bigger than the the minimal degree in the variable x of the Kauffman polynomial:

$$tb \leq -\max\text{-}\deg_x K.$$

The contribution of the author is to apply this to links of topological unknots. We had to be careful because the two groups of authors [**1, 3**] used different Kauffman polynomials and thus obtain slightly different inequalities: here we work with the Dubrovnik–version Chmutov and Goryunov used. Let us first recall the definition of the Thurston–Bennequin number:

DEFINITION 1. Let L be an oriented Legendrian link given by a wave front projection. Then the *Thurston–Bennequin number* of L, $tb(L)$, is the number of

2000 *Mathematics Subject Classification.* Primary 57R17; Secondary 57M27.
Key words and phrases. Legendrian links, Kauffman–polynomial.

sideward crossings minus the number of up– or downward crossings minus half the number of cusps.

From that the following observation is immediate

LEMMA 2. *Let $L = \coprod_i L_i$ be a Legendrian link such that, for $i \neq j$, the linking number of L_i and L_j is zero. Then the Thurston–Bennequin number of that link is simply given by the sum of those of the components*

$$tb(L) = \sum_i tb(L_i).$$

In particular, it does not depend on the orientation of the components.

To investigate Eliashberg's question we took the simplest examples we knew: The Borromean rings B and the Whitehead link W (see Figure 3 for which of the two possible Whitehead links we investigate here). The Kauffman polynomials are given by

$$\begin{aligned} K_W(x,y) = {} & yx^5 - 2x^4 - (2y^3+6y)x^3 + (-y^4-y^2+6+y^{-2})x^2 + (3y^3+9y+2y^{-1})x \\ & + (y^4+y^2-5-2y^{-2}) - (y^3+4y-2y^{-1})x^{-1} + (2+y^{-2})x^{-2} \end{aligned}$$

and

$$\begin{aligned} K_B(x,t) = {} & y^2x^4 + (-4y+y^{-3})x^3 + (-3y^4-10y^2+3y^{-2})x^2 \\ & + (-2y^5-2y^3+14y+3y^{-1}-3y^{-3})x + (6y^4+18y^2+1-6y^{-2}) \\ & - (-2y^5-2y^3+14y+3y^{-1}-3y^{-3})x^{-1} + (-3y^4-10y^2+3y^{-2})x^{-2} \\ & - (-4y+y^{-3})x^{-3} + y^2x^4. \end{aligned}$$

From that we easily deduce the main result of this note

PROPOSITION 3. *(1) For any Legendrian representation of the Borromean rings we have*

$$tb \leq -4.$$

(2) For any Legendrian representation of the Whitehead link we have

$$tb \leq -5.$$

Thus for both not all components may be Legendrian unknots.

REMARK 4. (1) In the cases of the Borromean rings and the Whitehead link the inequality is sharp as Figures 2 and 3 show.

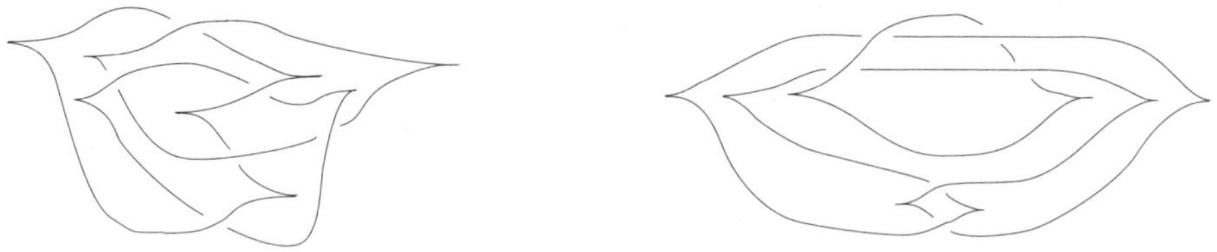

FIGURE 2. Legendrian Borromean rings with $tb = -4$ and Legendrian Whitehead mirror image consisting of Legendrian unknots

(2) In the case of the Whitehead link W the Kauffman polynomial gives no further obstruction for its mirror image. Indeed it can be represented as a Legendrian link of Legendrian unknots (see figure above).

On the other hand it is possible to give two different Legendrian Whithead links with $tb = -5$. One consists of components with Thurston–Bennequin numbers -4 and -1 the other with -3 and -2 as shown in the following figure.

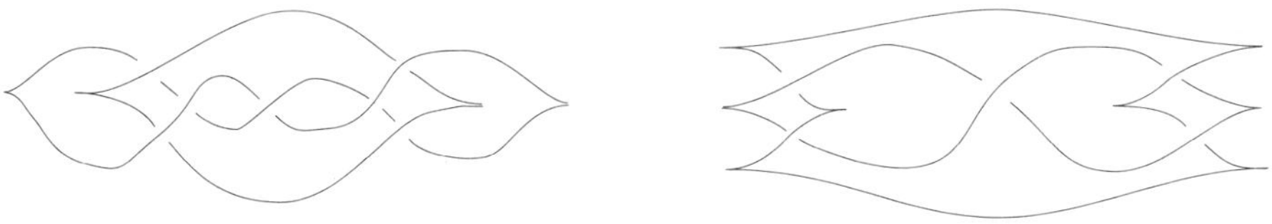

FIGURE 3. Two Legendrian Whitehead links with $tb = -5$

QUESTIONS 5. (1) Are there sharper bounds on the Thurston–Bennequin number than that given in [**1, 3**]?
(2) Are there sharper bounds if one imposes additional conditions? E.g. consider Brunnian links, i.e. links of unknots which fall apart if one removes one component.
(3) Is it at least true that a link of the type described in (2) may be represented as a Legendrian link consisting of Legendrian unknots iff the maximal degree in x of its Kauffman polynomial is equal to the number of components of the link (note that this degree is never less than the number of components)?
(4) Are there further restrictions for the distribution of the Thurston–Bennequin number on the components of a Legendrian link?

Acknowledgements. I would like to thank Uwe Kaiser and Alexander Pilz for patiently checking my computations.

References

1. S. Chmutov and V. Goryunov. Polynomial invariants of Legendrian links and their fronts. In Shin'chin Suzuki, editor, *Proceedings of Knots 96*, pages 239–256. World Scientific Publishing Co., 1997.
2. Yakov Eliashberg and Maia Fraser. Classification of topologically trivial Legendrian knots. In *Geometry, topology, and dynamics* (Montreal, PQ, 1995), 17–51, CRM Proc. Lecture Notes, 15, Amer. Math. Soc., Providence, RI, 1998.
3. Dmitry Fuks and Serge Tabachnikov. Invariants of Legendrian and transverse knots in the standard contact space. *Topology*, 36(5):1025–1053, 1997.
4. Lee Rudolph. A congruence between link polynomials. *Math. Proc. Camb. Phil. Soc.*, 107:319–327, 1990.

UNIVERSITÄT–GH SIEGEN, FACHBEREICH 6,WALTER-FLEX-STR.3, 57068 SIEGEN, GERMANY
Current address: Stanford University, Dept. of Math., Stanford, CA 94305
E-mail address: `mohnke@math.stanford.edu`

Contemporary Mathematics
Volume **279**, 2001

ALGEBRAIC POINCARÉ COBORDISM

ANDREW RANICKI

ABSTRACT. An introduction to the cobordism theory of algebraic Poincaré complexes and some its applications to manifolds, vector bundles and quadratic forms.

Introduction

This paper gives a reasonably leisurely account of the algebraic Poincaré cobordism theory of Ranicki [16], [17] and the further development due to Weiss [20], along with some of the applications to manifolds, vector bundles and quadratic forms. It is a companion paper to Ranicki [18], which is an introduction to algebraic surgery using forms and formations.

Algebraic Poincaré cobordism is modelled on the bordism groups $\Omega_*(X)$ of maps $f : M \to X$ from manifolds to a space X. The Wall [19] surgery obstruction groups $L_*(A)$ of a ring with involution A were expressed in [16] as the cobordism groups of A-module chain complexes C with a quadratic Poincaré duality

$$\psi \;:\; H^{n-*}(C) \;\cong\; H_*(C)\;,$$

and the surgery obstruction $\sigma_*(f,b) \in L_n(\mathbb{Z}[\pi_1(X)])$ of an n-dimensional normal map $(f,b) : M \to X$ was expressed as the cobordism class (C,ψ) of an n-dimensional f.g. free $\mathbb{Z}[\pi_1(X)]$-module chain complex C such that

$$H_*(C) \;=\; \ker(f_* : H_*(\widetilde{M}) \to H_*(\widetilde{X}))$$

together with an n-dimensional quadratic Poincaré duality ψ. The passage from the bundle map $b : \nu_M \to \nu_X$ to ψ used an equivariant chain level version of the relationship established by Thom between the Wu classes of the normal bundle ν_M of a manifold M and the action of the Steenrod algebra on the Thom class of ν_M.

A chain bundle (C,γ) over a ring with involution A is an A-module chain complex C together with a Tate $\mathbb{Z}_2$-hypercohomology class $\gamma \in \widehat{H}^0(\mathbb{Z}_2; C^* \otimes C^*)$. The L-groups $L^*(C,\gamma)$ of [20] are the cobordism groups of symmetric Poincaré

2000 *Mathematics Subject Classification.* Primary 57R67; Secondary 18G35.
Key words and phrases. Chain complexes, Poincaré duality, surgery.

complexes over A with a chain bundle map to (C,γ), which are related to the quadratic L-groups by an exact sequence of abelian groups

$$\cdots \to L_n(A) \to L^n(C,\gamma) \to Q_n(C,\gamma) \to L_{n-1}(A) \to \ldots$$

with the Q-groups $Q_*(C,\gamma)$ defined purely homologically. The surgery obstruction groups $L_*(A)$ of [19] and the symmetric L-groups $L^*(A)$ of Mishchenko [14] are particular examples of the generalized L-groups $L^*(C,\gamma)$. The main novelty of this paper is an explicit formula obtained in §7 for the addition of elements in $Q_*(C,\gamma)$. The Wu classes $v_*(\nu) \in H^*(X;\mathbb{Z}_2)$ of a $(k-1)$-spherical fibration ν over a space X (e.g. the sphere bundle of a k-plane bundle) determine a chain bundle $(C(\widetilde{X}),\gamma(\nu))$ over $\mathbb{Z}[\pi_1(X)]$, with $\widetilde{X}$ the universal cover of X, and with a morphism

$$\pi_{n+k}(T(\nu)) \to Q_n(C(\widetilde{X}),\gamma) \ .$$

For a k-plane bundle ν the morphism factors through the flexible signature map of [20]

$$\Omega_n(X,\nu) \ = \ \pi_{n+k}(T(\nu)) \to L^n(C(\widetilde{X}),\gamma(\nu)) \ .$$

with $\Omega_n(X,\nu)$ the bordism group of normal maps $(f : M \to X, b : \nu_M \to \nu)$ from n-dimensional manifolds.

In subsequent joint work with Frank Connolly a computation of $Q_*(C,\gamma)$ will be used to compute the Cappell Unil-groups in certain special cases.

The titles of the sections are

1. Rings with involution
2. Chain complexes
3. Symmetric, quadratic and hyperquadratic structures
4. Algebraic Wu classes
5. Algebraic Poincaré complexes
6. Chain bundles
7. Normal complexes
8. Normal cobordism
9. Normal Wu classes
10. Forms
11. An example.

§1. Rings with involution

In §1 we show how an involution $a \mapsto \bar{a}$ on a ring A determines a duality involution functor

$$\text{(f.g. projective left } A\text{-modules)} \to \text{(f.g. projective left } A\text{-modules)} \ .$$

More generally, duality can be defined using an antistructure on A in the sense of Wall [19], and the L-theory results described in this paper all have versions for rings with antistructure.

Let A be an associative ring with 1, together with an involution

$$\bar{\ } : A \to A \ ; \ a \mapsto \bar{a} \ ,$$

that is a function satisfying

$$\overline{a+b} = \bar{a} + \bar{b} \ , \ \bar{\bar{a}} = a \ , \ \overline{ab} = \bar{b}.\bar{a} \ , \ \bar{1} = 1 \in A \ (a, b \in A) \ .$$

In the topological applications $A = \mathbb{Z}[\pi]$ is a group ring, for some group π equipped with a morphism $w : \pi \to \mathbb{Z}_2 = \{+1, -1\}$, and the involution is defined by

$$\bar{\ } : A \to A \ ; \ \sum_{g \in \pi} n_g g \mapsto \sum_{g \in \pi} n_g w(g) g^{-1} \ .$$

We take A-modules to be left A-modules, unless a right A-action is expressly specified. Given an A-module M there is defined a right A-module M^t with the same additive group and

$$M^t \times A \to M^t \ ; \ (x, a) \mapsto \bar{a} x \ .$$

The *dual* of an A-module M is the A-module with additive group

$$M^* = \operatorname{Hom}_A(M, A)$$

and A acting by

$$A \times M^* \to M^* \ ; \ (a, f) \mapsto (x \mapsto f(x).\bar{a}) \ .$$

The *dual* of an A-module morphism $f \in \operatorname{Hom}_A(M, N)$ is the A-module morphism

$$f^* : N^* \to M^* \ ; \ g \mapsto (x \mapsto g(f(x))) \ .$$

For a f.g. (finitely generated) projective A-module M the natural A-module isomorphism

$$M \to M^{**} \ ; \ x \mapsto (f \mapsto \overline{f(x)})$$

will be used to identify

$$M^{**} = M \ .$$

§2. Chain complexes

In order to adequately deal with the quadratic nature of the function $C \to C^t \otimes_A C$ sending an A-module chain complex C to the $\mathbb{Z}$-module chain complex $C^t \otimes_A C$ it is necessary to use the equivalence of Dold [7] and Kan [8] (cf. May [10, §22]) between positive $\mathbb{Z}$-module chain complexes and simplicial $\mathbb{Z}$-modules. This is now recalled, along with some other properties of chain complexes that we shall require.

An A-module chain complex

$$C : \cdots \to C_{r+1} \xrightarrow{d} C_r \xrightarrow{d} C_{r-1} \to \ldots \ (r \in \mathbb{Z})$$

is *n-dimensional* if each C_r $(0 \leq r \leq n)$ is a f.g. projective A-module and $C_r = 0$ for $r < 0$ and $r > n$. By an abuse of terminology chain complexes of the chain homotopy type of an n-dimensional chain complex will also be called n-dimensional.

The *suspension* of an A-module chain complex C is the A-module chain complex defined by

$$d_{SC} = d_C : SC_r = C_{r-1} \to SC_{r-1} = C_{r-2} \ .$$

If C is n-dimensional then SC is $(n+1)$-dimensional.

Given an A-module chain complex C let

$$C^r = (C_r)^* \ (r \in \mathbb{Z}) .$$

The *dual A-module* chain complex C^{-*} is defined by

$$d_{C^{-*}} = (d_C)^* : (C^{-*})_r = C^{-r} \to (C^{-*})_{r-1} = C^{-r+1} .$$

The *n-dual A-module chain complex* C^{n-*} is defined by

$$d_{C^{n-*}} = (-1)^r (d_C)^* : (C^{n-*})_r = C^{n-r} \to (C^{n-*})_{r-1} = C^{n-r+1} .$$

The n-fold suspension of the dual $S^n C^{-*}$ is related to the n-dual C^{n-*} by the isomorphism

$$S^n C^{-*} \to C^{n-*} ; \ x \mapsto (-1)^{r(r-1)/2} x \ (x \in C^{n-r}) .$$

In particular, C^{0-*} is isomorphic to (but not identical to) C^{-*}.

A *chain map up to sign* between A-module chain complexes

$$f : C \to D$$

is a collection of A-module morphisms

$$\{f \in \mathrm{Hom}_A(C_r, D_r) \,|\, r \in \mathbb{Z}\}$$

such that

$$d_D f = \pm f d_C : C_r \to D_{r-1} \ (r \in \mathbb{Z}) .$$

If the sign is always $+$ this is just a *chain map* $f : C \to D$, as usual.

Given A-module chain complexes C, D let $C^t \otimes_A D$, $\mathrm{Hom}_A(C, D)$ be the $\mathbb{Z}$-module chain complexes defined by

$$(C^t \otimes_A D)_n = \sum_{p+q=n} C_p \otimes_A D_q ,$$

$$d_{C^t \otimes_A D}(x \otimes y) = x \otimes d_D(y) + (-1)^q d_C(x) \otimes y ,$$

$$\mathrm{Hom}_A(C, D)_n = \sum_{q-p=n} \mathrm{Hom}_A(C_p, D_q) ,$$

$$d_{\mathrm{Hom}_A(C,D)}(f)(x) = d_D(f(x)) + (-1)^q f(d_C(x)) .$$

A cycle $f \in \mathrm{Hom}_A(C, D)_n$ is a chain map up to sign $f : S^n C \to D$, and

$$H_n(\mathrm{Hom}_A(C, D)) = H_0(\mathrm{Hom}_A(S^n C, D))$$

is the $\mathbb{Z}$-module of chain homotopy classes of chain maps $S^n C \to D$.

For finite-dimensional C the *slant isomorphism* of $\mathbb{Z}$-module chain complexes

$$C^t \otimes_A D \to \mathrm{Hom}_A(C^{-*}, D) ; \ x \otimes y \mapsto (f \mapsto \overline{f(x)} \,.\, y)$$

will be used to identify

$$C^t \otimes_A D = \mathrm{Hom}_A(C^{-*}, D) .$$

A cycle

$$f \in (C^t \otimes_A D)_n = \mathrm{Hom}_A(C^{-*}, D)_n$$

is a chain map $f : C^{n-*} \to D$. Thus

$$H_n(C^t \otimes_A D) \;=\; H_n(\mathrm{Hom}_A(C^{-*}, D)) \;=\; H_0(\mathrm{Hom}_A(C^{n-*}, D))$$

is the $\mathbb{Z}$-module of chain homotopy classes of chain maps $C^{n-*} \to D$.

The *algebraic mapping cone* $C(f)$ of an A-module chain map $f : C \to D$ is the A-module chain complex defined as usual by

$$d_{C(f)} \;=\; \begin{pmatrix} d_D & (-1)^{r-1}f \\ 0 & d_C \end{pmatrix} \;:\; C(f)_r \;=\; D_r \oplus C_{r-1} \to C(f)_{r-1} \;=\; D_{r-1} \oplus C_{r-2} \; .$$

The relative homology A-modules

$$H_n(f) \;=\; H_n(C(f))$$

are such that there is defined an exact sequence

$$\dots \to H_n(C) \xrightarrow{f_*} H_n(D) \to H_n(f) \to H_{n-1}(C) \to \dots \; .$$

Let $C(\Delta^n)$ denote the cellular chain complex of the standard n-simplex Δ^n with the standard cell structure consisting of $\binom{n+1}{r+1}$ r-cells $(0 \leq r \leq n)$.

Given a $\mathbb{Z}$-module chain complex C let $K(C)$ denote the simplicial $\mathbb{Z}$-module defined by the Dold-Kan construction, with one n-simplex for each chain map $C(\Delta^n) \to C$ and the evident face and degeneracy maps d_i, s_i, such that

$$\pi_n(K(C)) \;=\; H_n(C) \; (n \geq 0) \; .$$

Given a chain $y \in C_n$ and cycles $x_i \in \ker(d : C_{n-1} \to C_{n-2})$ $(0 \leq i \leq n)$ such that

$$dy \;=\; \sum_{i=0}^{n} (-1)^i x_i \in C_{n-1}$$

let $(y; x_0, \dots, x_n)$ denote the n-simplex of $K(C)$ defined by the chain map

$$f \;:\; C(\Delta^n) \to C$$

with

$$\begin{aligned} f \;:\; C(\Delta^n)_n \;&=\; \mathbb{Z} \to C_n \;;\; 1 \mapsto y \; , \\ d_i f \;:\; C(\Delta^{n-1})_{n-1} \;&=\; \mathbb{Z} \to C_{n-1} \;;\; 1 \mapsto x_i \; (0 \leq i \leq n) \; . \end{aligned}$$

The chain $x \in C_n$ is identified with the n-simplex $(x; dx, 0, \dots, 0) \in K(C)^{(n)}$.

Given $\mathbb{Z}$-module chain complexes C, D and a simplicial map

$$f \;:\; K(C) \to K(D)$$

(which need not preserve the $\mathbb{Z}$-module structure) there is defined a function

$$f \;:\; C_n \to D_n \;;\; x \mapsto f(x) \;=\; f(x; dx, 0, \dots, 0)$$

such that

$$df(x) \;=\; f(dx) \in D_{n-1} \; (x \in C_n) \; .$$

In general, $f : C_n \to D_n$ is only additive on the level of homology, with

$$f(x + x') - f(x) - f(x') \;=\; d[x, x']_f \in D_n \; (x, x' \in \ker(d : C_n \to C_{n-1})$$

where the function

$$[\ ,\ ]_f\ :\ \ker(d:C_n\to C_{n-1})\times\ker(d:C_n\to C_{n-1})\to D_{n+1}$$

is defined by

$$[x,x']_f\ =\ f(0;x+x',x,-x',0,\dots,0)-f(0;x,x,0,\dots,0)-f(0;x',0,-x',0,\dots,0)\ .$$

The induced functions

$$f_*\ :\ H_n(C)\to H_n(D)\ ;\ x\mapsto f(x)$$

are thus morphisms of abelian groups, which fit into an exact sequence

$$\dots\to H_{n+1}(f)\to H_n(C)\xrightarrow{f_*}H_n(D)\to H_n(f)\to H_{n-1}(C)\to\dots\ ,$$

with the relative group $H_n(f)$ $(=\pi_n(f))$ the set of equivalence classes of pairs $(x,y)\in C_n\times D_{n+1}$ such that

$$dx\ =\ 0\in C_{n-1}\ ,\ f(x)\ =\ dy\in D_n\ ,$$

subject to the equivalence relation

$(x,y)\sim(x',y')$ if there exist $(u,v)\in C_{n+1}\times D_{n+2}$ such that

$$x-x'\ =\ du\in C_n\ ,\ y-y'\ =\ f(u;x,x',0,\dots,0)+dv\in D_{n+1}\ ,$$

and addition by

$$(x,y)+(x',y')\ =\ (x+x',y+y'+[x,x']_f)\in H_n(f)\ .$$

If $f:K(C)\to K(D)$ does preserve the $\mathbb{Z}$-module structure (so that $[\ ,\ ]_f=0$) then f is essentially just a chain map $f:C\to D$, and the relative homology groups $H_*(f)$ are just the homology groups $H_*(C(f))$ of the algebraic mapping cone $C(f)$, as usual.

§3. Symmetric, quadratic and hyperquadratic structures

An n-dimensional $\begin{cases}\text{symmetric}\\ \text{quadratic}\\ \text{hyperquadratic}\end{cases}$ structure on an A-module chain complex C is a cycle representing an element of the $\begin{cases}\mathbb{Z}_2\text{-hypercohomology}\\ \mathbb{Z}_2\text{-hyperhomology}\\ \text{Tate }\mathbb{Z}_2\text{-hypercohomology}\end{cases}$ group $\begin{cases}H^n(\mathbb{Z}_2;C^t\otimes_A C)\\ H_n(\mathbb{Z}_2;C^t\otimes_A C)\\ \widehat{H}^n(\mathbb{Z}_2;C^t\otimes_A C)\end{cases}$ in the sense of Cartan and Eilenberg [6].

Given an A-module chain complex C let the generator $T\in\mathbb{Z}_2$ act on $C^t\otimes_A C$ by the *transposition involution*

$$T\ :\ C^t\otimes_A C\to C^t\otimes_A C\ ;\ x\otimes y\ \mapsto\ (-1)^{pq}y\otimes x\ (x\in C_p,y\in C_q)\ .$$

For finite-dimensional C use the slant isomorphism to identify

$$C^t\otimes_A C\ =\ \mathrm{Hom}_A(C^{-*},C)\ .$$

Under this identification the transposition involution corresponds to the duality involution on $\mathrm{Hom}_A(C^{-*}, C)$

$$T \;:\; \mathrm{Hom}_A(C^{-*}, C) \to \mathrm{Hom}_A(C^{-*}, C) \;;\; \phi \mapsto (-1)^{pq}\phi^* \;\; (\phi \in \mathrm{Hom}_A(C^p, C_q)) \;.$$

A cycle $\phi \in \mathrm{Hom}_A(C^{-*}, C)_n = (C^t \otimes_A C)_n$ is a chain map $\phi : C^{n-*} \to C$, and $H_n(\mathrm{Hom}_A(C^{-*}, C))$ is the $\mathbb{Z}$-module of chain homotopy classes of A-module chain maps $C^{n-*} \to C$. Let W be the standard free $\mathbb{Z}[\mathbb{Z}_2]$-module resolution of $\mathbb{Z}$

$$W \;:\; \cdots \to \mathbb{Z}[\mathbb{Z}_2] \xrightarrow{1-T} \mathbb{Z}[\mathbb{Z}_2] \xrightarrow{1+T} \mathbb{Z}[\mathbb{Z}_2] \xrightarrow{1-T} \mathbb{Z}[\mathbb{Z}_2] \to 0$$

and let $\widehat{W}$ be the complete resolution

$$\widehat{W} \;:\; \cdots \to \mathbb{Z}[\mathbb{Z}_2] \xrightarrow{1-T} \mathbb{Z}[\mathbb{Z}_2] \xrightarrow{1+T} \mathbb{Z}[\mathbb{Z}_2] \xrightarrow{1-T} \mathbb{Z}[\mathbb{Z}_2] \to \dots \;.$$

The $\left\{\begin{array}{l} \mathbb{Z}_2\textit{-hypercohomology} \\ \mathbb{Z}_2\textit{-hyperhomology} \\ \textit{Tate } \mathbb{Z}_2\textit{-hypercohomology} \end{array}\right.$ *groups* of a $\mathbb{Z}[\mathbb{Z}_2]$-module chain complex C are defined by

$$\left\{\begin{array}{rcl} H^n(\mathbb{Z}_2; C) & = & H_n(\mathrm{Hom}_{\mathbb{Z}[\mathbb{Z}_2]}(W, C)) \\ H_n(\mathbb{Z}_2; C) & = & H_n(W \otimes_{\mathbb{Z}[\mathbb{Z}_2]} C) \\ \widehat{H}^n(\mathbb{Z}_2; C) & = & H_n(\mathrm{Hom}_{\mathbb{Z}[\mathbb{Z}_2]}(\widehat{W}, C)) \;. \end{array}\right.$$

The evident short exact sequence of $\mathbb{Z}[\mathbb{Z}_2]$-module chain complexes

$$0 \to SW^{-*} \to \widehat{W} \to W \to 0$$

induces a long exact sequence of abelian groups

$$\cdots \to H_n(\mathbb{Z}_2; C) \xrightarrow{1+T} H^n(\mathbb{Z}_2; C) \xrightarrow{J} \widehat{H}^n(\mathbb{Z}_2; C) \xrightarrow{H} H_{n-1}(\mathbb{Z}_2; C) \to \dots \;.$$

An element $\left\{\begin{array}{l} \phi \in H^n(\mathbb{Z}_2; C) \\ \psi \in H_n(\mathbb{Z}_2; C) \\ \theta \in \widehat{H}^n(\mathbb{Z}_2; C) \end{array}\right.$ is represented by an n-cycle of $\left\{\begin{array}{l} \mathrm{Hom}_{\mathbb{Z}[\mathbb{Z}_2]}(W, C) \\ W \otimes_{\mathbb{Z}[\mathbb{Z}_2]} C \\ \mathrm{Hom}_{\mathbb{Z}[\mathbb{Z}_2]}(\widehat{W}, C) \end{array}\right.$ which is just a collection of chains of C $\left\{\begin{array}{l} \{\phi_s \in C_{n+s} \,|\, s \geq 0\} \\ \{\psi_s \in C_{n-s} \,|\, s \geq 0\} \\ \{\theta_s \in C_{n+s} \,|\, s \in \mathbb{Z}\} \end{array}\right.$ such that

$$\left\{\begin{array}{rcl} d_C(\phi_s) + (-1)^{n+s-1}(\phi_{s-1} + (-1)^s T\phi_{s-1}) & = & 0 \in C_{n+s-1} \;\; (s \geq 0, \phi_{-1} = 0) \\ d_C(\psi_s) + (-1)^{n-s-1}(\psi_{s+1} + (-1)^{s+1} T\psi_{s+1}) & = & 0 \in C_{n-s-1} \;\; (s \geq 0) \\ d_C(\theta_s) + (-1)^{n+s-1}(\theta_{s-1} + (-1)^s T\theta_{s-1}) & = & 0 \in C_{n+s-1} \;\; (s \in \mathbb{Z}) \end{array}\right.$$

with

$$1+T \;:\; H_n(\mathbb{Z}_2;C) \to H^n(\mathbb{Z}_2;C) \;;$$
$$\psi \;=\; \{\psi_s \,|\, s \geq 0\} \mapsto \{(1+T)\psi_s = \begin{cases} (1+T)\psi_0 & \text{if } s=0 \\ 0 & \text{if } s \geq 1 \end{cases}\} ,$$
$$J \;:\; H^n(\mathbb{Z}_2;C) \to \widehat{H}^n(\mathbb{Z}_2;C) \;;$$
$$\phi \;=\; \{\phi_s \,|\, s \geq 0\} \mapsto \{J\phi_s = \begin{cases} \phi_s & \text{if } s \geq 0 \\ 0 & \text{if } s \leq -1 \end{cases}\} ,$$
$$H \;:\; \widehat{H}^n(\mathbb{Z}_2;C) \to H_{n-1}(\mathbb{Z}_2;C) \;;$$
$$\theta \;=\; \{\theta_s \,|\, s \in \mathbb{Z}\} \mapsto H\theta \;=\; \{H\theta_s = \theta_{s-1} \,|\, s \geq 0\} .$$

Given an A-module chain complex C use the action of $T \in \mathbb{Z}_2$ on $C^t \otimes_A C$ by the transposition involution to define the $\mathbb{Z}$-module chain complex

$$\begin{aligned} W^{\%}C &= \mathrm{Hom}_{\mathbb{Z}[\mathbb{Z}_2]}(W, C^t \otimes_A C) \\ W_{\%}C &= W \otimes_{\mathbb{Z}[\mathbb{Z}_2]} (C^t \otimes_A C) \\ \widehat{W}^{\%}C &= \mathrm{Hom}_{\mathbb{Z}[\mathbb{Z}_2]}(\widehat{W}, C^t \otimes_A C) . \end{aligned}$$

We shall be mainly concerned with finite-dimensional C, using the slant isomorphism to identify

$$C^t \otimes_A C \;=\; \mathrm{Hom}_A(C^{-*}, C)$$

and

$$\begin{aligned} W^{\%}C &= \mathrm{Hom}_{\mathbb{Z}[\mathbb{Z}_2]}(W, \mathrm{Hom}_A(C^{-*}, C)) \\ W_{\%}C &= W \otimes_{\mathbb{Z}[\mathbb{Z}_2]} \mathrm{Hom}_A(C^{-*}, C) \\ \widehat{W}^{\%}C &= \mathrm{Hom}_{\mathbb{Z}[\mathbb{Z}_2]}(\widehat{W}, \mathrm{Hom}_A(C^{-*}, C)) . \end{aligned}$$

An n-*dimensional* $\begin{cases} \textit{symmetric} \\ \textit{quadratic} \\ \textit{hyperquadratic} \end{cases}$ *structure* on a finite-dimensional A-module chain complex C is a cycle $\begin{cases} \phi \in (W^{\%}C)_n \\ \psi \in (W_{\%}C)_n \\ \theta \in (\widehat{W}^{\%}C)_n , \end{cases}$ which is just a collection of A-module morphisms

$$\begin{cases} \{\phi_s \in \mathrm{Hom}_A(C^{n-r+s}, C_r) \,|\, r \in \mathbb{Z}, s \geq 0\} \\ \{\psi_s \in \mathrm{Hom}_A(C^{n-r-s}, C_r) \,|\, r \in \mathbb{Z}, s \geq 0\} \\ \{\theta_s \in \mathrm{Hom}_A(C^{n-r+s}, C_r) \,|\, r \in \mathbb{Z}, s \in \mathbb{Z}\} \end{cases}$$

such that

$$\begin{cases} d\phi_s + (-1)^r \phi_s d^* + (-1)^{n+s-1}(\phi_{s-1} + (-1)^s T\phi_{s-1}) \;=\; 0 \\ \qquad\qquad : C^{n-r+s-1} \to C_r \ (s \geq 0, \phi_{-1} = 0) \\ d\psi_s + (-1)^r \psi_s d^* + (-1)^{n-s-1}(\psi_{s+1} + (-1)^{s+1} T\psi_{s+1}) \;=\; 0 \\ \qquad\qquad : C^{n-r-s-1} \to C_r \ (s \geq 0) \\ d\theta_s + (-1)^r \theta_s d^* + (-1)^{n+s-1}(\theta_{s-1} + (-1)^s T\theta_{s-1}) \;=\; 0 \\ \qquad\qquad : C^{n-r+s-1} \to C_r \ (s \in \mathbb{Z}) . \end{cases}$$

An *equivalence* $\begin{cases} \xi : \phi \to \phi' \\ \chi : \psi \to \psi' \\ \nu : \theta \to \theta' \end{cases}$ of n-dimensional $\begin{cases} \text{symmetric} \\ \text{quadratic} \\ \text{hyperquadratic} \end{cases}$ structures on C is a chain $\begin{cases} \xi \in (W^{\%}C)_{n+1} \\ \chi \in (W_{\%}C)_{n+1} \\ \nu \in (\widehat{W}^{\%}C)_{n+1} \end{cases}$ such that

$$\begin{cases} \phi' - \phi \ = \ d(\xi) \in (W^{\%}C)_n \\ \psi' - \psi \ = \ d(\chi) \in (W_{\%}C)_n \\ \theta' - \theta \ = \ d(\nu) \in (\widehat{W}^{\%}C)_n \ . \end{cases}$$

The n-*dimensional* $\begin{cases} \textit{symmetric} \\ \textit{quadratic} \\ \textit{hyperquadratic} \end{cases}$ *structure group* $\begin{cases} Q^n(C) \\ Q_n(C) \\ \widehat{Q}^n(C) \end{cases}$ of a chain complex C is the abelian group of equivalence classes of n-dimensional $\begin{cases} \text{symmetric} \\ \text{quadratic} \\ \text{hyperquadratic} \end{cases}$ structures on C, that is

$$\begin{cases} Q^n(C) \ = \ H^n(\mathbb{Z}_2; C^t \otimes_A C) \ = \ H_n(W^{\%}C) \\ Q_n(C) \ = \ H_n(\mathbb{Z}_2; C^t \otimes_A C) \ = \ H_n(W_{\%}C) \\ \widehat{Q}^n(C) \ = \ \widehat{H}^n(\mathbb{Z}_2; C^t \otimes_A C) \ = \ H_n(\widehat{W}^{\%}C) \ . \end{cases}$$

The Q-groups are related by a long exact sequence

$$\dots \to Q_n(C) \xrightarrow{1+T} Q^n(C) \xrightarrow{J} \widehat{Q}^n(C) \xrightarrow{H} Q_{n-1}(C) \to \dots$$

involving the morphisms induced in homology by the $\mathbb{Z}$-module chain maps

$$1+T \ : \ W_{\%}C \to W^{\%}C \ , \ J \ : \ W^{\%}C \to \widehat{W}^{\%}C \ , \ H \ : \ \widehat{W}^{\%}C \to S(W_{\%}C)$$

defined by

$$\begin{aligned} &1+T \ : \ (W_{\%}C)_n \to (W^{\%}C)_n \ ; \\ &\qquad \{\psi_s \in (C^t \otimes_A C)_{n-s} \,|\, s \geq 0\} \mapsto \{((1+T)\psi)_s = \begin{cases} (1+T)\psi_0 & \text{if } s = 0 \\ 0 & \text{if } s \geq 1 \end{cases}\} \ , \\ &J \ : \ (W^{\%}C)_n \to (\widehat{W}^{\%}C)_n \ ; \\ &\qquad \{\phi_s \in (C^t \otimes_A C)_{n+s} \,|\, s \geq 0\} \mapsto \{(J\phi)_s = \begin{cases} \phi_s & \text{if } s \geq 0 \\ 0 & \text{if } s \leq -1 \end{cases}\} \ , \\ &H \ : \ (\widehat{W}^{\%}C)_n \to (W_{\%}C)_{n-1} \ ; \\ &\qquad \{\theta_s \in (C^t \otimes_A C)_{n+s} \,|\, s \in \mathbb{Z}\} \ \mapsto \ \{(H\theta)_s = \theta_{-s-1} \,|\, s \geq 0\} \ . \end{aligned}$$

An n-dimensional symmetric structure $\phi \in (W^{\%}C)_n$ is equivalent to the symmetrization $(1+T)\psi$ of an n-dimensional quadratic structure $\psi \in (W_{\%}C)_n$ if and only if the n-dimensional hyperquadratic structure $J(\phi) \in (\widehat{W}^{\%}C)_n$ is equivalent to 0. An A-module chain map $f : C \to D$ induces a $\mathbb{Z}[\mathbb{Z}_2]$-module chain map

$$f^t \otimes_A f \ : \ C^t \otimes_A C \to D^t \otimes_A D \ ; \ x \otimes y \mapsto f(x) \otimes f(y)$$

and hence $\mathbb{Z}$-module chain maps

$$\begin{aligned} f^{\%} &: W^{\%}C \to W^{\%}D\ , \\ f_{\%} &: W_{\%}C \to W_{\%}D\ , \\ \widehat{f}^{\%} &: \widehat{W}^{\%}C \to \widehat{W}^{\%}D\ . \end{aligned}$$

An A-module chain homotopy

$$g\ :\ f \simeq f'\ :\ C \to D$$

determines $\mathbb{Z}$-module chain homotopies

$$\begin{aligned} (g;f,f')^{\%} &: f^{\%} \simeq f'^{\%}\ :\ W^{\%}C \to W^{\%}D \\ (g;f,f')_{\%} &: f_{\%} \simeq f'_{\%}\ :\ W_{\%}C \to W_{\%}D \\ \widehat{(g;f,f')}^{\%} &: \widehat{f}^{\%} \simeq \widehat{f}'^{\%}\ :\ \widehat{W}^{\%}C \to \widehat{W}^{\%}D \end{aligned}$$

with

$$\begin{aligned} (g;f,f')^{\%}\ :\ (W^{\%}C)_n &= \sum_{s=0}^{\infty}(C^t \otimes_A C)_{n+s} \\ \to (W^{\%}D)_{n+1} &= \sum_{s=0}^{\infty}\sum_q D^t_{n-q+s+1} \otimes_A D_q\ ; \end{aligned}$$

$$\sum_{s=0}^{\infty}\phi_s \mapsto \sum_{s=0}^{\infty}((f^t \otimes_A g + g^t \otimes_A f')(\phi_s) + (-1)^{q+s-1}(g^t \otimes_A g)(T\phi_{s-1}))$$

and similarly for $(g;f,f')_{\%}$, $\widehat{(g;f,f')}^{\%}$. Thus the induced morphisms in the Q-groups

$$\begin{aligned} f^{\%} &: Q^n(C) \to Q^n(D) \\ f_{\%} &: Q_n(C) \to Q_n(D) \\ \widehat{f}^{\%} &: \widehat{Q}^n(C) \to \widehat{Q}^n(D) \end{aligned}$$

depend only on the chain homotopy class of f, and are isomorphisms if f is a chain equivalence. For finite-dimensional C, D the slant isomorphisms are used to identify $f^t \otimes_A f : C^t \otimes_A C \to D^t \otimes_A D$ with

$$\mathrm{Hom}_A(f^*, f)\ :\ \mathrm{Hom}_A(C^{-*}, C) \to \mathrm{Hom}_A(D^*, D)\ ;\ \theta \mapsto f\theta f^*\ ,$$

and similarly for $f^{\%}, f_{\%}, \widehat{f}^{\%}$ and $(g;f,f')^{\%}, (g;f,f')_{\%}, \widehat{(g;f,f')}^{\%}$.

Although all the Q-groups are chain homotopy invariant, only the hyperquadratic Q-groups $\widehat{Q}^*(C)$ are additive. The sum of A-module chain maps $f, g : C \to D$ is an A-module chain map $f + g : C \to D$ such that

$$\begin{aligned} (f+g)^{\%} - f^{\%} - g^{\%}\ &:\ Q^n(C) \to H_n(C^t \otimes_A C) \xrightarrow{f^t \otimes_A g} H_n(D^t \otimes_A D) \to Q^n(D)\ , \\ (f+g)_{\%} - f_{\%} - g_{\%}\ &:\ Q_n(C) \to H_n(C^t \otimes_A C) \xrightarrow{f^t \otimes_A g} H_n(D^t \otimes_A D) \to Q_n(D)\ , \\ \widehat{(f+g)}^{\%} - \widehat{f}^{\%} - \widehat{g}^{\%}\ &=\ 0\ :\ \widehat{Q}^n(C) \to \widehat{Q}^n(D) \end{aligned}$$

with

$$\begin{aligned}
&Q^n(C) \to H_n(C^t \otimes_A C)\ ;\ \phi\ =\ \{\phi_s \,|\, s \geq 0\} \mapsto \phi_0\ ,\\
&Q_n(C) \to H_n(C^t \otimes_A C)\ ;\ \psi\ =\ \{\psi_s \,|\, s \geq 0\} \mapsto (1+T)\psi_0\ ,\\
&H_n(D^t \otimes_A D) \to Q^n(D)\ ;\ \theta \mapsto \{\phi_s = \begin{cases} (1+T)\theta & \text{if } s=0 \\ 0 & \text{if } s \geq 1 \end{cases}\}\ ,\\
&H_n(D^t \otimes_A D) \to Q_n(D)\ ;\ \theta \mapsto \{\psi_s = \begin{cases} \theta & \text{if } s=0 \\ 0 & \text{if } s \geq 1 \end{cases}\}\ .
\end{aligned}$$

Given a finite-dimensional A-module chain complex C and $n \geq 0$ define the *n-fold suspension chain isomorphism*

$$\begin{aligned}
S^n\ &: S^n(\widehat{W}^{\%}C) \to \widehat{W}^{\%}(S^nC)\ ;\\
&\theta\ =\ \{\theta_s \in \mathrm{Hom}_A(C^r, C_{m-r+s}) \,|\, s \in \mathbb{Z}\}\\
&\mapsto\ S^n\theta\ =\ \{(S^n\theta)_s = \theta_{s-n} \in \mathrm{Hom}_A(C^r, C_{m-n+r+s}) \,|\, s \in \mathbb{Z}\}\ .
\end{aligned}$$

For any (finite-dimensional) A-module chain complexes C, D there is defined a simplicial map

$$I\ :\ K(\mathrm{Hom}_A(C,D)) \to K(\mathrm{Hom}_{\mathbb{Z}}(\widehat{W}^{\%}C, \widehat{W}^{\%}D))$$

sending a cycle $f \in \mathrm{Hom}_A(C,D)_n$ (= a chain map up to sign $f : S^nC \to D$) to the $\mathbb{Z}$-module chain map up to sign

$$I(f)\ =\ \widehat{f}^{\%}S^n\ :\ S^n(\widehat{W}^{\%}C) \xrightarrow{S^n} \widehat{W}^{\%}S^nC \xrightarrow{\widehat{f}^{\%}} W^{\%}D\ .$$

An n-simplex $(g; f, f', 0, \dots, 0) \in K(\mathrm{Hom}_A(C,D))^{(n)}$ (= an A-module chain homotopy up to sign $g : f \simeq f' : S^nC \to D$) is sent to the $\mathbb{Z}$-module chain homotopy up to sign

$$I(g; f, f')\ =\ (g; f, f')^{\%}S^n\ :\ I(f)\ \simeq\ I(f')\ :\ S^n(\widehat{W}^{\%}C) \to \widehat{W}^{\%}D\ .$$

The failure of I to be linear on chain maps up to sign $f : S^nC \to D$ is given by the chain homotopy up to sign

$$[f, f']\ :\ (\widehat{f+f'})^{\%}\ \simeq\ \widehat{f}^{\%} + \widehat{f'}^{\%}\ :\ \widehat{W}^{\%}(S^nC) \to \widehat{W}^{\%}D$$

defined by

$$\begin{aligned}
&[f, f']\ :\ (S^n\widehat{W}^{\%}C)_m \to (\widehat{W}^{\%}D)_{m+n+1}\ ;\\
&\theta\ =\ \{\theta_s \in \mathrm{Hom}_A(C^r, C_{m-r+s}) \,|\, s \in \mathbb{Z}\}\\
&\mapsto [f, f']\theta\ =\ \{T^{n+1}f\theta_{s-n+1}f'^* \in \mathrm{Hom}_A(D^r, D_{m-n+r+s+1}) \,|\, s \in \mathbb{Z}\}\ .
\end{aligned}$$

§4. Algebraic Wu classes

The algebraic Wu classes are the fundamental invariants of a duality structure on a chain complex C, which are obtained by an algebraic analogue of the Steenrod squares in the cohomology groups of a topological space. In the topological applications the algebraic Wu classes are closely related to the topological Wu classes, as explained in Ranicki [16].

Let S^rA $(r \in \mathbb{Z})$ denote the A-module chain complex

$$S^rA\ :\ \dots \to 0 \to A \to 0 \to \dots$$

concentrated in degree r. For any A-module chain complex C there are defined natural isomorphisms

$$H_0(\mathrm{Hom}_A(C,S^rA)) \to H^r(C)\ ;\ (f : C_r \to A) \mapsto f^*(1)\ .$$

An element $f \in H^r(C)$ is just a chain homotopy class of chain maps $f : C \to S^rA$. The Wu classes of a quadratic structure on C are the invariants of the equivalence class defined by sending an element $f \in H^r(C)$ to the induced equivalence class of quadratic structures on S^rA. The quadratic structure groups of the elementary complexes S^rA are identified with subquotients of the ground ring A.

An A-*group* M is an abelian group together with an A-action

$$A \times M \to M\ ;\ (a,x) \mapsto ax$$

such that

$$a(x+y)\ =\ ax+ay\ ,\ a(bx)\ =\ (ab)x\ ,\ 1x\ =\ x\ (a,b \in A, x,y \in M)\ .$$

An A-module is an A-group M such that also

$$(a+b)x\ =\ ax+bx \in M\ .$$

An A-*morphism* of A-groups

$$f\ :\ M \to N$$

is a morphism of abelian groups such that

$$f(ax)\ =\ af(x) \in N\ (x \in M, a \in A)\ .$$

The set of A-group morphisms $f : M \to N$ defines an abelian group $\mathrm{Hom}_A(M,N)$, with addition by

$$(f+g)(x)\ =\ f(x)+g(x) \in N\ .$$

For A-modules M,N the A-morphisms $f : M \to N$ coincide with A-module morphisms.

For $\epsilon = \pm 1$ let the generator $T \in \mathbb{Z}_2$ act on A by the ϵ-involution

$$T_\epsilon\ :\ A \to A\ ;\ a \mapsto \epsilon\bar{a}\ .$$

Define the $\begin{cases} \mathbb{Z}_2\textit{-cohomology} \\ \mathbb{Z}_2\textit{-homology} \\ \textit{Tate } \mathbb{Z}_2\textit{-cohomology} \end{cases}$ A-*groups* $\begin{cases} H^*(\mathbb{Z}_2;A,\epsilon) \\ H_*(\mathbb{Z}_2;A,\epsilon) \\ \widehat{H}^*(\mathbb{Z}_2;A,\epsilon) \end{cases}$ by

$$\begin{cases} H^r(\mathbb{Z}_2;A,\epsilon)\ =\ \begin{cases} \ker(1-T_\epsilon : A \to A) & \text{if } r=0 \\ \widehat{H}^r(\mathbb{Z}_2;A,\epsilon) & \text{if } r \geq 1 \\ 0 & \text{if } r<0, \end{cases} \\ H_r(\mathbb{Z}_2;A,\epsilon)\ =\ \begin{cases} \mathrm{coker}(1-T_\epsilon : A \to A) & \text{if } r=0 \\ \widehat{H}^{r+1}(\mathbb{Z}_2;A,\epsilon) & \text{if } r \geq 1 \\ 0 & \text{if } r<0, \end{cases} \\ \widehat{H}^r(\mathbb{Z}_2;A,\epsilon)\ =\ \ker(1-(-1)^rT_\epsilon : A \to A)/\operatorname{im}(1+(-1)^rT_\epsilon : A \to A)\ (r \in \mathbb{Z})\ . \end{cases}$$

The A-action

$$A \times \widehat{H}^r(\mathbb{Z}_2;A,\epsilon) \to \widehat{H}^r(\mathbb{Z}_2;A,\epsilon)\ ;\ (a,x) \mapsto ax\bar{a}$$

defines an A-module structure on $\widehat{H}^r(\mathbb{Z}_2; A, \epsilon)$. The A-actions

$$A \times H^0(\mathbb{Z}_2; A, \epsilon) \to H^0(\mathbb{Z}_2; A, \epsilon) \ ; \ (a, x) \mapsto ax\bar{a}$$
$$A \times H_0(\mathbb{Z}_2; A, \epsilon) \to H_0(\mathbb{Z}_2; A, \epsilon) \ ; \ (a, x) \mapsto ax\bar{a}$$

are not linear in A, and so do not define A-module structures.

For $\epsilon = +1$ the groups $\begin{cases} H^*(\mathbb{Z}_2; A, \epsilon) \\ H_*(\mathbb{Z}_2; A, \epsilon) \\ \widehat{H}^*(\mathbb{Z}_2; A, \epsilon) \end{cases}$ are denoted by $\begin{cases} H^*(\mathbb{Z}_2; A) \\ H_*(\mathbb{Z}_2; A) \\ \widehat{H}^*(\mathbb{Z}_2; A). \end{cases}$

The natural $\mathbb{Z}$-module isomorphisms

$$Q^n(S^r A) \to H^{2r-n}(\mathbb{Z}_2; A, (-1)^r) \ ; \ \phi \mapsto \phi_{2r-n}(1)(1)$$
$$Q_n(S^r A) \to H_{n-2r}(\mathbb{Z}_2; A, (-1)^r) \ ; \ \psi \mapsto \psi_{n-2r}(1)(1)$$
$$\widehat{Q}^n(S^r A) \to \widehat{H}^{r-n}(\mathbb{Z}_2; A, (-1)^r) \ ; \ \theta \mapsto \theta_{2r-n}(1)(1)$$

will be used as identifications.

The *Wu classes* of a $\begin{cases} \text{symmetric} \\ \text{quadratic} \\ \text{hyperquadratic} \end{cases}$ structure $\begin{cases} \phi \in (W^{\%}C)_n \\ \psi \in (W_{\%}C)_n \\ \theta \in (\widehat{W}^{\%}C)_n \end{cases}$ are the invariants of the equivalence class of structures defined by the A-morphisms

$$\begin{cases} v_r(\phi) \ : \ H^{n-r}(C) \ = \ H_0(\mathrm{Hom}_A(C, S^{n-r}A)) \\ \qquad \to Q^n(S^{n-r}A) \ = \ H^{n-2r}(\mathbb{Z}_2; A, (-1)^{n-r}) \ ; \ f \mapsto (f \otimes f)(\phi_{n-2r}) \\ v^r(\psi) \ : \ H^{n-r}(C) \ = \ H_0(\mathrm{Hom}_A(C, S^{n-r}A)) \\ \qquad \to Q_n(S^{n-r}A) \ = \ H_{2r-n}(\mathbb{Z}_2; A, (-1)^{n-r}) \ ; \ f \mapsto (f \otimes f)(\psi_{2r-n}) \\ \widehat{v}_r(\theta) \ : \ H^{n-r}(C) \ = \ H_0(\mathrm{Hom}_A(C, S^{n-r}A)) \\ \qquad \to \widehat{Q}^n(S^{n-r}A) \ = \ \widehat{H}^r(\mathbb{Z}_2; A) \ ; \ f \mapsto (f \otimes f)(\theta_{n-2r}) \ . \end{cases}$$

§5. **Algebraic Poincaré complexes**

An algebraic Poincaré complex is a chain complex with Poincaré duality, such as arises from a compact n-manifold or a normal map.

An *n-dimensional* $\begin{cases} \textit{symmetric} \\ \textit{quadratic} \end{cases}$ *(Poincaré) complex over* A $\begin{cases} (C, \phi) \\ (C, \psi) \end{cases}$ is an n-dimensional A-module chain complex C together with an n-dimensional $\begin{cases} \text{symmetric} \\ \text{quadratic} \end{cases}$ structure $\begin{cases} \phi \in (W^{\%}C) \\ \psi \in (W_{\%}C)_n \end{cases}$ (such that $\begin{cases} \phi_0 \\ (1+T)\psi_0 \end{cases} : C^{n-*} \to C$ is a chain equivalence).

An *$(n+1)$-dimensional* $\begin{cases} \textit{symmetric} \\ \textit{quadratic} \\ \textit{(symmetric, quadratic)} \end{cases}$ *(Poincaré) pair over* A

$$\left(f \ : \ C \to D \ , \ \begin{cases} (\delta\phi, \phi) \\ (\delta\psi, \psi) \\ (\delta\phi, \psi) \end{cases} \right)$$

consists of an n-dimensional A-module chain complex C, an $(n+1)$-dimensional A-module chain complex D, a chain map $f : C \to D$ and a cycle

$$\begin{cases} (\delta\phi, \phi) \in C(f^{\%} : W^{\%}C \to W^{\%}D)_{n+1} = (W^{\%}D)_{n+1} \oplus (W^{\%}C)_n \\ (\delta\psi, \psi) \in C(f_{\%} : W_{\%}C \to W_{\%}D)_{n+1} = (W_{\%}D)_{n+1} \oplus (W_{\%}C)_n \\ (\delta\phi, \psi) \in C((1+T)f_{\%} : W_{\%}C \to W^{\%}D)_{n+1} = (W^{\%}D)_{n+1} \oplus (W_{\%}C)_n \end{cases}$$

(such that the A-module chain map $D^{n+1-*} \to C(f)$ defined by

$$\begin{cases} (\delta\phi, \phi)_0 = \begin{pmatrix} \delta\phi_0 \\ \phi_0 f^* \end{pmatrix} \\ (1+T)(\delta\psi, \psi)_0 = \begin{pmatrix} (1+T)\delta\psi_0 \\ (1+T)\psi_0 f^* \end{pmatrix} \\ (\delta\phi, (1+T)\psi)_0 = \begin{pmatrix} \delta\phi_0 \\ (1+T)\psi_0 f^* \end{pmatrix} \end{cases} : D^{n+1-r} \to C(f)_r = D_r \oplus C_{r-1}$$

is a chain equivalence). The *boundary* of the pair is the n-dimensional $\begin{cases} \text{symmetric} \\ \text{quadratic} \\ \text{quadratic} \end{cases}$ (Poincaré) complex $\begin{cases} (C, \phi) \\ (C, \psi) \\ (C, \psi). \end{cases}$

A *homotopy equivalence* of n-dimensional $\begin{cases} \text{symmetric} \\ \text{quadratic} \end{cases}$ complexes

$$\begin{cases} (f, \chi) : (C, \phi) \to (C', \phi') \\ (f, \xi) : (C, \psi) \to (C', \psi') \end{cases}$$

is a chain equivalence $f : C \to C'$ together with an equivalence of $\begin{cases} \text{symmetric} \\ \text{quadratic} \end{cases}$ structures on C' $\begin{cases} \chi : f^{\%}(\phi) \to \phi' \\ \xi : f_{\%}(\psi) \to \psi' . \end{cases}$ There is a similar notion of homotopy equivalence for pairs.

An n-dimensional $\begin{cases} \text{symmetric} \\ \text{quadratic} \end{cases}$ complex $\begin{cases} (C, \phi) \\ (C, \psi) \end{cases}$ is *connected* if

$$\begin{cases} H_0(\phi_0 : C^{n-*} \to C) = 0 \\ H_0((1+T)\psi_0 : C^{n-*} \to C) = 0 . \end{cases}$$

It was shown in Ranicki [16] that there is a natural one-one correspondence between the homotopy equivalence classes of connected n-dimensional $\begin{cases} \text{symmetric} \\ \text{quadratic} \end{cases}$ complexes over A and the homotopy equivalence classes of n-dimensional $\begin{cases} \text{symmetric} \\ \text{quadratic} \end{cases}$ Poincaré pairs over A. A connected n-dimensional $\begin{cases} \text{symmetric} \\ \text{quadratic} \end{cases}$

complex $\begin{cases} (C,\phi) \\ (C,\psi) \end{cases}$ determines the n-dimensional $\begin{cases} \text{symmetric} \\ \text{quadratic} \end{cases}$ Poincaré pair

$$(i_C : \partial C \to C^{n-*}, \begin{cases} (0,\partial\phi) \\ (0,\partial\psi) \end{cases})$$

defined by

$$\begin{aligned}
i_C \; &= \; (0 \quad 1) \; : \; \partial C_r \; = \; C_{r+1} \oplus C^{n-r} \to C^{n-r} \; , \\
d_{\partial C} \; &= \; \begin{cases} \begin{pmatrix} d_C & (-1)^r \phi_0 \\ 0 & (-1)^r d_C^* \end{pmatrix} : \\ \begin{pmatrix} d_C & (-1)^r (1+T)\psi_0 \\ 0 & (-1)^r d_C^* \end{pmatrix} : \end{cases} \\
&\qquad \partial C_r \; = \; C_{r+1} \oplus C^{n-r} \to \partial C_{r-1} \; = \; C_r \oplus C^{n-r+1} \; , \\
&\begin{cases} \partial\phi_0 \; = \; \begin{pmatrix} (-1)^{n-r-1} T\phi_1 & (-1)^{r(n-r-1)} \\ 1 & 0 \end{pmatrix} : \\ \partial\psi_0 \; = \; \begin{pmatrix} 0 & 0 \\ 1 & 0 \end{pmatrix} : \end{cases} \\
&\qquad \partial C^{n-r-1} \; = \; C^{n-r} \oplus C_{r+1} \to \partial C_r \; = \; C_{r+1} \oplus C^{n-r} \; , \\
&\begin{cases} \partial\phi_s \; = \; \begin{pmatrix} (-1)^{n-r+s-1} T\phi_{s+1} & 0 \\ 0 & 0 \end{pmatrix} : \\ \qquad \partial C^{n-r+s-1} \; = \; C^{n-r+s} \oplus C_{r-s+1} \to \partial C_r \; = \; C_{r+1} \oplus C^{n-r} \; (s \geq 1) \; , \\ \partial\psi_s \; = \; \begin{pmatrix} (-1)^{n-r-s} T\psi_{s-1} & 0 \\ 0 & 0 \end{pmatrix} : \\ \qquad \partial C^{n-r-s-1} \; = \; C^{n-r-s} \oplus C_{r+s+1} \to \partial C_r \; = \; C_{r+1} \oplus C^{n-r} \; (s \geq 1) \; . \end{cases}
\end{aligned}$$

The $(n-1)$-dimensional $\begin{cases} \text{symmetric} \\ \text{quadratic} \end{cases}$ Poincaré complex

$$\begin{cases} \partial(C,\phi) \; = \; (\partial C, \partial\phi) \\ \partial(C,\psi) \; = \; (\partial C, \partial\psi) \end{cases}$$

is the *boundary* of the connected n-dimensional $\begin{cases} \text{symmetric} \\ \text{quadratic} \end{cases}$ complex $\begin{cases} (C,\phi) \\ (C,\psi). \end{cases}$ The connected complex $\begin{cases} (C,\phi) \\ (C,\psi) \end{cases}$ is a Poincaré complex if and only if the boundary $\begin{cases} \partial(C,\phi) \\ \partial(C,\psi) \end{cases}$ is contractible (= homotopy equivalent to 0). A Poincaré complex $\begin{cases} (C,\phi) \\ (C,\psi) \end{cases}$ is the boundary of an $(n+1)$-dimensional $\begin{cases} \text{symmetric} \\ \text{quadratic} \end{cases}$ Poincaré pair $\begin{cases} (f : C \to D, (\delta\phi, \phi)) \\ (f : C \to D, (\delta\psi, \psi)) \end{cases}$ if and only if it is homotopy equivalent to the boundary of a connected $(n+1)$-dimensional $\begin{cases} \text{symmetric} \\ \text{quadratic} \end{cases}$ complex.

The n-dimensional $\begin{cases} \text{symmetric} \\ \text{quadratic} \end{cases}$ Poincaré complexes $\begin{cases} (C,\phi) \\ (C,\psi) \end{cases}$, $\begin{cases} (C',\phi') \\ (C',\psi') \end{cases}$ are *cobordant* if $\begin{cases} (C,\phi)\oplus(C',-\phi') \\ (C,\psi)\oplus(C',-\psi') \end{cases}$ is the boundary of an $(n+1)$-dimensional $\begin{cases} \text{symmetric} \\ \text{quadratic} \end{cases}$ Poincaré pair $\begin{cases} ((f\ f') : C\oplus C' \to D, (\delta\phi, \phi\oplus -\phi')) \\ ((f\ f') : C\oplus C' \to D, (\delta\psi, \psi\oplus -\psi'))\ . \end{cases}$ Homotopy equivalent Poincaré complexes are cobordant.

The $\begin{cases} \textit{symmetric} \\ \textit{quadratic} \end{cases}$ *L-groups* $\begin{cases} L^n(A) \\ L_n(A) \end{cases}$ $(n\geq 0)$ are the cobordism groups of n-dimensional $\begin{cases} \text{symmetric} \\ \text{quadratic} \end{cases}$ Poincaré complexes over A. The quadratic L-groups $L_*(A)$ are 4-periodic, with isomorphisms

$$L_n(A)\to L_{n+4}(A)\ ;\ (C,\psi)\mapsto (S^2C,\psi)\ (n\geq 0)\ ,$$

and are just the surgery obstruction groups of Wall [19]. The symmetric L-groups $L^*(A)$ were introduced by Mishchenko [14]. The corresponding maps in the symmetric L-groups

$$L^n(A)\to L^{n+4}(A)\ ;\ (C,\phi)\mapsto (S^2C,\phi)\ (n\geq 0)$$

are not isomorphisms in general. The symmetric and quadratic L-groups are related by an exact sequence

$$\dots\to L_n(A)\overset{1+T}{\to} L^n(A)\to \widehat{L}^n(A)\to L_{n-1}(A)\to\dots$$

with

$$1+T\ :\ L_n(A)\to L^n(A)\ ;\ (C,\psi)\mapsto (C,(1+T)\psi)$$

and $\widehat{L}^n(A)$ the relative cobordism group of n-dimensional (symmetric, quadratic) Poincaré pairs over A. The relative L-groups $\widehat{L}^*(A)$ are 8-torsion, so that the symmetrization maps $1+T : L_n(A)\to L^n(A)$ are isomorphisms modulo 8-torsion. If $\widehat{H}^*(\mathbb{Z}_2;A)=0$ (e.g. if $1/2\in A$) then $\widehat{L}^*(A)=0$ and the symmetrization maps are isomorphisms.

The *symmetric construction* of Ranicki [16] is the natural chain map

$$\phi_X\ =\ 1\otimes\Delta\ :\ C(X)\to W^{\%}C(\widetilde{X})\ =\ \mathrm{Hom}_{\mathbb{Z}[\mathbb{Z}_2]}(W,C(\widetilde{X})\otimes_{\mathbb{Z}[\pi]}C(\widetilde{X}))$$

induced by an Alexander-Whitney-Steenrod diagonal chain approximation Δ, for any space X and any regular cover $\widetilde{X}$, with π the group of covering translations. For $\widetilde{X}=X$ the mod 2 reduction of the composite

$$H_n(X)\xrightarrow{\phi_X} Q^n(C(X))\xrightarrow{v_r}\mathrm{Hom}_{\mathbb{Z}}(H^{n-r}(X),Q^n(S^{n-r}\mathbb{Z}))$$

is given by the rth Steenrod square

$$v_r(\phi_X(x))(y)\ =\ \langle Sq^r(y),x\rangle\in\mathbb{Z}_2\ .$$

The symmetric signature of Mishchenko [14] is defined for any n-dimensional geometric Poincaré complex X to be the symmetric Poincaré cobordism class

$$\sigma^*(X)\ =\ (C(\widetilde{X}),\phi_X([X]))\in L^n(\mathbb{Z}[\pi_1(X)])\ .$$

The symmetric L-groups of $\mathbb{Z}$ were computed in Ranicki [16] to be

$$L^n(\mathbb{Z}) = \begin{cases} \mathbb{Z} \text{ (signature)} & \text{if } n \equiv 0 (\bmod 4) \\ \mathbb{Z}_2 \text{ (deRham invariant)} & \text{if } n \equiv 1 (\bmod 4) \\ 0 & \text{if } n \equiv 2 (\bmod 4) \\ 0 & \text{if } n \equiv 3 (\bmod 4) \;. \end{cases}$$

The symmetric signature map on geometric Poincaré bordism

$$\sigma^* \;:\; \Omega_n^P(X) \to L^n(\mathbb{Z}[\pi]) \; (\pi = \pi_1(X))$$

factors through the visible symmetric L-groups $VL^*(\mathbb{Z}[\pi])$ of Weiss [21] and the closely related L-groups $L^*_{RL}(\mathbb{Z}[\pi])$ of Milgram [12]. The essential difference between $VL^*(\mathbb{Z}[\pi])$ and $L^*(\mathbb{Z}[\pi])$ is that in a visible symmetric Poincaré complex (C, ϕ) over $\mathbb{Z}[\pi]$ the Wu classes are restricted to be such that

$$\widehat{v}_r(\phi) \;:\; H_r(C) \to \widehat{H}^r(\mathbb{Z}_2; \mathbb{Z}) \subseteq \widehat{H}^r(\mathbb{Z}_2; \mathbb{Z}[\pi]) \;.$$

The symmetric signature map also factors through the visible L-groups $VL^*(X)$ of the space X, defined in Ranicki [17, §14] using sheaves over X of symmetric Poincaré complexes over $\mathbb{Z}$.

The *quadratic construction* of Ranicki [16] associates to any stable π-equivariant map $F : \Sigma^\infty \widetilde{X}_+ \to \Sigma^\infty \widetilde{Y}_+$ a natural chain map

$$\psi_F \;:\; C(X) \to W_\% C(\widetilde{Y}) \;=\; W \otimes_{\mathbb{Z}[\mathbb{Z}_2]} (C(\widetilde{Y}) \otimes_{\mathbb{Z}[\pi]} C(\widetilde{Y}))$$

such that

$$(1+T)\psi_F \;=\; F^\% \phi_X - \phi_Y F_* \;:\; C(X) \to W^\% C(\widetilde{Y})$$

with $\widetilde{X}$ a regular cover of X with group of covering translations π, $\widetilde{X}_+ = \widetilde{X} \cup \{\text{pt.}\}$, and similarly for Y. For $\widetilde{X} = X$, $\widetilde{Y} = Y$, $\pi = \{1\}$ the mod 2 reduction of the composite

$$H_n(X) \xrightarrow{\psi_F} Q_n(C(Y)) \xrightarrow{v^r} \operatorname{Hom}_{\mathbb{Z}}(H^{n-r}(Y), Q_n(S^{n-r}\mathbb{Z}))$$

is given by the $(r+1)$th functional Steenrod square

$$v^r(\psi_F(x))(y) \;=\; \langle Sq^{r+1}_{(\Sigma^\infty y)F}(\Sigma^\infty \iota), \Sigma^\infty x \rangle \in \mathbb{Z}_2$$

with $\iota \in H^{n-r}(K(\mathbb{Z}_2, n-r); \mathbb{Z}_2) = \mathbb{Z}_2$ the generator.

The Wall [19] surgery obstruction of an n-dimensional normal map $(f, b) : M \to X$ was expressed in [16] as the quadratic Poincaré cobordism class

$$\sigma_*(f, b) \;=\; (C(f^!), e_\% \psi_F([X])) \in L_n(\mathbb{Z}[\pi_1(X)])$$

with $F : \Sigma^\infty \widetilde{X}_+ \to \Sigma^\infty \widetilde{M}_+$ a $\pi_1(X)$-equivariant S-dual of $T(\widetilde{b}) : T(\nu_{\widetilde{M}}) \to T(\nu_{\widetilde{X}})$ inducing the Umkehr chain map

$$f^! \;:\; C(\widetilde{X}) \;\simeq\; C(\widetilde{X})^{n-*} \xrightarrow{f^*} C(\widetilde{M})^{n-*} \;\simeq\; C(\widetilde{M})$$

and $e : C(\widetilde{M}) \to C(f^!)$ the inclusion in the algebraic mapping cone. The quadratic L-groups of $\mathbb{Z}$ are given by

$$L_n(\mathbb{Z}) \;=\; \begin{cases} \mathbb{Z}\ (\text{signature}/8) & \text{if } n \equiv 0 (\text{mod } 4) \\ 0 & \text{if } n \equiv 1 (\text{mod } 4) \\ \mathbb{Z}_2\ (\text{Arf invariant}) & \text{if } n \equiv 2 (\text{mod } 4) \\ 0 & \text{if } n \equiv 3 (\text{mod } 4)\ . \end{cases}$$

§6. **Chain bundles**

A *bundle* over a finite-dimensional A-module chain complex C is a 0-dimensional hyperquadratic structure on C^{0-*}, that is a cycle

$$\gamma \in (\widehat{W}^{\%} C^{0-*})_0\ ,$$

as defined by a collection of A-module morphisms

$$\{\gamma_s \in \mathrm{Hom}_A(C_{r-s}, C^{-r}) \,|\, r, s \in \mathbb{Z}\}$$

such that

$$(-1)^{r+1} d^*_C \gamma_s + (-1)^s \gamma_s d_C + (-1)^{s-1}(\gamma_{s-1} + (-1)^s T\gamma_{s-1}) \;=\; 0 \;:\; C_{r-s+1} \to C^{-r}\ .$$

An *equivalence* of bundles over C

$$\chi \;:\; \gamma \to \gamma'$$

is an equivalence of hyperquadratic structures, as defined by a collection of A-module morphisms

$$\{\chi_s \in \mathrm{Hom}_A(C_{r-s-1}, C^{-r}) \,|\, r, s \in \mathbb{Z}\}$$

such that

$$\gamma'_s - \gamma_s \;=\; (-1)^{r+1} d^*_C \chi_s + (-1)^s \chi_s d_C + (-1)^s(\chi_{s-1} + (-1)^s T\chi_{s-1}) \;:\; C_{r-s} \to C^{-r}\ .$$

Thus

$$\widehat{Q}^0(C^{0-*}) \;=\; H_0(\widehat{W}^{\%} C^{0-*})$$

is the abelian group of equivalence classes of bundles over C.

A *chain bundle* over A (C, γ) is a finite-dimensional A-module chain complex C together with a bundle $\gamma \in (\widehat{W}^{\%} C^{0-*})_0$.

Given a chain bundle (C, γ) over A and an A-module chain map $f : B \to C$ define the *pullback* chain bundle $(B, f^*\gamma)$ using the image of γ under the $\mathbb{Z}$-module chain map

$$\widehat{f^*} \;:\; \widehat{W}^{\%} C^{0-*} \to \widehat{W}^{\%} B^{0-*}$$

induced by the dual A-module chain map $f^* : C^{0-*} \to B^{0-*}$. The equivalence class of the pullback bundle $f^*\gamma$ depends only on the chain homotopy class of the chain map f, by the chain homotopy invariance of the Q-groups.

A *map* of chain bundles over A

$$(f, \chi) \;:\; (C, \gamma) \to (C', \gamma')$$

is a chain map $f : C \to C'$ together with an equivalence of bundles over C

$$\chi \;:\; \gamma \to f^*\gamma'\ .$$

The *composite* of chain bundle maps

$$(f,\chi)\ :\ (C,\gamma)\to(C',\gamma')\ ,\ (f',\chi')\ :\ (C',\gamma')\to(C'',\gamma'')$$

is the chain bundle map

$$(f',\chi')(f,\chi)\ =\ (f'f,\chi+\widehat{f^*}^{\%}(\chi'))\ :\ (C,\gamma)\to(C'',\gamma'')\ .$$

A *homotopy of chain bundle maps*

$$(g,\eta)\ :\ (f,\chi)\ \simeq\ (f',\chi')\ :\ (C,\gamma)\to(C',\gamma')$$

is a chain homotopy

$$g\ :\ f\ \simeq\ f'\ :\ C\to C'$$

together with an equivalence of 1-dimensional hyperquadratic structures on C^{0-*}

$$\eta\ :\ \chi-\chi'+(g^*;f^*,f'^*)^{\%}(\gamma')\to 0\ .$$

A map of chain bundles $(f,\chi):(C,\gamma)\to(C',\gamma')$ is an *equivalence* if there exists a homotopy inverse. This happens precisely when $f:C\to C'$ is a chain equivalence, in which case any chain homotopy inverse

$$f'\ =\ f^{-1}\ :\ C'\to C$$

can be used to define a homotopy inverse

$$(f',\chi')\ :\ (C',\gamma')\to(C,\gamma)\ .$$

Given a chain bundle (B,β) over A and a finite-dimensional A-module chain complex C use the pullback construction to define abelian group morphisms

$$I_\beta\ :\ H_n(B^t\otimes_A C)\to\widehat{Q}^n(C)\ ;\ f\mapsto\widehat{f}^{\%}(S^n\beta)\ ,$$

using the slant isomorphism

$$B^t\otimes_A C\to\mathrm{Hom}_A(B^{-*},C)\ ;\ x\otimes y\mapsto(f\mapsto\overline{f(x)}\,.\,y)$$

to identify a cycle $f\in(B^t\otimes_A C)_n$ with a chain map $f:B^{n-*}\to C$. Weiss [20] developed an algebraic analogue of the representation theorem of Brown [3] to obtain for any ring with involution A the existence of a directed system $\{(B(r),\beta(r))\,|\,r\geq 0\}$ of chain bundles over A and chain bundle maps

$$(B(r),\beta(r))\to(B(r+1),\beta(r+1))$$

such that the abelian group morphisms

$$\varinjlim_r I_{\beta_r}\ :\ \varinjlim_r H_n(B(r)^t\otimes_A C)\to\widehat{Q}^n(C)$$

are isomorphisms for any finite-dimensional A-module chain complex C. In general, the direct limit A-module chain complex

$$B(\infty)\ =\ \varinjlim_r B(r)$$

is not finite-dimensional. As in [20] we shall ignore this inconvenience and treat $B(\infty)$ as if it were finite-dimensional, so that there is defined the *universal chain bundle* over A

$$(B(\infty),\beta(\infty))\ =\ \varinjlim_r(B(r),\beta(r))\ ,$$

with the universal property that for any finite-dimensional A-module chain complex C the abelian group morphisms

$$I_{\beta(\infty)} \ : \ H_n(B(\infty)^t \otimes_A C) \to \widehat{Q}^n(C)$$

are isomorphisms. In particular, there are defined isomorphisms

$$H_0(\mathrm{Hom}_A(C, B(\infty))) \to \widehat{Q}^0(C^{0-*}) \ ; \ f \mapsto f^*(\beta(\infty))$$

for any finite-dimensional C. Thus every chain bundle (C, γ) has a classifying map

$$(f, \chi) \ : \ (C, \gamma) \to (B(\infty), \beta(\infty))$$

and the equivalence classes of bundles $\gamma \in (\widehat{W}^{\%}C^{0-*})_0$ over C are in one-one correspondence with the chain homotopy classes of chain maps $f : C \to B(\infty)$.

The *Wu classes* of a chain bundle (C, γ) are the Wu classes of γ, the A-module morphisms

$$\widehat{v}_r(\gamma) \ : \ H_r(C) \to \widehat{H}^r(\mathbb{Z}_2; A) \ ; \ x \mapsto \langle \gamma_{-2r}, x \otimes x \rangle \ .$$

The universal chain bundle $(B(\infty), \beta(\infty))$ is characterized by the property that the Wu classes define A-module isomorphisms

$$\widehat{v}_r(\gamma) \ : \ H_r(C) \to \widehat{H}^r(\mathbb{Z}_2; A) \ (r \geq 0) \ .$$

For example, if $A = \mathbb{Z}$ the chain bundle $(B(\infty), \beta(\infty))$ defined by

$$d_{B(\infty)} = \begin{cases} 2 & \text{if } r \text{ is odd} \\ 0 & \text{if } r \text{ is even} \end{cases} : B(\infty)_r = \mathbb{Z} \to B(\infty)_{r-1} = \mathbb{Z} \ ,$$

$$\beta(\infty)_s = \begin{cases} 1 & \text{if } 2r = s \\ 0 & \text{otherwise} \end{cases} : B(\infty)_{r-s} = \mathbb{Z} \to B(\infty)^{-r} = \mathbb{Z}$$

is universal, with the Wu classes defining isomorphisms

$$\widehat{v}_r(\beta(\infty)) \ : \ H_r(B(\infty)) \to \widehat{H}^r(\mathbb{Z}_2; \mathbb{Z}) = \begin{cases} \mathbb{Z}_2 & \text{if } r \text{ is even} \\ 0 & \text{if } r \text{ is odd} \ . \end{cases}$$

The symmetric and quadratic constructions of Ranicki [16] were extended in Weiss [20] and Ranicki [17, 2.5] : a spherical fibration $\nu : X \to BG(k)$ determines a chain bundle $(C(\widetilde{X}), \gamma)$ over $\mathbb{Z}[\pi_1(X)]$, and there is defined a natural transformation of exact sequences from the certain exact sequence of Whitehead [22]

$$\begin{array}{ccccccccc}
\cdots \longrightarrow & \Gamma_{n+k+1}(T(\nu)) & \longrightarrow & \pi_{n+k}(T(\nu)) & \xrightarrow{h} & \dot{H}_{n+k}(T(\nu)) & \longrightarrow & \Gamma_{n+k}(T(\nu)) & \longrightarrow \cdots \\
& \downarrow & & \downarrow & & \downarrow \phi_X U & & \downarrow & \\
\cdots \longrightarrow & \widehat{Q}^{n+1}(C(\widetilde{X})) & \longrightarrow & Q_n(C(\widetilde{X}), \gamma) & \longrightarrow & Q^n(C(\widetilde{X})) & \xrightarrow{J_\gamma} & \widehat{Q}^n(C(\widetilde{X})) & \longrightarrow \cdots
\end{array}$$

with h the Hurewicz map and $U : \dot{H}_{n+k}(T(\nu)) \to H_n(X)$ the Thom isomorphism. The topological Wu classes of ν are the algebraic Wu classes of the induced chain bundle $(C(X; \mathbb{Z}_2), 1 \otimes \gamma)$ over $\mathbb{Z}_2$

$$v_*(\nu) = \widehat{v}_*(1 \otimes \gamma) \in H^*(X; \mathbb{Z}_2) = \mathrm{Hom}_{\mathbb{Z}_2}(H_*(X; \mathbb{Z}_2), \mathbb{Z}_2) \ .$$

§7. Normal complexes

An n-dimensional normal space (X, ν_X, ρ_X) (Quinn [15]) is a finite n-dimensional CW complex X together with a $(k-1)$-spherical fibration $\nu_X : X \to BG(k)$ and a map $\rho_X : S^{n+k} \to T(\nu_X)$ to the Thom space of ν_X. An n-dimensional geometric Poincaré complex X has a unique equivalence class of normal structures (ν_X, ρ_X), with ν_X the Spivak normal fibration and ρ_X representing the fundamental class $[X] \in H_n(X)$. A normal complex is the algebraic analogue of a normal space, consisting of a symmetric complex with normal chain bundle.

A *normal structure* (γ, θ) on an n-dimensional symmetric complex (C, ϕ) is a bundle $\gamma \in (\widehat{W}^{\%}C^{0-*})_0$ together with an equivalence of n-dimensional hyperquadratic structures on C

$$\theta \;:\; J(\phi) \to (\widehat{\phi}_0)^{\%}(S^n\gamma) \;,$$

as defined by a chain $\theta \in (\widehat{W}^{\%}C)_{n+1}$ such that

$$J(\phi) - (\widehat{\phi}_0)^{\%}(S^n\gamma) \;=\; d\theta \in (\widehat{W}^{\%}C)_n \;.$$

The Wu classes of ϕ and γ are then related by a commutative diagram

$$\begin{array}{ccc}
H^{n-r}(C) & \xrightarrow{\quad \phi_0 \quad} & H_r(C) \\
\Big\downarrow{\scriptstyle v_r(\phi)} & & \Big\downarrow{\scriptstyle \widehat{v}_r(\gamma)} \\
H^{n-2r}(\mathbb{Z}_2; A, (-1)^{n-r}) & \xrightarrow{\quad J \quad} & \widehat{H}^r(Z_2; A) \;.
\end{array}$$

An *equivalence* of n-dimensional normal structures on (C, ϕ)

$$(\chi, \eta) \;:\; (\gamma, \theta) \to (\gamma', \theta')$$

is an equivalence of bundles $\chi : \gamma \to \gamma'$ together with an equivalence of $(n+1)$-dimensional hyperquadratic structures on C

$$\eta \;:\; \theta - \theta' + (\widehat{\phi}_0)^{\%}(S^n\chi) \to 0 \;.$$

An n-dimensional symmetric Poincaré complex $(C, \phi \in (W^{\%}C)_n)$ has a unique equivalence class of normal structures (γ, θ), with the equivalence class of bundles $[\gamma] \in \widehat{Q}^0(C^{0-*})$ the image of the equivalence class of symmetric structures $[\phi] \in Q^n(C)$ under the composite

$$Q^n(C) \stackrel{J}{\to} \widehat{Q}^n(C) \stackrel{((\phi_0)^{\%})^{-1}}{\to} \widehat{Q}^n(C^{n-*}) \stackrel{(S^n)^{-1}}{\to} \widehat{Q}^0(C^{0-*}) \;.$$

If (γ, θ), (γ', θ') are two such normal structures on (C, ϕ) there exists an equivalence of bundles $\chi : \gamma \to \gamma'$. As $\phi_0 : C^{n-*} \to C$ is a chain equivalence the cycle

$$\theta - \theta' + (\phi_0)^{\%}(S^n\chi) \in (\widehat{W}^{\%}C)_{n+1}$$

is such that there exist a cycle $\lambda \in \widehat{W}^{\%}C^{n-*})_{n+1}$ and a chain $\mu \in (\widehat{W}^{\%}C)_{n+2}$ such that

$$\theta - \theta' + (\phi_0)^{\%}(S^n\chi) \;=\; (\phi_0)^{\%}(\lambda) + d\mu \in (\widehat{W}^{\%}C)_{n+1} \;.$$

There is now defined an equivalence of normal structures on (C, ϕ)

$$(\chi - (S^n)^{-1}(\lambda), \mu) \;:\; (\gamma, \theta) \to (\gamma', \theta') \;.$$

An *n-dimensional normal* (*Poincaré*) *complex* over A (C,ϕ,γ,θ) is an n-dimensional symmetric (Poincaré) complex (C,ϕ) together with a normal structure (γ,θ). Symmetric Poincaré complexes are regarded as normal Poincaré complexes by choosing a normal structure in the unique equivalence class.

An n-dimensional normal complex (C,ϕ,γ,θ) is *connected* if the n-dimensional symmetric complex (C,ϕ) is connected, that is

$$H_0(\phi_0 : C^{n-*} \to C) \;=\; 0 \;.$$

The correspondence described in §5 between the homotopy equivalence classes of connected n-dimensional $\left\{\begin{array}{l}\text{symmetric}\\ \text{quadratic}\end{array}\right.$ complexes and those of n-dimensional $\left\{\begin{array}{l}\text{symmetric}\\ \text{quadratic}\end{array}\right.$ Poincaré pairs has the following generalization to connected normal complexes and (symmetric, quadratic) Poincaré pairs.

A connected n-dimensional normal complex (C,ϕ,γ,θ) determines the n-dimensional (symmetric, quadratic) Poincaré pair

$$(i_C : \partial C \to C^{n-*}, (\delta\phi,\psi))$$

defined by

$$\begin{array}{l}
i_C \;=\; (0\;1) \;:\; \partial C_r \;=\; C_{r+1}\oplus C^{n-r} \to C^{n-r} \;,\\
d_{\partial C} \;=\; \begin{pmatrix} d_C & (-1)^r\phi_0 \\ 0 & (-1)^r d_C^* \end{pmatrix} \;: \\
\qquad \partial C_r \;=\; C_{r+1}\oplus C^{n-r} \to \partial C_{r-1} \;=\; C_r \oplus C^{n-r+1} \;,\\
\psi_0 \;=\; \begin{pmatrix} \chi_0 & 0 \\ 1+\gamma_{-n}\phi_0^* & \gamma_{-n-1}^* \end{pmatrix} \;: \\
\qquad \partial C^r \;=\; C^{r+1}\oplus C_{n-r} \to \partial C_{n-r-1} \;=\; C_{n-r}\oplus C^{r+1} \;,\\
\psi_s \;=\; \begin{pmatrix} \chi_{-s} & 0 \\ \gamma_{-n-s}\phi_0^* & \gamma_{-n-s-1}^* \end{pmatrix} \;: \\
\qquad \partial C^r \;=\; C^{r+1}\oplus C_{n-r} \to \partial C_{n-r-s-1} \;=\; C_{n-r-s}\oplus C^{r+s+1} \;(s\geq 1) \;,\\
\delta\phi_s \;=\; \gamma_{-n-s} \;:\; C_r \to C^{n-r+s} \;(s \geq 0) \;.
\end{array}$$

The *quadratic boundary* of a connected n-dimensional normal complex (C,ϕ,γ,θ) is the $(n-1)$-dimensional quadratic Poincaré complex

$$\partial(C,\phi,\gamma,\theta) \;=\; (\partial C,\psi) \;,$$

generalizing the definition in §6 of the boundary $(n-1)$-dimensional $\left\{\begin{array}{l}\text{symmetric}\\ \text{quadratic}\end{array}\right.$ Poincaré complex $\left\{\begin{array}{l}\partial(C,\phi)\\ \partial(C,\psi)\end{array}\right.$ of a connected n-dimensional $\left\{\begin{array}{l}\text{symmetric}\\ \text{quadratic}\end{array}\right.$ complex $\left\{\begin{array}{l}(C,\phi)\\ (C,\psi)\end{array}\right.$. Conversely, given an n-dimensional (symmetric, quadratic) Poincaré pair

$(f: C \to D, (\delta\phi, \psi))$ there is defined a connected n-dimensional normal complex $(C(f), \phi, \gamma, \theta)$ with the symmetric structure

$$\phi_s = \begin{cases} \begin{pmatrix} \delta\phi_0 & 0 \\ (1+T)\psi_0 f^* & 0 \end{pmatrix} & \text{if } s = 0 \\ \begin{pmatrix} \delta\phi_1 & 0 \\ 0 & (1+T)\psi_0 \end{pmatrix} & \text{if } s = 1 \\ \begin{pmatrix} \delta\phi_s & 0 \\ 0 & 0 \end{pmatrix} & \text{if } s \geq 2 \end{cases}$$

$$: C(f)^r = D^r \oplus C^{r-1} \to C(f)_{n-r+s} = D_{n-r+s} \oplus C_{n-r+s-1} .$$

The normal structure (γ, χ) is determined up to equivalence by the Poincaré duality, with $\gamma \in \widehat{Q}^0(D^{-*})$ the image of $(\delta\phi/(1+T)\psi) \in Q^n(C(f))$ under the composite

$$Q^n(C(f)) \xrightarrow{((\delta\phi_0, (1+T)\psi_0)^{\%})^{-1}} Q^n(D^{n-*}) \xrightarrow{J} \widehat{Q}^n(D^{n-*}) \xrightarrow{S^{-n}} \widehat{Q}^0(D^{-*}) .$$

The composite isomorphism

$$\widehat{Q}^0(C(f)^{0-*}) \xrightarrow{S^n} \widehat{Q}^n(C(f)^{n-*}) \xrightarrow{(\delta\phi_0, (1+T)\psi_0)^{\%}} \widehat{Q}^n(D)$$

sends the equivalence class $[\gamma] \in \widehat{Q}^0(C(f)^{0-*})$ to the element $\alpha \in \widehat{Q}^n(D)$ represented by

$$\alpha_s = \begin{cases} \delta\phi_s & \text{if } s \geq 0 \\ f\psi_{-s-1}f^* & \text{if } s \leq -1 \end{cases} : D^r \to D_{n-r+s} .$$

There is thus a natural one-one correspondence between the homotopy equivalence classes of connected n-dimensional normal complexes over A and the homotopy equivalence classes of n-dimensional (symmetric, quadratic) Poincaré pairs over A. In §8 below this correspondence will be used to identify the cobordism group $\widehat{L}^n(A)$ of n-dimensional (symmetric, quadratic) Poincaré pairs over A with the cobordism group of n-dimensional normal complexes over A.

Let (B, β) be a chain bundle over A. A *normal* (B, β)*-structure* $(\gamma, \theta, f, \chi)$ on an n-dimensional symmetric complex (C, ϕ) over A is a normal structure (γ, θ) on (C, ϕ) together with a chain bundle map

$$(f, \chi) : (C, \gamma) \to (B, \beta) .$$

There are also the corresponding relative notions of *normal* (B, β)-structure on symmetric and (symmetric, quadratic) pairs. For the universal chain bundle $(B(\infty), \beta(\infty))$ over A a normal $(B(\infty), \beta(\infty))$-structure $(\gamma, \theta, f, \chi)$ on a symmetric complex (C, ϕ) is to all intents and purposes the same as a normal structure (γ, θ).

A normal (0,0)-structure $(\gamma, \theta, 0, \chi)$ on an n-dimensional symmetric complex (C, ϕ) determines an equivalence to 0 of the hyperquadratic structure $J(\phi) \in (\widehat{W}^{\%}C)_n$

$$\xi = \theta + \phi_0^{\%}(S^n\chi) : J(\phi) \to 0 .$$

Such an equivalence $\xi : J(\phi) \to 0$ consists of a quadratic structure $\psi \in (W_{\%}C)_n$ and an equivalence of symmetric structures

$$\eta \, : \, (1+T)\psi \to \phi \, ,$$

with

$$\begin{aligned} \psi_s &= \xi_{-s-1} \in \mathrm{Hom}_A(C^*, C)_{n-s} \ (s \geq 0) \ , \\ \eta_s &= \xi_s \in \mathrm{Hom}_A(C^*, C)_{n+s+1} \ (s \geq 0) \ . \end{aligned}$$

Thus a normal (0,0)-structure on a symmetric complex (C, ϕ) is to all intents and purposes an equivalence of the symmetric structure ϕ to $(1+T)\psi$ for some quadratic structure ψ on C.

An *n-dimensional (B, β)-normal (Poincaré) complex* $(C, \phi, \gamma, \theta, f, \chi)$ is an n-dimensional symmetric (Poincaré) complex (C, ϕ) together with a normal (B, β)-structure $(\gamma, \theta, \phi, \chi)$.

In §8 below the cobordism group $\widehat{L}\langle B, \beta\rangle^n(A)$ of n-dimensional (B, β)-normal complexes over A will be identified with the twisted quadratic group $Q_n(B, \beta)$ (introduced by Weiss [20]) of equivalence classes of pairs (ϕ, θ) such that $(B, \phi, \beta, \theta, 1, 0)$ is an n-dimensional (B, β)-normal complex.

An *n-dimensional symmetric structure* (ϕ, θ) on a chain bundle (C, γ) is an n-dimensional symmetric structure $\phi \in (W^{\%}C)_n$ together with an equivalence of n-dimensional hyperquadratic structures on C

$$\theta \, : \, J(\phi) \to (\phi_0)^{\%}(S^n\gamma) \ ,$$

as defined by a chain $\theta \in (\widehat{W}^{\%}C)_{n+1}$ such that

$$J(\phi) - (\phi_0)^{\%}(S^n\gamma) \ = \ d(\theta) \in (\widehat{W}^{\%}C)_n \ .$$

Thus (C, ϕ) is an n-dimensional symmetric complex with normal structure (γ, θ).

An *equivalence* of n-dimensional symmetric structures on (C, γ)

$$(\xi, \eta) \, : \, (\phi, \theta) \to (\phi', \theta')$$

is defined by an equivalence of symmetric structures $\xi : \phi \to \phi'$ together with an equivalence of hyperquadratic structures on C

$$\eta \, : \, \theta - \theta' + J(\xi) + (\xi_0; \phi_0, \phi_0')^{\%}(S^n\gamma) \to 0 \ ,$$

as defined by chains $\xi \in (W^{\%}C)_{n+1}$, $\eta \in (\widehat{W}^{\%}C)_{n+2}$ such that

$$\begin{aligned} \phi' - \phi &= d(\xi) \in (W^{\%}C)_n C^{-*} \\ \theta' - \theta + J(\xi) + (\xi_0; \phi_0, \phi_0')^{\%}(S^n\gamma) &= d(\eta) \in (\widehat{W}^{\%}C)_{n+1} \ . \end{aligned}$$

The *twisted quadratic Q-group* $Q_n(C, \gamma)$ is the abelian group of equivalence classes of n-dimensional symmetric structures on a chain bundle (C, γ), with addition by

$$(\phi, \theta) + (\phi', \theta') \ = \ (\phi + \phi', \theta + \theta' + [\phi_0, \phi_0'](S^n\gamma)) \in Q_n(C, \gamma) \ .$$

The twisted quadratic Q-groups $Q_*(C, \gamma)$ fit into an exact sequence of abelian groups

$$\dots \to \widehat{Q}^{n+1}(C) \xrightarrow{H_\gamma} Q_n(C, \gamma) \xrightarrow{N_\gamma} Q^n(C) \xrightarrow{J_\gamma} \widehat{Q}^n(C) \to \dots$$

with the morphisms

$$\begin{aligned} H_\gamma &: \widehat{Q}^{n+1}(C) \to Q_n(C,\gamma) \; ; \; \theta \mapsto (0,\theta) \; , \\ N_\gamma &: Q_n(C,\gamma) \to Q^n(C) \; ; \; (\phi,\theta) \mapsto \phi \; , \\ J_\gamma &: Q^n(C) \to \widehat{Q}^n(C) \; ; \; \phi \mapsto J(\phi) - (\phi_0)^{\%}(S^n\gamma) \end{aligned}$$

induced by simplicial maps. In the untwisted case $\gamma = 0$ there is defined an isomorphism of exact sequences

$$\begin{array}{ccccccccc} \cdots \longrightarrow & \widehat{Q}^{n+1}(C) & \xrightarrow{H} & Q_n(C) & \xrightarrow{1+T} & Q^n(C) & \xrightarrow{J} & \widehat{Q}^n(C) & \longrightarrow \cdots \\ & \| & & \downarrow \cong & & \| & & \| & \\ \cdots \longrightarrow & \widehat{Q}^{n+1}(C) & \xrightarrow{H_0} & Q_n(C,0) & \xrightarrow{N_0} & Q^n(C) & \xrightarrow{J_0} & \widehat{Q}^n(C) & \longrightarrow \cdots \end{array}$$

with

$$Q_n(C) \to Q_n(C,0) \; ; \; \psi \mapsto ((1+T)\psi, \theta) \; , \; \theta_s \; = \; \begin{cases} \psi_{-s-1} & \text{if } s \leq -1 \\ 0 & \text{if } s \geq 0. \end{cases}$$

The twisted quadratic groups $Q_*(C,\gamma)$ are covariant in (C,γ). Given a map of chain bundles $(f,\chi) : (C,\gamma) \to (C',\gamma')$ and an n-dimensional symmetric structure (ϕ,θ) on (C,γ) define an n-dimensional symmetric structure on (C',γ')

$$(f,\chi)_{\%}(\phi,\theta) \; = \; (f^{\%}(\phi), \widehat{f}^{\%}(\theta) + (f\phi_0)^{\%}(S^n\chi)) \; .$$

The resulting morphisms of the twisted quadratic Q-groups

$$(f,\chi)_{\%} \; : \; Q_n(C,\gamma) \to Q_n(C',\gamma')$$

depend only on the homotopy class of (f,χ). There is defined a morphism of exact sequences

$$\begin{array}{ccccccccc} \cdots \longrightarrow & \widehat{Q}^{n+1}(C) & \xrightarrow{H_\gamma} & Q_n(C,\gamma) & \xrightarrow{N_\gamma} & Q^n(C) & \xrightarrow{J_\gamma} & \widehat{Q}^n(C) & \longrightarrow \cdots \\ & \downarrow \widehat{f}^{\%} & & \downarrow (f,\chi)_{\%} & & \downarrow f^{\%} & & \downarrow \widehat{f}^{\%} & \\ \cdots \longrightarrow & \widehat{Q}^{n+1}(C') & \xrightarrow{H_{\gamma'}} & Q_n(C',\gamma') & \xrightarrow{N_{\gamma'}} & Q^n(C') & \xrightarrow{J_{\gamma'}} & \widehat{Q}^n(C') & \longrightarrow \cdots \end{array}$$

which is an isomorphism if (f,χ) is an equivalence.

The *characteristic element* of an n-dimensional (B,β)-normal complex $(C,\phi,\gamma,\theta,f,\chi)$ is defined by

$$(f,\chi)_{\%}(\phi,\theta) \in Q_n(B,\beta) \; .$$

In §8 the cobordism class of a (B,β)-normal complex will be identified with the characteristic element.

A *map of n-dimensional* $\begin{cases} \textit{normal} \\ (B,\beta)\textit{-normal} \end{cases}$ *complexes*

$$\begin{cases} (f,\xi,\chi,\eta) \; : \; (C,\phi,\gamma,\theta) \to (C',\phi',\gamma',\theta') \\ (f,\xi,\chi,\eta,h,\mu) \; : \; (C,\phi,\gamma,\theta,g,\lambda) \to (C',\phi',\gamma',\theta',g',\lambda') \end{cases}$$

consists of

(i) a chain map $f : C \to C'$,
(ii) an equivalence $\xi : f^{\%}(\phi) \to \phi'$ of n-dimensional symmetric structures on C',
(iii) an equivalence $\chi : \gamma \to f^*\gamma'$ of bundles on C,
(iv) an equivalence of $(n+1)$-dimensional hyperquadratic structures on C'

$$\eta \ : \ J(\xi) + \theta' - \widehat{f^{\%}}(\theta) + (\xi_0; f^{\%}\phi_0, \phi_0')^{\%}(S^n\gamma') + (f\phi_0)^{\%}(S^n\chi) \to 0 \ ,$$

and in the (B,β)-normal case also

(v) a homotopy of bundle maps

$$(h,\mu) \ : \ (g,\lambda) \ \simeq \ (g',\lambda')(f,\chi) \ : \ (C,\gamma) \to (B,\beta) \ .$$

Note that $(C,\phi,\gamma,\theta,g,\lambda)$ and $(C',\phi',\gamma',\theta',g',\lambda')$ have the same characteristic element

$$(g,\lambda)_{\%}(\phi,\theta) \ = \ (g',\lambda')_{\%}(\phi',\theta') \in Q_n(B,\beta) \ .$$

It is convenient for computational purposes to describe the behaviour of the twisted quadratic groups under direct sum. The direct sum of chain bundles (C,γ), (C',γ') is the chain bundle

$$(C,\gamma) \oplus (C',\gamma') \ = \ (C \oplus C', \gamma \oplus \gamma') \ .$$

Let

$$i \ = \ \begin{pmatrix} 1 \\ 0 \end{pmatrix} \ : \ C \to C \oplus C' \ , \ i' \ = \ \begin{pmatrix} 0 \\ 1 \end{pmatrix} \ : \ C' \to C \oplus C' \ ,$$
$$j \ = \ (1 \quad 0) \ : \ C \oplus C' \to C \ , \ j' \ = \ (0 \quad 1) \ : \ C \oplus C' \to C' \ .$$

The twisted quadratic groups of the direct sum are such that there is defined a long exact sequence

$$\begin{aligned} \dots \to Q_n(C,\gamma) \oplus Q_n(C',\gamma') &\xrightarrow{i_*} Q_n(C \oplus C', \gamma \oplus \gamma') \xrightarrow{j_*} H_n(C^t \otimes_A C') \\ &\xrightarrow{k_*} Q_{n-1}(C,\gamma) \oplus Q_{n-1}(C',\gamma') \xrightarrow{i_*} Q_{n-1}(C \oplus C', \gamma \oplus \gamma') \to \dots \end{aligned}$$

with

$$\begin{aligned} i_* \ &= \ (i_{\%} \quad i'_{\%}) \ : \ Q_n(C,\gamma) \oplus Q_n(C',\gamma') \to Q_n(C \oplus C', \gamma \oplus \gamma') \ , \\ j_* \ &: \ Q_n(C \oplus C', \gamma \oplus \gamma') \to H_n(C^t \otimes_A C') \ ; \ (\phi,\theta) \mapsto (j \otimes j')\phi_0 \ , \\ k_* \ &: \ H_n(C^t \otimes_A C') \to Q_{n-1}(C,\gamma) \oplus Q_{n-1}(C',\gamma') \ ; \\ &\quad (f : C^{n-*} \to C') \mapsto ((0, \widehat{f^{\%}}(S^n\gamma')), (0, -\widehat{f^{*\%}}(S^n\gamma))) \ . \end{aligned}$$

For $\gamma = 0$ and $\gamma' = 0$ the long exact sequence collapses into split exact sequences of the untwisted quadratic Q-groups

$$0 \to Q_n(C) \oplus Q_n(C') \to Q_n(C \oplus C') \to H_n(C^t \otimes_A C') \to 0 \ .$$

§8. Normal cobordism

Given a k-plane vector bundle $\nu : X \to BO(k)$ over a space X let $\Omega_n(X,\nu)$ $(n \geq 0)$ denote the bordism groups of bundle maps

$$(f,b) \ : \ (M^n, \nu_M) \to (X,\nu)$$

with M^n a smooth closed n-manifold and $\nu_M : M \to BO(k)$ the normal bundle of an embedding $M^n \subset S^{n+k}$ (Lashof [9]). The Thom space of ν_M is given by

$$T(\nu_M) \;=\; E(\nu_M)/\partial E(\nu_M)$$

with $E(\nu_M)$ the tubular neighbourhood of M^n in S^{n+k}, so that there is defined a collapse map

$$\rho_M \;:\; S^{n+k} \to S^{n+k}/(S^{n+k}\backslash E(\nu_M)) \;=\; E(\nu_M)/\partial E(\nu_M) \;=\; T(\nu_M)\;.$$

The Pontrjagin-Thom isomorphism

$$\begin{aligned}&\Omega_n(X,\nu) \to \pi_{n+k}(T(\nu))\;;\\ &(f : M^n \to X, b : \nu_M \to \nu) \mapsto (T(b)(\rho_M) : S^{n+k} \overset{\rho_M}{\to} T(\nu_M) \overset{T(b)}{\to} T(\nu))\end{aligned}$$

has inverse

$$\begin{aligned}&\pi_{n+k}(T(\nu)) \to \Omega_n(X,\nu)\;;\\ &(\rho : S^{n+k} \to T(\nu)) \mapsto (f = \rho| : M^n = \rho^{-1}(X) \to X, b : \nu_M \to \nu)\;,\end{aligned}$$

using smooth transversality to choose a representative ρ transverse regular at the zero section $X \subset T(\nu)$.

Given a $(k-1)$-spherical fibration $\nu : X \to BG(k)$ over a space X let $\Omega^N_n(X,\nu)$ (resp. $\Omega^P_n(X,\nu)$) denote the bordism group of fibration maps

$$(f,b) \;:\; (M^n,\nu_M) \to (X,\nu)$$

with $(M^n, \nu_M : M \to BG(k), \rho_M : S^{n+k} \to T(\nu_M))$ an n-dimensional normal space (resp. geometric Poincaré complex with Spivak normal structure). According to the theory of Quinn [15] there is a geometric theory of transversality for normal spaces, so that by analogy with the Pontrjagin-Thom isomorphism for smooth bordism there is defined an isomorphism

$$\begin{aligned}&\Omega^N_n(X,\nu) \to \pi_{n+k}(T(\nu))\;;\\ &(f : M^n \to X, b : \nu_M \to \nu) \mapsto (T(b)(\rho_M) : S^{n+k} \overset{\rho_M}{\to} T(\nu_M) \overset{T(b)}{\to} T(\nu))\;,\end{aligned}$$

with inverse

$$\begin{aligned}&\pi_{n+k}(T(\nu)) \to \Omega^N_n(X,\nu)\;;\\ &(\rho : S^{n+k} \to T(\nu)) \mapsto (f = \rho| : M^n = \rho^{-1}(X) \to X, b : \nu_M \to \nu)\;.\end{aligned}$$

The geometric Poincaré and normal bordism groups for $n \geq 5$ are related by the Levitt-Jones-Quinn exact sequence

$$\dots \to L_n(\mathbb{Z}[\pi_1(X)]) \to \Omega^P_n(X,\nu) \to \Omega^N_n(X,\nu) \to L_{n-1}(\mathbb{Z}[\pi_1(X)]) \to \dots\;.$$

If $\nu : X \to BG(k)$ admits a TOP reduction $\widetilde{\nu} : X \to BTOP(k)$ the forgetful maps from manifold to normal space bordism $\Omega_n(X,\nu) \to \Omega^N_n(X,\nu)$ are isomorphisms, and

$$\Omega^P_n(X,\nu) \;=\; L_n(\mathbb{Z}[\pi_1(X)]) \oplus \Omega^N_n(X,\nu)\;.$$

A map of n-dimensional normal spaces

$$(f,b,c) : (M^n,\nu_M,\rho_M) \to (X^n,\nu_X,\rho_X)$$

is defined by a map of fibrations $(f,b) : (M,\nu_M) \to (X,\nu_X)$ together with a homotopy

$$c \ : \ T(b)\rho_M \ \simeq \ \rho_X \ : \ S^{n+k} \to T(\nu_X) \ .$$

The mapping cylinder of f

$$M(f) \ = \ M \times [0,1] \cup X/\{(x,1) = f(x) \,|\, x \in M\}$$

defines a cobordism $(M(f); M, X)$ of normal spaces, identifying

$$M \ = \ M \times \{0\} \subset M(f) \ .$$

If M^n and X^n are Poincaré complexes the corresponding element of the relative bordism group is just the surgery obstruction

$$(M(f); M \cup -X) \ = \ \sigma_*(f,b) \in \Omega^{N,P}_{n+1}(X,\nu_X) \ = \ L_n(\mathbb{Z}[\pi_1(X)]) \ .$$

Ignoring questions of finite-dimensionality (or assuming that X is a finite n-dimensional CW complex) it is therefore possible to define the inverse isomorphism to $\Omega^N_n(X,\nu) \to \pi_{n+k}(T(\nu))$ by

$$\pi_{n+k}(T(\nu)) \to \Omega^N_n(X,\nu) \ ; \ \rho \mapsto (X,\nu,\rho) \ ,$$

without an appeal to the transversality of normal spaces. The group $\pi_{n+k}(T(\nu))$ consists of the equivalence classes of normal structures $(\nu_X : X \to BG(k), \rho_X : S^{n+k} \to T(\nu_X))$ on X with $\nu_X = \nu$.

Following Weiss [20] we shall now identify the algebraic normal bordism groups $\widehat{L}\langle B,\beta\rangle^n(A)$ with the twisted quadratic groups $Q_n(B,\beta)$, the algebraic analogues of the homotopy groups of the Thom space $\pi_{n+k}(T(\nu))$.

A *cobordism* of n-dimensional normal complexes (C,ϕ,γ,θ), $(C',\phi',\gamma',\theta')$ is defined by an $(n+1)$-dimensional symmetric pair

$$((f\ f') : C \oplus C' \to D, (\delta\phi, \phi \oplus -\phi'))$$

together with bundle maps

$$(f,\zeta) \ : \ (C,\gamma) \to (D,\delta\gamma) \ , \ (f',\zeta') \ : \ (C',\gamma') \to (D,\delta\gamma)$$

and an equivalence of hyperquadratic structures on D

$$\delta\theta \ : \ J(\delta\phi) - (\delta\phi_0; f\phi_0 f^*, f'\phi'_0 f'^*)^{\%}(S^n\delta\gamma) + f; (\phi'_0)^{\%}(S^n\zeta')) \to 0 \ .$$

Similarly for the cobordism of (B,β)-normal complexes.

The *symmetric (B,β)-structure L-groups of A* $L\langle B,\beta\rangle^n(A)$ $(n \geq 0)$ of Weiss [20] are the cobordism groups of n-dimensional (B,β)-normal Poincaré complexes over A $(C,\phi,\gamma,\theta,f,\chi)$. For the $\begin{cases} \text{universal} \\ \text{zero} \end{cases}$ chain bundle $\begin{cases} (B(\infty),\beta(\infty)) \\ (0,0) \end{cases}$ over A these are just the $\begin{cases} \text{symmetric} \\ \text{quadratic} \end{cases}$ L-groups

$$\begin{cases} L\langle B(\infty),\beta(\infty)\rangle^n(A) \ = \ L^n(A) \\ L\langle 0,0\rangle^n(A) \ = \ L_n(A) \ . \end{cases}$$

The *symmetric* (B,β)*-structure* $\widehat{L}$*-groups* $\widehat{L}\langle B,\beta\rangle^n(A)$ $(n\geq 0)$ are the cobordism groups of n-dimensional (B,β)-normal complexes over A. For the $\begin{cases}\text{universal}\\ \text{zero}\end{cases}$ chain bundle $\begin{cases}(B(\infty),\beta(\infty))\\ (0,0)\end{cases}$ over A these are just the

$$\begin{cases}\widehat{L}\langle B(\infty),\beta(\infty)\rangle^n(A) \;=\; \widehat{L}^n(A)\\ \widehat{L}\langle 0,0\rangle^n(A) \;=\; 0\ .\end{cases}$$

Algebraic surgery was used in Ranicki [16] to prove that every n-dimensional quadratic Poincaré complex (C,ψ) is cobordant to a highly-connected complex (C',ψ'), with

$$H_r(C') \;=\; 0\ (2r\leq n-2)\ .$$

The boundary of an n-dimensional normal complex (C,ϕ,γ,θ) is an $(n-1)$-dimensional quadratic Poincaré complex $(\partial C,\psi)$. Glueing on to (C,ϕ,γ,θ) the trace of the surgery making $(\partial C,\psi)$ highly-connected there is obtained an n-dimensional normal complex $(C',\phi',\gamma',\theta')$ which is cobordant to (C,ϕ,γ,θ) and which has a highly-connected boundary, with

$$H_r(\partial C') \;=\; H_{r+1}(\phi'_0 : C'^{n-*}\to C') \;=\; 0\ (2r\leq n-3)\ .$$

In particular, this shows that every normal complex is cobordant to a connected complex. Thus $\widehat{L}\langle B,\beta\rangle^n(A)$ is also the cobordism group of connected n-dimensional (B,β)-normal complexes over A. The one-one correspondence established in §7 between connected n-dimensional normal complexes and n-dimensional (symmetric, quadratic) Poincaré pairs generalizes to a one-one correspondence between connected n-dimensional (B,β)-normal complexes over A and n-dimensional (symmetric, quadratic) (B,β)-normal Poincaré pairs over A, for any chain bundle (B,β) over A. It follows that $\widehat{L}\langle B,\beta\rangle^n(A)$ can be identified with the cobordism group of n-dimensional (symmetric, quadratic) (B,β)-normal Poincaré pairs, and that there is defined an exact sequence

$$\dots\to L_n(A)\to L\langle B,\beta\rangle^n(A)\to \widehat{L}\langle B,\beta\rangle^n(A)\xrightarrow{\partial} L_{n-1}(A)\to\dots\ ,$$

with ∂ defined by the quadratic boundary

$$\partial\ :\ \widehat{L}\langle B,\beta\rangle^n(A)\to L_{n-1}(A)\ ;\ (C,\phi,\gamma,\theta,f,\chi)\mapsto \partial(C,\phi,\gamma,\theta)\ .$$

A map of n-dimensional normal complexes

$$(f,\xi,\chi,\eta)\ :\ (C,\phi,\gamma,\theta)\to(C',\phi',\gamma',\theta')$$

determines an $(n+1)$-dimensional symmetric pair $((f\ 1) : C\oplus C'\to C',(\xi,\phi\oplus-\phi'))$, bundle maps

$$(f,\chi)\ :\ (C,\gamma)\to(C',\gamma')\ ,\ (1,0)\ :\ (C',\gamma')\to(C',\gamma')$$

and an equivalence of hyperquadratic structures on C'

$$\eta : J(\xi)-(\xi_0;f\phi_0f^*,\phi'_0)^{\%}(S^n\gamma')+f^{\%}(\theta-(\phi_0)^{\%}(S^n\gamma))-\theta'\to 0\ ,$$

defining a cobordism between (C,ϕ,γ,θ) and $(C',\phi',\gamma',\theta')$ by analogy with the mapping cylinder construction of geometric normal bordisms. Similarly for maps of (B,β)-normal complexes. It follows that the abelian group morphisms

$$\begin{aligned} &\widehat{L}\langle B,\beta\rangle^n(A) \to Q_n(B,\beta)\ ;\ (C,\phi,\gamma,\theta,f,\chi) \mapsto (f,\chi)_\%(\phi,\theta)\ ,\\ &Q_n(B,\beta) \to \widehat{L}\langle B,\beta\rangle^n(A)\ ;\ (\phi,\theta) \mapsto (B,\phi,\beta,\theta,1,0)\end{aligned}$$

are inverse isomorphisms.

For example, if $A=\mathbb{Z}$ and $(B(\infty),\beta(\infty))$ is the universal chain bundle over $\mathbb{Z}$ (as constructed at the end of §6) then

$$L\langle B(\infty),\beta(\infty)\rangle^n(\mathbb{Z}) \ =\ L^n(\mathbb{Z}) \ =\ \begin{cases} \mathbb{Z} & \text{if } n\equiv 0(\text{mod } 4)\\ \mathbb{Z}_2 & \text{if } n\equiv 1(\text{mod } 4)\\ 0 & \text{if } n\equiv 2,3(\text{mod } 4)\ ,\end{cases}$$

$$L_n(\mathbb{Z}) \ =\ \begin{cases} \mathbb{Z} & \text{if } n\equiv 0(\text{mod } 4)\\ \mathbb{Z}_2 & \text{if } n\equiv 2(\text{mod } 4)\\ 0 & \text{if } n\equiv 1,3(\text{mod } 4)\ ,\end{cases}$$

$$\widehat{L}\langle B(\infty),\beta(\infty)\rangle^n(\mathbb{Z}) \ =\ Q_n(B(\infty),\beta(\infty)) \ =\ \begin{cases} \mathbb{Z}_8 & \text{if } n\equiv 0(\text{mod } 4)\\ \mathbb{Z}_2 & \text{if } n\equiv 1,3(\text{mod } 4)\\ 0 & \text{if } n\equiv 2(\text{mod } 4)\ .\end{cases}$$

A spherical fibration $\nu : X \to BG(k)$ determines a chain bundle $(C(\widetilde{X}),\gamma)$ over $\mathbb{Z}[\pi_1(X)]$ ([16], [20]) and there is defined a natural transformation of exact sequences from the Levitt-Jones-Quinn Poincaré bordism sequence

$$\begin{array}{ccccccccc} \dots \to & L_n(\mathbb{Z}[\pi_1(X)]) & \longrightarrow & \Omega^P_n(X,\nu) & \longrightarrow & \pi_{n+k}(T(\nu)) & \to & L_{n-1}(\mathbb{Z}[\pi_1(X)]) & \to \dots\\ & \| & & \downarrow & & \downarrow & & \| & \\ \dots \to & L_n(\mathbb{Z}[\pi_1(X)]) & \to & L^n(C(\widetilde{X}),\gamma) & \to & Q_n(C(\widetilde{X}),\gamma) & \to & L_{n-1}(\mathbb{Z}[\pi_1(X)]) & \to \dots \end{array}$$

with $\Omega^P_n(X,\nu) \to L^n(C(\widetilde{X}),\gamma)$ a generalized symmetric signature map.

§9. Normal Wu classes

The Wu classes of the symmetric structure ϕ and the bundles β,γ in an n-dimensional (B,β)-normal complex $(C,\phi,\gamma,\theta,f,\chi)$ are related by a commutative diagram

$$\begin{array}{ccccc} H^{n-r}(C) & \xrightarrow{\ \phi_0\ } & H_r(C) & \xrightarrow{\ f_*\ } & H_r(B)\\ \Big\downarrow{\scriptstyle v_r(\phi)} & & \Big\downarrow{\scriptstyle \widehat{v}_r(\gamma)} & \swarrow{\scriptstyle \widehat{v}_r(\beta)} & \\ H^{n-2r}(\mathbb{Z}_2;A,(-1)^{n-r}) & \xrightarrow{\ J\ } & \widehat{H}^r(\mathbb{Z}_2;A) & & \end{array}$$

For any chain bundle (B,β) and any chain complex C we shall now define symmetric (B,β)-structure groups $Q\langle B,\beta\rangle^n(C)$ $(n\geq 0)$ to fit into an exact sequence

$$\dots \to Q\langle B,\beta\rangle^n(C) \to Q^n(C)\oplus H_n(B^t\otimes_A C) \to \widehat{Q}^n(C) \to Q\langle B,\beta\rangle^{n-1}(C) \to \dots\ .$$

The Wu classes $v_r(\phi)$ of a symmetric complex (C,ϕ) will then be refined to the normal Wu classes of a (B,β)-normal complex $(C,\phi,\gamma,\theta,f,\chi)$

$$v_r \ = \ v_r(\phi,\gamma,\theta,f,\chi) \ : \ H^{n-r}(C) \to Q\langle B,\beta\rangle^n(S^{n-r}A) \ ,$$

with

$$v_r(\phi) : H^{n-r}(C) \xrightarrow{v_r} Q\langle B,\beta\rangle^n(S^{n-r}A) \to Q^n(S^{n-r}A) = H^{n-2r}(\mathbb{Z}_2;A,(-1)^{n-r}).$$

In §11 below the normal Wu classes will be used to define a $\mathbb{Z}_4$-valued quadratic function on $H^n(C)$ for a $2n$-dimensional symmetric Poincaré complex (C,ϕ) over $\mathbb{Z}_2$ with normal $(v_{n+1}=0)$-structure, as required to define the $\mathbb{Z}_8$-valued invariant of Brown [4].

Let (B,β) be a chain bundle over A, and let C be a finite-dimensional A-module chain complex. An *n-dimensional symmetric (B,β)-structure on C (ϕ,θ,f)* is defined by an n-dimensional symmetric structure $\phi \in (W^{\%}C)_n$ together with a chain $\theta \in (\widehat{W}^{\%}C)_{n+1}$ and a chain map $f : B^{n-*} \to C$ such that

$$J(\phi) - \widehat{f}^{\%}(S^n\beta) \ = \ d\theta \in (\widehat{W}^{\%}C)_n \ .$$

An n-dimensional (B,β)-normal structure $(\phi,\gamma,\theta,g,\chi)$ on C determines the n-dimensional symmetric (B,β)-structure $(\phi,\theta+(\phi_0)^{\%}(S^n\chi),\phi_0 g^*)$ on C. Conversely, if $f^* : C^{n-*} \to B$ is a composite

$$f^* \ : \ C^{n-*} \xrightarrow{\phi_0} C \xrightarrow{g} B$$

(as is always the case up to chain homotopy if (C,ϕ) is a Poincaré complex) the symmetric (B,β)-structure (ϕ,θ,f) determines the n-dimensional (B,β)-normal structure $(\phi,g^*\gamma,\theta,g,0)$.

An *n-dimensional symmetric (B,β)-structure (Poincaré) complex over A* (C,ϕ,θ,f) is an n-dimensional A-module chain complex C together with an n-dimensional symmetric (B,β)-structure (ϕ,θ,f) (such that $\phi_0 : C^{n-*} \to C$ is a chain equivalence). As for symmetric (Poincaré) pairs there is also the analogous notion of *symmetric (B,β)-structure (Poincaré) pair.* There is essentially no difference between symmetric (B,β)-structure Poincaré complexes and (B,β)-normal Poincaré complexes, so that the L-groups $L\langle B,\beta\rangle^n(A)$ $(n \geq 0)$ can also be regarded as the cobordism groups of n-dimensional symmetric (B,β)-structure Poincaré complexes over A.

An *equivalence* of n-dimensional symmetric (B,β)-structures on C

$$(\xi,\eta,g) \ : \ (\phi,\theta,f) \to (\phi',\theta',f')$$

is defined by an equivalence of symmetric structures $\xi : \phi \to \phi'$ together with a chain $\eta \in (\widehat{W}^{\%}C)_{n+2}$ and a chain homotopy $g : f \simeq f' : B^{n-*} \to C$ such that

$$J(\xi) - (g;f,f')^{\%}(S^n\beta) - \theta' + \theta \ = \ d\eta \in (\widehat{W}^{\%}C)_{n+1} \ .$$

The *n-dimensional symmetric (B,β)-structure group of C $Q\langle B,\beta\rangle^n(C)$* is the abelian group of equivalence classes of n-dimensional symmetric (B,β)-symmetric structures on C, with addition by

$$(\phi,\theta,f)+(\phi',\theta',f') \ = \ (\phi+\phi',\theta+\theta'+[f,f'](S^n\beta),f+f') \in Q\langle B,\beta\rangle^n(C) \ .$$

There is also a more economical description of $Q\langle B,\beta\rangle^n(C)$ as the abelian group of equivalence classes of pairs (ψ, f) defined by an n-dimensional quadratic structure $\psi \in (W_\% C)_n$ and a chain map $f : B^{n-*} \to C$ such that

$$f_\% H(S^n\beta) \;=\; d\psi \in (W_\% C)_{n-1}\ ,$$

so that up to signs

$$f\beta_{-n-s-1}f^* \;=\; d\psi_s + \psi_s d^* + \psi_{s+1} + \psi^*_{s+1} \in \mathrm{Hom}_A(C^{-*},C)_{n-s}\ (s \geq 0)\ ,$$

subject to the equivalence relation

$$\begin{array}{ll}(\psi,f) \sim (\psi',f') & \text{if there exist a chain homotopy } g : f \simeq f' : B^{n-*} \to C \\ & \text{and an equivalence of quadratic structures} \\ & \chi\ :\ \psi' - \psi \to (g;f,f')_\% H(S^n\beta)\ ,\end{array}$$

with addition by

$$(\psi,f) + (\psi',f') \;=\; (\psi + \psi' + H([f,f'](S^n\beta)), f + f')\ .$$

The pair (ψ, f) determines the triple (ϕ, θ, f) with

$$\phi_s \;=\; \begin{cases} f\beta_{s-n}f^* & \text{if } s \geq 1 \\ f\beta_{-n}f^* + (1+T)\psi_0 & \text{if } s = 0\ , \end{cases}$$

$$\theta_s \;=\; \begin{cases} 0 & \text{if } s \geq 0 \\ \psi_{-s-1} & \text{if } s \leq -1\ . \end{cases}$$

Conversely, a triple (ϕ, θ, f) determines the pair (ψ, f) with

$$\psi_s \;=\; \theta_{-s-1}\ (s \geq 0)\ .$$

Given an n-dimensional symmetric (B,β)-structure (ϕ,θ,f) on C, a chain bundle map $(g,\chi) : (B,\beta) \to (B',\beta')$ and a chain map $h : C \to C'$ define the *pushforward* n-dimensional symmetric (B',β')-structure on C'

$$\langle g,\chi\rangle(h)^\%(\phi,\theta,f) \;=\; (h^\%(\phi), h^\%(\theta + S^n(\widehat{f}^\%\chi)), hfg^*)\ .$$

Thus the groups $Q\langle B,\beta\rangle^*(C)$ are covariant in both (B,β) and C, with pushforward abelian group morphisms

$$\begin{array}{l}\langle g,\chi\rangle(h)^\%\ : Q\langle B,\beta\rangle^n(C) \to Q\langle B',\beta'\rangle^n(C')\ ; \\ \qquad (\phi,\theta,f) \mapsto (h^\%(\phi), \widehat{h}^\%(\theta + \widehat{f}^\%(S^n\chi)), hfg^*)\end{array}$$

depending only on the homotopy classes of (g,χ) and h.

An n-dimensional symmetric (B,β)-structure (ϕ,θ,f) on C determines an n-dimensional symmetric structure $\phi \in (W^\% C)_n$ on C, so that there is defined a forgetful map

$$s\ :\ Q\langle B,\beta\rangle^n(C) \to Q^n(C)\ ;\ (\phi,\theta,f) \mapsto \phi\ .$$

An n-dimensional quadratic structure $\psi \in (W_\% C)_n$ on C determines an n-dimensional symmetric (B,β)-structure $((1+T)\psi, \theta, 0)$ on C for any (B,β), with

$$\theta_s \;=\; \begin{cases} \psi_{-s-1} & \text{if } s \leq -1 \\ 0 & \text{if } s \geq 0\ . \end{cases}$$

Thus there are also defined forgetful maps

$$s \;:\; Q_n(C) \to Q\langle B,\beta\rangle^n(C) \;;\; \psi \mapsto ((1+T)\psi, \theta, 0) \;,$$

and $sr = 1 + T : Q_n(C) \to Q^n(C)$.

Let $P\langle B,\beta\rangle^n(C)$ be the abelian group of equivalence classes of n-dimensional symmetric (B,β)-structures (ϕ,θ,f) with $\phi = 0$, to be denoted (θ,f), subject to the equivalence relation

$$\begin{aligned}(\theta,f) \sim (\theta',f') \text{ if there exists an equivalence of } (B,\beta)\text{-structures} \\ (0,\eta,g) : (0,\theta,f) \to (0,\theta',f').\end{aligned}$$

The symmetric (B,β)-structure groups $Q\langle B,\beta\rangle^*(C)$ and the groups $P\langle B,\beta\rangle^*(C)$ are related by a commutative braid of exact sequences of abelian groups

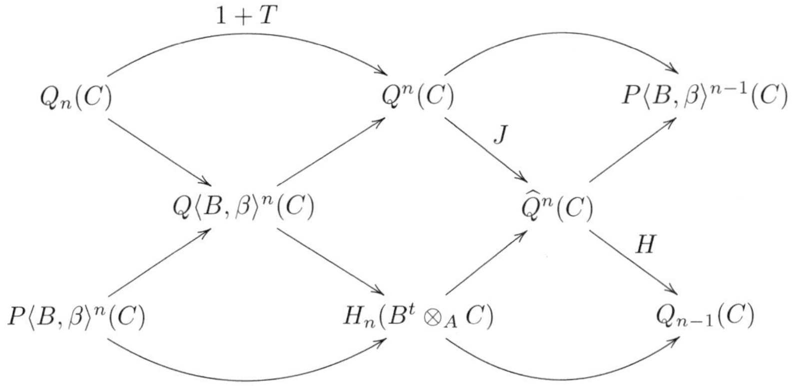

If (B,β) is the $\begin{cases}\text{universal} \\ \text{zero}\end{cases}$ chain bundle $\begin{cases}(B(\infty),\beta(\infty)) \\ (0,0)\end{cases}$ the forgetful map

$$\begin{cases} Q\langle B(\infty),\beta(\infty)\rangle^n(C) \to Q^n(C) \;;\; (\phi,\theta,f) \mapsto \phi \\ Q_n(C) \to Q\langle 0,0\rangle^n(C) \;;\; \psi \mapsto ((1+T)\psi,\theta) \end{cases}$$

is an isomorphism and

$$\begin{cases} P\langle B(\infty),\beta(\infty)\rangle^n(C) \;=\; 0 \\ P\langle 0,0\rangle^n(C) \;=\; \widehat{Q}^{n+1}(C) \;. \end{cases}$$

The *Wu classes* of an n-dimensional symmetric (B,β)-structure (ϕ,θ,f) on C are the A-morphisms

$$\begin{aligned} v_r(\phi,\theta,f) \;:\; H^{n-r}(C) \to Q\langle B,\beta\rangle^n(S^{n-r}A) \;; \\ (x : C \to S^{n-r}A) \mapsto \langle 1,0\rangle(x)^{\%}(\phi,\theta,f) \;. \end{aligned}$$

Now

$$Q\langle B,\beta\rangle^n(S^{n-r}A) \;=$$

$$\begin{cases} H_r(B) & \text{if } 2r < n \\ \{(a,b) \in A \oplus B_r \,|\, db = 0 \in B_{r-1}\}/\sim & \text{if } 2r = n \\ \{(a,b) \in A \oplus B_r \,|\, a + (-1)^r\overline{a} + \beta_{-2r}(b)(b) = 0 \in A, db = 0 \in B_{r-1}\}/\sim & \\ & \text{if } 2r > n \end{cases}$$

with the equivalence relation $\sim$ defined by

$$\begin{aligned}(a,b)\sim(a',b') \text{ if there exists } & (x,y)\in A\oplus B_{r+1} \text{ such that}\\ & a'-a = x+(-1)^{r+1}\overline{x}+\beta_{-2r-2}(y)(y)\\ &\qquad\qquad +\beta_{-2r-1}(y)(b)+\beta_{-2r-1}(b')(y)\in A,\\ & b'-b = dy\in B_r\end{aligned}$$

and addition by

$$(a,b)+(a',b') = (a+a'+\beta_{-2r}(b)(b'),b+b') .$$

The map to the symmetric Q-group is given by

$$Q\langle B,\beta\rangle^n(S^{n-r}A)\to Q^n(S^{n-r}A) = H^{n-2r}(\mathbb{Z}_2;A,(-1)^{n-r}) ;$$
$$\begin{cases} b\mapsto \beta_{-2r}(b)(b) & \text{if } 2r<n\\ (a,b)\mapsto a+(-1)^r\overline{a}+\beta_{-2r}(b)(b) & \text{if } 2r=n\\ 0 & \text{if } 2r>n .\end{cases}$$

The Wu classes are given by

$$v_r(\phi,\theta,f) : H^{n-r}(C)\to Q\langle B,\beta\rangle^n(S^{n-r}A) ;$$
$$z\mapsto\begin{cases} f^*(z) & \text{if } 2r<n\\ (\theta_{n-2r-1}(z)(z),f^*(z)) & \text{if } 2r\geq n\end{cases} \quad (z\in C^{n-r},d^*z=0) .$$

§10. **Forms**

In Ranicki [16] the even-dimensional $\begin{cases}\text{symmetric}\\ \text{quadratic}\end{cases}$ L-groups $\begin{cases}L^{2n}(A)\\ L_{2n}(A)\end{cases}$ $(n\geq 0)$ were related to the Witt groups $\begin{cases}W^{(-1)^n}(A)\\ W_{(-1)^n}(A)\end{cases}$ of nonsingular $\begin{cases}(-1)^n\text{-symmetric}\\ (-1)^n\text{-quadratic}\end{cases}$ forms over A. In particular, it was shown that

$$\begin{cases} L^0(A) = W^{+1}(A)\\ L_{2n}(A) = W_{(-1)^n}(A)\end{cases} (n\geq 0) .$$

This relationship between L-groups and Witt groups will now be generalized to the even-dimensional symmetric (B,β)-structure L-groups $L\langle B,\beta\rangle^{2n}(A)$ and the Witt groups $W_{Q(n)}(A)$ of nonsingular $Q(n)$-quadratic forms over A, with (B,β) any chain bundle over A and

$$Q(n) = Q\langle B,\beta\rangle^{2n}(S^nA) .$$

Let $\epsilon=\pm 1$. An *ϵ-symmetric form over A* (M,λ) is a f.g. projective A-module M together with an element $\lambda\in\mathrm{Hom}_A(M,M^*)$ such that

$$\epsilon\lambda^* = \lambda : M\to M^* .$$

Equivalently, the form is defined by a pairing

$$\lambda : M\times M\to A ; (x,y)\to\lambda(x,y) = \lambda(x)(y)$$

such that

$$\begin{aligned}\lambda(ax,by) &= b\lambda(x,y)\overline{a} ,\\ \lambda(x+x',y) &= \lambda(x,y)+\lambda(x',y) ,\\ \epsilon\overline{\lambda(y,x)} &= \lambda(x,y)\ (x,y\in M,a,b\in A) .\end{aligned}$$

Let $Q(\epsilon)$ be an A-group together with A-morphisms

$$r \ : \ Q(\epsilon) \to H^0(\mathbb{Z}_2; A, \epsilon) \ = \ \{a \in A \mid \epsilon\bar{a} = a\} \ ,$$
$$s \ : \ H_0(\mathbb{Z}_2; A, \epsilon) \ = \ A/\{b - \epsilon\bar{b} \mid b \in A\} \to Q(\epsilon)$$

such that

$$rs \ = \ 1 + T_\epsilon \ : \ H_0(\mathbb{Z}_2; A, \epsilon) \to H^0(\mathbb{Z}_2; A, \epsilon) \ .$$

A *$Q(\epsilon)$-quadratic form over A* (M, λ, μ) is an ϵ-symmetric form (M, λ) together with an A-morphism $\mu : M \to Q(\epsilon)$ such that

$$r(\mu(x)) \ = \ \lambda(x, x) \in H^0(\mathbb{Z}_2; A, \epsilon) \ ,$$
$$\mu(x + y) - \mu(x) - \mu(y) \ = \ s(\lambda(x, y)) \in Q(\epsilon) \ (x, y \in M) \ .$$

There is an evident notion of *isomorphism* of $Q(\epsilon)$-quadratic forms.

A $Q(\epsilon)$-quadratic form (M, λ, μ) is *nonsingular* if $\lambda \in \mathrm{Hom}_A(M, M^*)$ is an isomorphism of A-modules.

A nonsingular $Q(\epsilon)$-quadratic form (M, λ, μ) is *hyperbolic* if there exists a direct summand $L \subset M$ such that

(i) the inclusion $j \in \mathrm{Hom}_A(L, M)$ fits into an exact sequence

$$0 \to L \xrightarrow{j} M \xrightarrow{j^*\lambda} L^* \to 0 \ ,$$

(ii) $\mu j \ = \ 0 \ : \ L \to Q(\epsilon)$.

The *$Q(\epsilon)$-quadratic Witt group of A* $W_{Q(\epsilon)}(A)$ is the abelian group of equivalence classes of nonsingular $Q(\epsilon)$-quadratic forms (M, λ, μ), subject to the equivalence relation

$(M, \lambda, \mu) \sim (M', \lambda', \mu')$ if there exists an isomorphism
$(M, \lambda, \mu) \oplus (N, \nu, \rho) \to (M', \lambda', \mu') \oplus (N', \nu', \rho')$
for some hyperbolic $Q(\epsilon)$-quadratic forms
(N, ν, ρ), (N', ν', ρ').

For $Q(\epsilon) = H^0(\mathbb{Z}_2; A, \epsilon)$, $r = 1$, $s = 1 + T_\epsilon$ a $Q(\epsilon)$-quadratic form (M, λ, μ) may be identified with the ϵ-symmetric form (M, λ), since λ determines μ by

$$\mu(x) \ = \ \lambda(x, x) \in H^0(\mathbb{Z}_2; A, \epsilon) \ (x \in M) \ .$$

The Witt group of ϵ-symmetric forms $W_{Q(\epsilon)}(A)$ is denoted by $W^\epsilon(A)$.

For $Q(\epsilon) = H_0(\mathbb{Z}_2; A, \epsilon)$, $r = 1 + T_\epsilon$, $s = 1$ a $Q(\epsilon)$-quadratic form (M, λ, μ) is just a ϵ-quadratic form in the sense of Wall [19]. The Witt group of ϵ-quadratic forms $W_{Q(\epsilon)}(A)$ is denoted by $W_\epsilon(A)$.

For $Q(\epsilon) = \mathrm{im}(1 + T_\epsilon : H_0(\mathbb{Z}_2; A, \epsilon) \to H^0(\mathbb{Z}_2; A, \epsilon))$, r = projection, s = injection a $Q(\epsilon)$-quadratic form (M, λ, μ) is just an ϵ-symmetric form (M, λ) for which there exists an ϵ-quadratic form $(M, \lambda, \mu : M \to H_0(\mathbb{Z}_2; A, \epsilon))$. Such an ϵ-symmetric form is *even*. The Witt group of even ϵ-symmetric forms $W_{Q(\epsilon)}(A)$ is denoted by $W\langle v_0\rangle^\epsilon(A)$.

For $\epsilon = +1$ $\begin{cases} \epsilon\text{-symmetric} \\ \epsilon\text{-quadratic} \end{cases}$ is abbreviated to $\begin{cases} \text{symmetric} \\ \text{quadratic} . \end{cases}$

A $2n$-dimensional $\begin{cases} \text{symmetric} \\ \text{quadratic} \end{cases}$ (Poincaré) complex over A $\begin{cases} (C,\phi) \\ (C,\psi) \end{cases}$ with $H^n(C)$ a f.g. projective A-module determines a (nonsingular) $\begin{cases} (-1)^n\text{-symmetric} \\ (-1)^n\text{-quadratic} \end{cases}$ form over A $\begin{cases} (H^n(C),\phi_0,v_n(\phi)) \\ (H^n(C),(1+T)\psi_0,v^n(\psi)) \end{cases}$ with

$$\begin{cases} v_n(\phi)\ :\ H^n(C)\to H^0(\mathbb{Z}_2;A,(-1)^n)\ ;\ x\mapsto \phi_0(x)(x) \\ v^n(\psi)\ :\ H^n(C)\to H_0(\mathbb{Z}_2;A,(-1)^n)\ ;\ x\mapsto \psi_0(x)(x)\ . \end{cases}$$

Conversely, a (nonsingular) $\begin{cases} (-1)^n\text{-symmetric} \\ (-1)^n\text{-quadratic} \end{cases}$ form $\begin{cases} (M,\lambda) \\ (M,\lambda,\mu) \end{cases}$ determines a $2n$-dimensional $\begin{cases} \text{symmetric} \\ \text{quadratic} \end{cases}$ (Poincaré) complex $\begin{cases} (C,\phi) \\ (C,\psi) \end{cases}$ such that

$$\begin{cases} \phi_0 \\ (1+T)\psi_0 \end{cases} =\ \lambda\ :\ C^n\ =\ M\to C_n\ =\ M^*\ ,\ C_r\ =\ 0\ (r\neq n)\ ,$$

$$v^n(\psi)\ =\ \mu\ :\ H^n(C)\ =\ M\to H_0(\mathbb{Z}_2;A,(-1)^n)\ .$$

The corresponding morphisms from the Witt groups to the L-groups

$$\begin{cases} W^{(-1)^n}(A)\to L^{2n}(A)\ ;\ (M,\lambda)\mapsto (C,\phi) \\ W_{(-1)^n}(A)\to L_{2n}(A)\ ;\ (M,\lambda,\mu)\mapsto (C,\psi) \end{cases}$$

were shown in Ranicki [16] to be isomorphisms for $n=0$ if A is any ring, and for all $n\geq 0$ if A is $\begin{cases} \text{a Dedekind} \\ \text{any} \end{cases}$ ring. For a Dedekind ring A the inverse isomorphism in symmetric L-theory is given by

$$L^{2n}(A)\to W^{(-1)^n}(A)\ ;\ (C,\phi)\mapsto (H^n(C)/(\text{torsion}),\phi_0)\ .$$

The inverse isomorphism in quadratic L-theory is given for any A by

$$L_{2n}(A)\to W_{(-1)^n}(A)\ ;$$

$$(C,\psi)\mapsto (\text{coker}(\begin{pmatrix} d^* & 0 \\ (1+T)\psi_0 & d \end{pmatrix} : C^{n-1}\oplus C_{n+2}\to C^n\oplus C_{n+1}), \begin{bmatrix} \psi_0 & d \\ 0 & 0 \end{bmatrix})\ .$$

If A is a field this isomorphism can also be expressed as

$$(C,\psi)\mapsto (H^n(C),(1+T)\psi_0,v^n(\psi))$$

but this is not the case in general – see Milgram and Ranicki [13, p.406].

Given A-groups M,N and a symmetric bilinear pairing

$$\phi\ :\ N\times N\to M$$

such that

$$\phi(ay,ay')\ =\ a\phi(y,y')\in M\ (a\in A, y,y'\in N)$$

let $M\times_\phi N$ be the A-group of pairs $(x\in M, y\in N)$, with addition by

$$(x,y)+(x',y')\ =\ (x+x'+\phi(y,y'),y+y')\in M\times_\phi N$$

and A acting by

$$A\times(M\times_\phi N)\to M\times_\phi N\ ;\ (a,(x,y))\mapsto (ax,ay)\ .$$

There is then defined a short exact sequence of A-groups and A-morphisms

$$0 \to M \to M \times_\phi N \to N \to 0$$

with

$$M \to M \times_\phi N \; ; \; x \mapsto (x, 0)$$
$$M \times_\phi N \to N \; ; \; (x, y) \mapsto y \; .$$

Given a chain bundle (B, β) over A define the A-group

$$\begin{aligned} Q(n) &= Q\langle B, \beta\rangle^{2n}(S^n A) \\ &= \{(a, b) \in A \oplus B_n \,|\, db = 0 \in B_{n-1}\}/\!\sim \; , \end{aligned}$$

where

$(a, b) \sim (a', b')$ if there exist $(x, y) \in A \oplus B_{n+1}$ such that

$$\begin{aligned} a' - a &= x + (-1)^{n+1}\overline{x} + \beta_{-2n-2}(y)(y) + \beta_{-2n-1}(y)(b) + \beta_{-2n-1}(b')(y) \; , \\ b' - b &= dy \; , \end{aligned}$$

with addition by

$$(a, b) + (a', b') = (a + a' + \beta_{-2n}(b)(b'), b + b')$$

and A-action by

$$A \times Q(n) \to Q(n) \; ; \; (x, (a, b)) \mapsto x(a, b) = (xa\overline{x}, xb) \; .$$

The A-morphisms

$$\begin{aligned} r \; &: \; Q_{2n}(S^n A) = H_0(\mathbb{Z}_2; A, (-1)^n) \to Q(n) \; ; \; a \mapsto (a, 0) \; , \\ s \; &: \; Q(n) \to Q^{2n}(S^n A) = H^0(\mathbb{Z}_2; A, (-1)^n) \; ; \\ & \qquad (a, b) \mapsto a + (-1)^n\overline{a} + \beta_{-2n}(b)(b) \end{aligned}$$

are such that there is defined a commutative braid of exact sequences

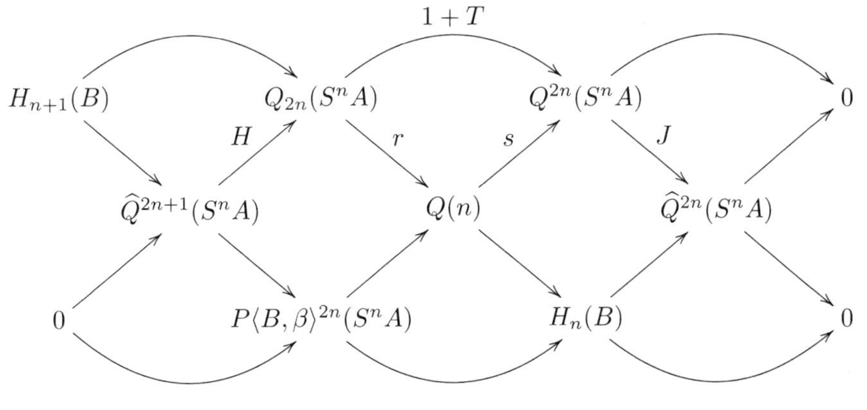

with

$$Q(n) \to H_n(B) \; ; \; (a, b) \mapsto b \; .$$

If (B, β) is such that for all $y \in B_{n+1}$ there exists $x \in A$ such that

$$(\beta_{-2n-2} + \beta_{-2n-1}d)(y)(y) = x + (-1)^n\overline{x} \in A$$

(e.g. if $d = 0 : B_{n+1} \to B_n$ and $v_{n+1}(\beta) = 0 : H_{n+1}(B) \to \widehat{H}^{n+1}(\mathbb{Z}_2; A)$) then there is a natural identification of A-groups

$$Q(n) \;=\; Q_{2n}(S^nA) \times_{\beta_{-2n}} H_n(B)$$

with

$$\beta_{-2n} \;:\; H_n(B) \times H_n(B) \to Q_{2n}(S^nA) \;;\; (b, b') \mapsto \beta_{-2n}(b)(b') \;.$$

For any chain bundle (B, β) and any $Q(n)$-quadratic form (M, λ, μ) there exist A-module morphisms

$$g \;:\; M \to B_n \;,\; \psi \;:\; M \to M^*$$

such that

$$\begin{aligned} dg \;&=\; 0 \;:\; M \to B_{n-1} \;, \\ \lambda - g^*\beta_{-2n}g \;&=\; \psi + (-1)^n\psi^* \;:\; M \to M^* \;, \\ \mu \;&:\; M \to Q(n) \;;\; x \mapsto (\psi(x)(x), g(x)) \;. \end{aligned}$$

If (M, λ, μ) is a nonsingular form there is thus defined a $2n$-dimensional symmetric (B, β)-structure Poincaré complex (C, ϕ, θ, f) with

$$\begin{aligned} \phi_0 \;&=\; \lambda \;:\; C^n \;=\; M \to C_n \;=\; M^* \;,\; C_r \;=\; 0 \; (r \neq n) \;, \\ \theta_{-1} \;&=\; \psi \;:\; C^n \;=\; M \to C_n \;=\; M^* \;, \\ f \;&=\; g\lambda^{-1} \;:\; C_n \;=\; M^* \xrightarrow{\lambda^{-1}} M \xrightarrow{g} B_n \;, \\ v_n(\phi, \theta, f) \;&=\; \mu \;:\; H^n(C) \;=\; M \to Q(n) \;. \end{aligned}$$

The construction defines a morphism of abelian groups

$$W_{Q(n)}(A) \to L\langle B, \beta\rangle^{2n}(A) \;;\; (M, \lambda, \mu) \mapsto (C, \phi, \theta, f) \;.$$

Conversely, if (C, ϕ, θ, f) is a $2n$-dimensional symmetric (B, β)-structure Poincaré complex such that $H^n(C)$ is a f.g. projective A-module there is defined a nonsingular $Q(n)$-quadratic form $(H^n(C), \phi_0, v_n(\phi, \theta, f))$, with

$$v_n(\phi, \theta, f) \;:\; H^n(C) \to Q(n) \;;\; x \mapsto (\theta_{-1}(x)(x), f(x)) \;.$$

It follows that for any ring A there is a natural identification of the 0-dimensional L-group with the Witt group

$$L\langle B, \beta\rangle^0(A) \;=\; W_{Q(0)}(A) \;.$$

For a field A the morphisms

$$W_{Q(n)}(A) \to L\langle B, \beta\rangle^{2n}(A) \; (n \geq 0)$$

are injections, which are split by

$$L\langle B, \beta\rangle^{2n}(A) \to W_{Q(n)}(A) \;;\; (C, \phi, \theta, f) \mapsto (H^n(C), \phi_0, v_n(\phi, \theta, f)) \; (n \geq 0) \;.$$

For any ring with involution A let $(B(\infty), \beta(\infty))$ be the universal chain bundle of Weiss [20] (cf. §6 above), with isomorphisms

$$\begin{aligned} \widehat{v}_m(\beta(\infty)) \;&:\; H_m(B(\infty)) \to \widehat{H}^m(\mathbb{Z}_2; A) \;, \\ L\langle B(\infty), \beta(\infty)\rangle^m(A) \;&\cong\; L^m(A) \end{aligned}$$

and an exact sequence

$$\cdots \to L_m(A) \xrightarrow{1+T} L^m(A) \to Q_m(B(\infty),\beta(\infty)) \xrightarrow{\partial} L_{m-1}(A) \to \dots .$$

The cokernel of the symmetrization map in the Witt groups

$$\mathrm{coker}(1+T : L_0(A) \to L^0(A)) \;=\; \mathrm{im}(L^0(A) \to Q_0(B(\infty),\beta(\infty)))$$

was computed for noetherian A by Carlsson [5] in terms of 'Wu invariants' prior to the general theory of Weiss [20].

For $n \geq 0$ let $(B\langle n+1\rangle, \beta\langle n+1\rangle)$ be the $(v_{n+1}=0)$*-universal chain bundle over* A, characterized up to equivalence by the properties

(i) $\widehat{v}_r(\beta\langle n+1\rangle) : H_r(B\langle n+1\rangle) \to \widehat{H}^r(\mathbb{Z}_2;A)$ is an isomorphism for $r \neq n+1$,
(ii) $H_{n+1}(B\langle n+1\rangle) = 0$.

The $(v_{n+1}=0)$*-symmetric L-groups of A* are defined by

$$L\langle v_{n+1}\rangle^m(A) \;=\; L\langle B\langle n+1\rangle, \beta\langle n+1\rangle\rangle^m(A) \;(m \geq 0) .$$

Define the A-group

$$Q\langle v_{n+1}\rangle \;=\; Q\langle B\langle n+1\rangle, \beta\langle n+1\rangle\rangle^{2n}(S^nA) ,$$

to fit into the commutative braid of exact sequences

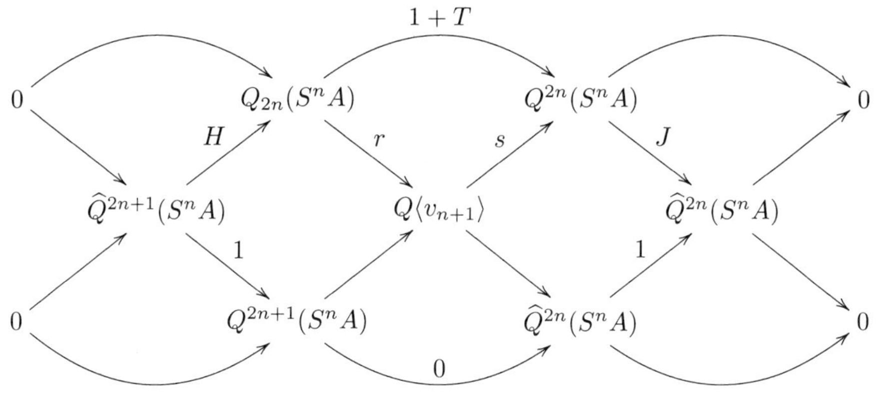

In §11 below we shall make use of the surjections

$$L\langle v_{n+1}\rangle^{2n}(A) \to W_{Q\langle v_{n+1}\rangle}(A) \; ; \; (C,\phi,\theta,f) \mapsto (H^n(C), \phi_0, v_n(\phi,\theta,f)) \; (n \geq 0)$$

defined for a field A.

§11. An example

As an illustration of the exact sequence of §8

$$\cdots \to L_n(A) \to L\langle B,\beta\rangle^n(A) \to Q_n(B,\beta) \xrightarrow{\partial} L_{n-1}(A) \to \dots$$

we compute the Witt groups $L^0(A)$, $L_0(A)$, $L\langle v_1\rangle^0(A)$ for A a perfect field of characteristic 2, without appealing to the theorem of Arf [1] on the classification of quadratic forms over such A (cf. Example 2.14 of Ranicki [17]).

For any field A $\begin{cases} L^{2n}(A) \\ L_{2n}(A) \end{cases}$ is the Witt group of nonsingular $\begin{cases} (-1)^n\text{-symmetric} \\ (-1)^n\text{-quadratic} \end{cases}$ forms over A, and $\begin{cases} L^{2n+1}(A) = 0 \\ L_{2n+1}(A) = 0 \end{cases}$ $(n \geq 0)$ – see Ranicki [16] for details.

Let then A be a perfect field of characteristic 2, so that squaring defines an automorphism

$$A \to A \; ; \; a \mapsto a^2 \; .$$

Let A have the identity involution

$$\bar{\ } : A \to A \; ; \; a \mapsto \bar{a} \; = \; a \; .$$

As an additive group

$$\widehat{H}^r(\mathbb{Z}_2; A) \; = \; A \; (r \in \mathbb{Z})$$

with A acting by

$$A \times \widehat{H}^r(\mathbb{Z}_2; A) \to \widehat{H}^r(\mathbb{Z}_2; A) \; ; \; (a, x) \mapsto a^2 x \; ,$$

and there is defined an isomorphism of A-modules

$$A \to \widehat{H}^r(\mathbb{Z}_2; A) \; ; \; a \mapsto a^2 \; .$$

The chain bundle over A $(B(\infty), \beta(\infty))$ defined by

$$\begin{aligned} d_{B(\infty)} \; &= \; 0 \; : \; B(\infty)_r \; = \; A \to B(\infty)_{r-1} \; = \; A \; , \\ \beta(\infty)_s \; &= \; \begin{cases} 1 \\ 0 \end{cases} : \; B(\infty)_r \; = \; A \to B(\infty)^{-r-s} \; = \; A \text{ if } \begin{cases} s = -2r \\ s \neq -2r \end{cases} \end{aligned}$$

is universal. The twisted quadratic groups of $(B(\infty), \beta(\infty))$ are given up to isomorphism by

$$\begin{aligned} Q_{2n}(B(\infty), \beta(\infty)) \; &= \; Q^{\bullet}(A) \; , \\ Q_{2n+1}(B(\infty), \beta(\infty)) \; &= \; Q_{\bullet}(A) \; , \end{aligned}$$

with the abelian groups $Q^{\bullet}(A)$, $Q_{\bullet}(A)$ defined by

$$\begin{aligned} Q^{\bullet}(A) \; &= \; \{a \in A \,|\, a + a^2 = 0\} \; = \; \mathbb{Z}_2 \; , \\ Q_{\bullet}(A) \; &= \; A/\{b + b^2 \,|\, b \in A\} \; , \end{aligned}$$

and isomorphisms defined by

$$\begin{aligned} & Q_{2n}(B(\infty), \beta(\infty)) \to Q^{\bullet}(A) \; ; \; (\phi, \theta) \mapsto \phi_0(1)(1) \; , \\ & \qquad \phi_0 \; : \; B(\infty)^n \; = \; A \to B(\infty)_n \; = \; A \; , \\ & Q_{2n+1}(B(\infty), \beta(\infty)) \to Q_{\bullet}(A) \; ; \; (\phi, \theta) \mapsto \theta_{-1}(1)(1) \; , \\ & \qquad \theta_{-1} \; : \; B(\infty)^{n+1} \; = \; A \to B(\infty)_{n+1} \; = \; A \; . \end{aligned}$$

A symmetric form over A (M, λ) is even if and only if

$$\lambda(x, x) \; = \; 0 \in A \; (x \in M) \; .$$

A nonsingular even symmetric form over A (M,λ) is hyperbolic, since for any $x\in M$ there exists $y\in M$ such that $\lambda(x)(y)=1\in A$, so that a hyperbolic summand may be split off (M,λ)

$$(M,\lambda) \;=\; (Ax\oplus Ay,\begin{pmatrix}0&1\\1&0\end{pmatrix})\oplus(M',\lambda')\ ,$$

with

$$\mathrm{rank}_A M' \;=\; (\mathrm{rank}_A M)-2\ .$$

Thus

$$L\langle v_0\rangle^0(A) \;=\; W\langle v_0\rangle(A) \;=\; 0$$

and the symmetrization maps

$$1+T\ :\ L_{2n}(A) \;=\; L_0(A)\to L\langle v_0\rangle^0(A)\to L^{2n}(A)$$

are zero. It is now immediate from the exact sequence

$$\dots\to L_m(A)\xrightarrow{1+T} L^m(A)\to Q_m(B(\infty),\beta(\infty))\xrightarrow{\partial} L_{m-1}(A)\to\dots$$

that

$$L^{2n}(A) \;=\; Q^\bullet(A)\ ,\ L_{2n}(A) \;=\; Q_\bullet(A)\ .$$

In the symmetric case there is defined an isomorphism

$$L^{2n}(A)\to Q^\bullet(A) \;=\; \mathbb{Z}_2\ ;\ (C,\phi)\mapsto\phi_0(v)(v) \;=\; \mathrm{rank}_A H^n(C)\ ,$$

sending a $2n$-dimensional symmetric Poincaré complex (C,ϕ) over A to the element $\phi_0(v)(v)\in Q^\bullet(A)$, with $v\in H^n(C)$ the unique cohomology class such that

$$\phi_0(x)(v) \;=\; \phi_0(x)(x)\in\widehat{H}^n(\mathbb{Z}_2;A)\ (x\in H^n(C))\ .$$

The inverse isomorphism

$$Q^\bullet(A) \;=\; \mathbb{Z}_2\to L^{2n}(A)$$

sends $1\in Q^\bullet(A)$ to the $2n$-dimensional symmetric Poincaré complex (C,ϕ) defined by

$$\phi_0 \;=\; 1\ :\ C^n \;=\; A\to C_n \;=\; A\ ,\ C_r \;=\; 0\ (r\neq 0)\ .$$

In the quadratic case there is defined an isomorphism

$$\partial\ :\ Q_\bullet(A)\to L_{2n}(A)\ ;\ a\mapsto(C,\psi)$$

with (C,ψ) the $2n$-dimensional quadratic Poincaré complex over A given by

$$\psi_0 \;=\; \begin{pmatrix}a&1\\0&1\end{pmatrix}\ :\ C^n \;=\; A\oplus A\to C_n \;=\; A\oplus A\ ,\ C_r \;=\; 0\ (r\neq n)\ .$$

The inverse isomorphism $L_{2n}(A)\to Q_\bullet(A)$ sends a $2n$-dimensional quadratic Poincaré complex (C,ψ) over A to the Arf invariant $c\in Q_\bullet(A)$ of the nonsingular quadratic form $(H^n(C),(1+T)\psi_0,v^n(\psi))$ over A, as defined by

$$c \;=\; \sum_{i=1}^{m} v^n(\psi)(x_{2i})v^n(\psi)(x_{2i+1})\in Q_\bullet(A)$$

with $\{x_i\,|\,1\leq i\leq m\}$ any basis for $H^n(C)$ such that

$$(1+T)\psi_0(x_i,x_j) \;=\; \begin{cases}1 & \text{if } (i,j)=(2r,2r+1)\text{ or }(2r+1,2r)\\ 0 & \text{otherwise}\ .\end{cases}$$

The chain bundle over A $(B\langle v_{n+1}\rangle, \beta\langle v_{n+1}\rangle)$ $(n \geq 0)$ defined by

$$\begin{aligned} B\langle v_{n+1}\rangle_r &= \begin{cases} A & \text{if } r \neq n+1 \\ 0 & \text{if } r = n+1 \end{cases}, \\ d &= 0 : B\langle v_{n+1}\rangle_r \to B\langle v_{n+1}\rangle_{r-1}, \\ \beta\langle v_{n+1}\rangle_s &= \begin{cases} 1 \\ 0 \end{cases} : B\langle v_{n+1}\rangle_r \to B\langle v_{n+1}\rangle^{-r-s} \text{ if } \begin{cases} s = -2r, r \neq n+1 \\ \text{otherwise} \end{cases} \end{aligned}$$

is $(v_{n+1} = 0)$-universal. Define symmetric bilinear pairings

$$\begin{aligned} \rho &: A \times A \to A ; (a,b) \mapsto ab , \\ \sigma &: Q^\bullet(A) \times Q^\bullet(A) \to A ; (1,1) \mapsto 1 \end{aligned}$$

such that

$$\begin{aligned} Q\langle B\langle v_{n+1}\rangle, \beta\langle v_{n+1}\rangle\rangle^{2n}(S^n A) &= A \times_\rho A , \\ Q_{2n}(B\langle v_{n+1}\rangle, \beta\langle v_{n+1}\rangle) &= A \times_\sigma Q^\bullet(A) , \\ Q_{2n+1}(B\langle v_{n+1}\rangle, \beta\langle v_{n+1}\rangle) &= 0 . \end{aligned}$$

Let

$$\begin{aligned} Q\langle v_1\rangle &= Q\langle B\langle v_1\rangle, \beta\langle v_1\rangle\rangle^0(A) \\ &= A \times_\rho A . \end{aligned}$$

Given a nonsingular $Q\langle v_1\rangle$-quadratic form (M, λ, μ) over A there exist $v \in M$, $\psi \in \operatorname{Hom}_A(M, M^*)$ such that

$$\begin{aligned} \lambda(x,y) &= \lambda(x,v)\lambda(y,v) + \psi(x)(y) + \psi(y)(x) \in A \ (x, y \in M) \\ \mu : M \to Q\langle v_1\rangle &= A \times_\rho A ; \ x \mapsto (\psi(x)(x), \lambda(x,v)) . \end{aligned}$$

The morphism

$$\begin{aligned} L\langle v_1\rangle^0(A) &= W_{Q\langle v_1\rangle}(A) \to Q_0(B\langle v_1\rangle, \beta\langle v_1\rangle) ; \\ (M, \lambda : M \times M \to A, \mu : M \to Q\langle v_1\rangle) \mapsto \mu(v) &= (\psi(v)(v), \lambda(v,v)) \end{aligned}$$

fits into a short exact sequence

$$0 \to L_0(A) \to L\langle v_1\rangle^0(A) \to Q_0(B\langle v_1\rangle, \beta\langle v_1\rangle) \to 0 .$$

The injection

$$L_{2n}(A) \to L\langle v_{n+1}\rangle^{2n}(A) \to W_{Q\langle v_1\rangle}(A) = L\langle v_1\rangle^0(A)$$

sends the cobordism class of a $2n$-dimensional quadratic Poincaré complex over A (C, ψ) to the Witt class of the nonsingular $Q\langle v_1\rangle$-quadratic form $(H^n(C), (1+T)\psi_0, iv^n(\psi))$, with i the canonical injection

$$i : H_0(\mathbb{Z}_2; A, (-1)^n) = A \to Q\langle v_1\rangle = A \times_\rho A ; \ a \mapsto (a, 0) .$$

In the special case $A = \mathbb{Z}_2$

$$\begin{aligned} L_{2n}(\mathbb{Z}_2) &= Q_\bullet(\mathbb{Z}_2) = \mathbb{Z}_2 , \ L^{2n}(\mathbb{Z}_2) = Q^\bullet(\mathbb{Z}_2) = \mathbb{Z}_2 , \\ Q\langle v_{n+1}\rangle &= \mathbb{Z}_4 , \ Q_0(B\langle v_1\rangle, \beta\langle v_1\rangle) = \mathbb{Z}_4 , \end{aligned}$$

with

$$L\langle v_{n+1}\rangle^{2n}(\mathbb{Z}_2) = W_{\mathbb{Z}_4}(\mathbb{Z}_2) = \mathbb{Z}_8$$

the Witt group of nonsingular $\mathbb{Z}_4$-valued quadratic forms over $\mathbb{Z}_2$. See Weiss [20, §11] for the algebraic Poincaré bordism interpretation of the work of Browder [2] and Brown [4] on the Kervaire invariant and its generalization, which applies also to the work of Milgram [11].

References

[1] Č. Arf, *Untersuchungen über quadratische Formen in Körpern der Charakteristik 2*, J. Reine Angew. Math. **183** (1941), 148–167.

[2] W. Browder, *The Kervaire invariant of framed manifolds and its generalization*, Ann. of Maths. **90** (1969), 157–186.

[3] E. H. Brown, *Cohomology theories*, Ann. of Maths. **75** (1962), 467–484.

[4] ———, *Generalisations of the Kervaire invariant*, Ann. of Maths. **95** (1972), 368–383.

[5] G. Carlsson, *Wu invariants of hermitian forms*, J. Algebra **65** (1980), 188–205.

[6] H. Cartan and S. Eilenberg, *Homological algebra*, Princeton, 1956.

[7] A. Dold, *Homology of symmetric products and other functors of complexes*, Ann. of Maths. **68** (1958), 54–80.

[8] D. M. Kan, *Functors involving c.s.s. complexes*, Trans. Amer. Math. Soc. **87** (1958), 330–346.

[9] R. Lashof, *Poincaré duality and cobordism*, Trans. Amer. Math. Soc. **109** (1963), 257–277.

[10] J. P. May, *Simplicial objects in algebraic topology*, Van Nostrand Mathematical Studies, vol. 11, Van Nostrand, 1967.

[11] R. J. Milgram, *Surgery with coefficients*, Ann. of Maths. **100** (1974), 194–248.

[12] ———, *Orientations for Poincaré duality spaces and applications*, Proc. 1986 Arcata Conference on Algebraic Topology, Lecture Notes in Mathematics, vol. 1370, Springer, 1989, pp. 293–324.

[13] ——— and A. A. Ranicki, *Some product formulae in non-simply-connected surgery*, Trans. Amer. Math. Soc. **297** (1986), 383–413.

[14] A. S. Mishchenko, *Homotopy invariants of non-simply connected manifolds III. Higher signatures*, Izv. Akad. Nauk SSSR, ser. mat. **35** (1971), 1316–1355.

[15] F. Quinn, *Surgery on Poincaré and normal spaces*, Bull. Amer. Math. Soc. **78** (1972), 262–267.

[16] A. Ranicki, *The algebraic theory of surgery*, Proc. Lond. Math. Soc. **40 (3)** (1980), I. 87–192, II. 193–287.

[17] ———, *Algebraic L-theory and topological manifolds*, Cambridge Tracts in Mathematics, vol. 102, Cambridge, 1992.

[18] ———, *An introduction to algebraic surgery*, Surveys on Surgery Theory, Vol. 2, Ann. of Maths. Studies 149, Princeton University Press, 2001, pp. 79–160.

[19] C. T. C. Wall, *Surgery on compact manifolds*, 2nd edition, Mathematical Surveys and Monographs, vol. 69, A.M.S., 1999.

[20] M. Weiss, *Surgery and the generalized Kervaire invariant*, Proc. Lond. Math. Soc. **51 (3)** (1985), I. 146–192, II. 193–230.

[21] ———, *Visible L-theory*, Forum Math. **4** (1992), 465–498.

[22] J. H. C. Whitehead, *A certain exact sequence*, Ann. of Maths. **52** (1950), 51–110.

Department of Mathematics and Statistics
University of Edinburgh
Edinburgh EH9 3JZ
Scotland, UK

E-mail address: aar@maths.ed.ac.uk